THE PSYCHOLOGY OF TECHNOLOGY

THE PSYCHOLOGY OF TECHNOLOGY

Social Science Research
in the Age of Big Data

Edited by SANDRA C. MATZ

 AMERICAN PSYCHOLOGICAL ASSOCIATION

Copyright © 2022 by the American Psychological Association. All rights reserved. Except as permitted under the United States Copyright Act of 1976, no part of this publication may be reproduced or distributed in any form or by any means, including, but not limited to, the process of scanning and digitization, or stored in a database or retrieval system, without the prior written permission of the publisher.

The opinions and statements published are the responsibility of the authors, and such opinions and statements do not necessarily represent the policies of the American Psychological Association.

Published by
American Psychological Association
750 First Street, NE
Washington, DC 20002
https://www.apa.org

Order Department
https://www.apa.org/pubs/books
order@apa.org

In the U.K., Europe, Africa, and the Middle East, copies may be ordered from Eurospan
https://www.eurospanbookstore.com/apa
info@eurospangroup.com

Typeset in Minion by Circle Graphics, Inc., Reisterstown, MD

Printer: Gasch Printing, Odenton, MD
Cover Designer: Gwen J. Grafft, Minneapolis, MN

Library of Congress Cataloging-in-Publication Data

Names: Matz, Sandra C., editor.
Title: The psychology of technology : social science research in the age of big data / edited by Sandra C. Matz.
Description: Washington, DC : American Psychological Association, [2022] | Includes bibliographical references and index.
Identifiers: LCCN 2022003017 (print) | LCCN 2022003018 (ebook) | ISBN 9781433836268 (paperback) | ISBN 9781433838477 (ebook)
Subjects: LCSH: Psychology--Research--Methodology. | Psychology--Technological innovations. | Technological innovations--Psychological aspects. | COVID-19 (Disease) in mass media. | Change (Psychology) | BISAC: PSYCHOLOGY / Research & Methodology | PSYCHOLOGY / Applied Psychology
Classification: LCC BF76.5 .P7923 2022 (print) | LCC BF76.5 (ebook) | DDC 150.72--dc23/eng/20220310
LC record available at https://lccn.loc.gov/2022003017
LC ebook record available at https://lccn.loc.gov/2022003018

https://doi.org/10.1037/0000290-000

Printed in the United States of America

10 9 8 7 6 5 4 3 2 1

Contents

Contributors	vii
Foreword. The Critical Role of Psychological Science in Understanding the Human–Technology Relationship *Juliana Schroeder and Nathanael Fast*	xi
Introduction: Welcome to the Psychology of Technology *Sandra C. Matz*	3
1. What Our Data Reveal About Our Minds: Predicting Psychological Characteristics From Digital Footprints *Poruz Khambatta*	9
2. Saying More Than We Know: How Language Provides a Window Into the Human Psyche *M. Asher Lawson and Sandra C. Matz*	45
3. The Big Data Toolkit for Psychologists: Data Sources and Methodologies *Heinrich Peters, Zachariah Marrero, and Samuel D. Gosling*	87
4. The Psychology of Mobile Technology and Daily Mobility *Morgan Quinn Ross, Sandrine R. Müller, and Joseph B. Bayer*	125
5. The Psychology of Virtual Reality *Marijn Mado and Jeremy Bailenson*	155

6.	Social Media and Psychological Well-Being *Jeffrey T. Hancock, Sunny Xun Liu, Mufan Luo, and Hannah Mieczkowski*	195
7.	How Social Media Contexts Affect the Expression of Moral Emotions *William J. Brady and Killian L. McLoughlin*	239
8.	Big Data in the Workplace *Peter J. Mancarella and Tara S. Behrend*	267
9.	Human–Robot Interaction Challenges in the Workplace *Guy Hoffman, Alap Kshirsagar, and Matthew V. Law*	305
10.	The Psychology of Big Data: Developing a "Theory of Machine" to Examine Perceptions of Algorithms *Jennifer M. Logg*	349
11.	Privacy and Ethics in the Age of Big Data *Sandra C. Matz, Ruth E. Appel, and Brian Croll*	379
12.	The Psychology of Technology: Where the Future Might Take Us *Moran Cerf and Sandra C. Matz*	421
Index		431
About the Editor		451

Contributors

Ruth E. Appel, MS, PhD Student, Department of Communication, Stanford University, Stanford, CA, United States

Jeremy Bailenson, PhD, Thomas More Storke Professor, Department of Communication; Founding Director, Virtual Human Interaction Lab, Stanford University, Stanford, CA, United States

Joseph B. Bayer, PhD, Assistant Professor, School of Communication; Core Faculty, Translational Data Analytics Institute, The Ohio State University, Columbus, OH, United States

Tara S. Behrend, PhD, Associate Professor of Industrial–Organizational Psychology, Department of Psychological Sciences; Director, Workplaces and Virtual Environments (WAVE) Lab, Purdue University, West Lafayette, IN, United States

William J. Brady, PhD, Postdoctoral Fellow, Department of Psychology, Yale University, New Haven, CT, United States

Moran Cerf, PhD, Associate Professor of Neuroscience and Marketing, Kellogg School of Management, Northwestern University, Chicago, IL, United States

Brian Croll, MS, Vice President (retired), Worldwide Product Marketing, Apple Inc., Cupertino, CA, United States

Nathanael Fast, PhD, Associate Professor of Management and Organization, Marshall School of Business, University of Southern California, Los Angeles, CA, United States

CONTRIBUTORS

Samuel D. Gosling, PhD, Professor, Department of Psychology, University of Texas at Austin, Austin, TX, United States

Jeffrey T. Hancock, PhD, Professor, Department of Communication; Founding Director, Stanford Social Media Lab, Stanford University, Stanford, CA, United States

Guy Hoffman, PhD, Associate Professor and Mills Family Faculty Fellow, Sibley School of Mechanical and Aerospace Engineering, Cornell University, Ithaca, NY, United States

Poruz Khambatta, PhD, Postdoctoral Fellow, Anderson School of Management, University of California, Los Angeles, CA, United States

Alap Kshirsagar, MS, PhD Candidate, Sibley School of Mechanical and Aerospace Engineering, Cornell University, Ithaca, NY, United States

Matthew V. Law, MS, PhD Candidate, Department of Information Science, Cornell University, Ithaca, NY, United States

M. Asher Lawson, BA, Doctoral Student, Fuqua School of Business, Duke University, Durham, NC, United States

Sunny Xun Liu, PhD, Social Science Research Scientist; Associate Director, Stanford Social Media Lab, Stanford University, Stanford, CA, United States

Jennifer M. Logg, PhD, Assistant Professor of Management, McDonough School of Business, Georgetown University, Washington, DC, United States

Mufan Luo, MA, PhD Candidate, Department of Communication, Stanford University, Stanford, CA, United States

Marijn Mado, MSc, Doctoral Student, Department of Communication, Stanford University, Stanford, CA, United States

Peter J. Mancarella, MS, Doctoral Student, Department of Psychological Sciences and Workplaces and Virtual Environments (WAVE) Lab, Purdue University, West Lafayette, IN, United States

Zachariah Marrero, MS, PhD Student, Department of Psychology, University of Texas at Austin, Austin, TX, United States

Sandra C. Matz, PhD, Professor of Management, Columbia University, New York, NY, United States

Killian L. McLoughlin, MS, PhD Student, Department of Psychology, Yale University, New Haven, CT, United States

CONTRIBUTORS

Hannah Mieczkowski, MA, PhD Candidate, Department of Communication, Stanford University, Stanford, CA, United States

Sandrine R. Müller, PhD, Postdoctoral Research Fellow, Data Science Institute, Columbia University, New York, NY, United States

Heinrich Peters, MS, PhD Student, Department of Psychology, University of Texas at Austin, Austin, TX, United States

Morgan Quinn Ross, BS, PhD Student, School of Communication, The Ohio State University, Columbus, OH, United States

Juliana Schroeder, PhD, Associate Professor, Haas School of Business, University of California, Berkeley, CA, United States

Foreword: The Critical Role of Psychological Science in Understanding the Human–Technology Relationship

Technological advances are changing every aspect of the human experience, including how people work (e.g., virtual conference calls, virtual assistants), play (e.g., virtual reality video games), socialize (e.g., social media), and make decisions (e.g., artificial intelligence, algorithmic decision making). Technology is also changing *perceptions* of the human experience, causing people to update their very understanding of what it means to be human and to lead a fulfilling life. However, the fact that many new technologies are introduced into society before people understand their consequences, including how they influence human psychology, has led not only to new opportunities for human flourishing (e.g., being able to connect instantly with family, friends, and colleagues living thousands of miles away) but also to a number of unanticipated dangers (e.g., harmful data breaches, the spread of misinformation, and increased societal polarization). Capitalizing on technology's benefits while avoiding its potential costs is one of the most critical tasks of our time. Doing so requires *trustworthy, independent knowledge* about when and why people adopt new technology and how doing so affects the human psyche.

In our opinion, the institution of science is best positioned to produce this kind of knowledge about the psychology of technology. While the scientific enterprise is not perfect, it is uniquely capable of providing the critical knowledge needed to give people confidence about how to consume technology and to elucidate technology designers on the consequences of their products. In contrast to media institutions—which have a conflict of interest because they are incentivized to attract public attention—and technology companies—which are incentivized to produce profit—scientists can be relatively objective in seeking the truth.

This is important because our current relationship to technology reminds us of the state of medicine in the early 1800s. At that point in history, our future health and advancement as a species depended on our ability to use science to understand and improve medicine. Imagine for a moment how misguided it would have been to turn our backs on science and its power to help us better understand medical treatments and their effects on human bodies. Health consumers would have been left on their own to try to blindly discern which drugs and procedures to use and which to avoid. Yet, this is precisely the predicament we face with technology today. Neither the users of our newest technologies nor the technologists who created them know what kinds of effects such tools will have on individual and societal well-being.

Thankfully, as the rapidly growing "psychology of technology" community and the insightful chapters in this book illustrate, scientific researchers are working hard to pave the way toward better research methodologies as well as beginning to shed more precise light on how different forms of technology influence the human psyche. For example, one key question, discussed by Logg in Chapter 10 of this book, is when and why people are willing to embrace the use of algorithms in everyday life. Some researchers have found that decision makers reject the use of algorithms as decision aids (Dietvorst et al., 2015; Yeomans et al., 2019), a tendency that is reduced when people are given some control over the algorithm (Dietvorst et al., 2018). Other research has found that employees resist the use of algorithms to make human resources decisions—even when they could reduce or eliminate human bias—as a result of "algorithmic reductionism" or the perceived inability of algorithms to form nuanced, contextualized judgments (Newman et al., 2020). Research in the health care domain reveals similar results; consumers resist the use of artificial intelligence to provide treatment recommendations (Longoni et al., 2019). On the flip side, however, researchers have found that people sometimes prefer algorithms over humans for certain types of recommendations (Logg et al., 2019) and functions (Raveendhran & Fast, 2021), even when doing so undermines their privacy (Fast & Jago, 2020). Clearly, the research in this area is ongoing, and Chapter 5 provides a helpful overview of what we currently know and where the field is headed.

Another critical question, discussed by Hancock, Liu, Luo, and Mieczkowski in Chapter 6, is how social media affects human well-being. While some scholars identified negative associations between social media use and psychological well-being (e.g., Rosen et al., 2014; Twenge & Campbell, 2019), others have contested these results, suggesting that these associations are not causal and, when rigorous causal inference techniques are used to analyze the data, there does not appear to be a strong association between digital media consumption and well-being (Orben & Przybylski, 2019). Moreover, the effect of social media on well-being appears to depend both on how people use social media (e.g., active vs. passive browsing influences social connections, Bapna et al., 2016; and whether digital technology displaces or supplements in-person connections, Lieberman & Schroeder, 2020; Waytz & Gray, 2018) and on who uses it (e.g., there is a modest negative impact of social media use for teen girls; Orben et al., 2019). How much one uses social media also matters: both too little use of social media and too much use are associated with reduced well-being, whereas moderate use is associated with beneficial effects (Przybylski & Weinstein, 2017). The debate in the literature surrounding social media use and well-being illustrates the value of the scientific approach; science scrutinizes its own findings and leaves no rock unturned. The work of these and many other researchers also shows that more work needs to be done to ascertain the direct and indirect effects of living in a society dominated by the use of social media.

Another debate on which science can shed light, discussed by Brady and McLoughlin in Chapter 7, is whether technology is enhancing or reducing polarization. Many have argued that echo chambers act as hermetically sealed ideological bubbles that prevent exposure to the other side (e.g., Mutz, 2006). By this account, polarization and radicalization occur because we simply don't know the other side's positions. But in the case of the political left and right, this is not generally true: In fact, social media use is associated with *more* exposure to materials from the opposing side of the political spectrum (Flaxman et al., 2016). Most politically engaged people do get exposed to the other side; it's just that this exposure happens in a context that incentivizes outrage and moral grandstanding rather than mutual understanding (e.g., Brady & Crockett, 2019). In fact, one study showed that more exposure to opposing views online actually increases polarization (Bail et al., 2018). This shows that our intuitions about both problems and solutions need to be tested empirically before being acted on and/or incorporated by policy makers.

Ongoing public debates like the ones highlighted above—attitudes about adopting artificial intelligence, the effect of social media on well-being, and the extent of polarization—would be better informed if there were greater attention paid to the scientific research. This book, edited by Dr. Sandra C. Matz and summarizing the research of many scientific experts, is a much-needed step toward collating and capitalizing upon scientific knowledge to address such questions about the psychology of technology. In particular, four exciting developments in psychological science offer promising opportunities to better understand how technology is changing humanity. First, researchers are increasingly studying human behavior by scrutinizing the digital footprints they leave behind, proving new methods and discoveries. Second, scientists are producing insight into the psychological underpinnings of human–machine interaction, such as attitudes about the use of algorithms to navigate daily life. Third, researchers are documenting the impact of technologies on individuals and society, allowing for more reasoned analysis (and typically less-extreme conclusions) about how technology is affecting humanity. And finally, a Big Data approach to studying the psychology of technology is both fascinating and critical in unlocking new ways of analyzing the costs and benefits of new technologies.

In order to realize the benefits of new technologies while avoiding the harms, we must capitalize on the scientific method and employ collaborative approaches to produce reliable, trustworthy knowledge about how technology affects human well-being. We applaud the editor of this book, along with the contributors, for their important efforts to do just that. Enjoy!

—Juliana Schroeder
University of California, Berkeley

—Nathanael Fast
University of Southern California
The authors are codirectors, Psychology of Technology Institute, and contributed equally to this foreword.

REFERENCES

Bail, C. A., Argyle, L. P., Brown, T. W., Bumpus, J. P., Chen, H., Hunzaker, M. B. F., Lee, J., Mann, M., Merhout, F., & Volfovsky, A. (2018). Exposure to opposing views on social media can increase political polarization. *Proceedings of the National Academy of Sciences of the United States of America, 115*(37), 9216–9221. https://doi.org/10.1073/pnas.1804840115

Bapna, R., Ramaprasad, J., Shmueli, G., & Umyarov, A. (2016). One-way mirrors in online dating: A randomized field experiment. *Management Science, 62*(11), 3100–3122. https://doi.org/10.1287/mnsc.2015.2301

Brady, W. J., & Crockett, M. J. (2019). How effective is online outrage? *Trends in Cognitive Sciences, 23*(2), 79–80. https://doi.org/10.1016/j.tics.2018.11.004

Dietvorst, B. J., Simmons, J. P., & Massey, C. (2015). Algorithm aversion: People erroneously avoid algorithms after seeing them err. *Journal of Experimental Psychology: General, 144*(1), 114–126. https://doi.org/10.1037/xge0000033

Dietvorst, B. J., Simmons, J. P., & Massey, C. (2018). Overcoming algorithm aversion: People will use imperfect algorithms if they can (even slightly) modify them. *Management Science, 64*(3), 1155–1170. https://doi.org/10.1287/mnsc.2016.2643

Fast, N. J., & Jago, A. S. (2020). Privacy matters . . . or does it? Algorithms, rationalization, and the erosion of concern for privacy. *Current Opinion in Psychology, 31*, 44–48. https://doi.org/10.1016/j.copsyc.2019.07.011

Flaxman, S., Goel, S., & Rao, J. M. (2016). Filter bubbles, echo chambers, and online news consumption. *Public Opinion Quarterly, 80*(S1), 298–320. https://doi.org/10.1093/poq/nfw006

Lieberman, A., & Schroeder, J. (2020). Two social lives: How differences between online and offline interaction influence social outcomes. *Current Opinion in Psychology, 31*, 16–21. https://doi.org/10.1016/j.copsyc.2019.06.022

Logg, J. M., Minson, J. A., & Moore, D. A. (2019). Algorithm appreciation: People prefer algorithmic to human judgment. *Organizational Behavior and Human Decision Processes, 151*, 90–103. https://doi.org/10.1016/j.obhdp.2018.12.005

Longoni, C., Bonezzi, A., & Morewedge, C. K. (2019). Resistance to medical artificial intelligence. *The Journal of Consumer Research, 46*(4), 629–650. https://doi.org/10.1093/jcr/ucz013

Mutz, D. C. (2006). *Hearing the other side*. Cambridge University Press. https://doi.org/10.1017/CBO9780511617201

Newman, D. T., Fast, N. J., & Harmon, D. J. (2020). When eliminating bias isn't fair: Algorithmic reductionism and procedural justice in human resource decisions. *Organizational Behavior and Human Decision Processes, 160*, 149–167. https://doi.org/10.1016/j.obhdp.2020.03.008

Orben, A., Dienlin, T., & Przybylski, A. K. (2019). Social media's enduring effect on adolescent life satisfaction. *Proceedings of the National Academy of Sciences of the United States of America, 116*(21), 10226–10228. https://doi.org/10.1073/pnas.1902058116

Orben, A., & Przybylski, A. K. (2019). The association between adolescent well-being and digital technology use. *Nature Human Behaviour, 3*(2), 173–182. https://doi.org/10.1038/s41562-018-0506-1

Przybylski, A. K., & Weinstein, N. (2017). A large-scale test of the Goldilocks hypothesis: Quantifying the relations between digital-screen use and the mental well-being of adolescents. *Psychological Science, 28*(2), 204–215. https://doi.org/10.1177/0956797616678438

Raveendhran, R., & Fast, N. J. (2021). Humans judge, algorithms nudge: The psychology of behavior tracking acceptance. *Organizational Behavior and Human Decision Processes, 164*, 11–26. https://doi.org/10.1016/j.obhdp.2021.01.001

Rosen, L. D., Lim, A. F., Felt, J., Carrier, L. M., Cheever, N. A., Lara-Ruiz, J. M., Mendoza, J. S., & Rokkum, J. (2014). Media and technology use predicts ill-being among children, preteens and teenagers independent of the negative health impacts of exercise and eating habits. *Computers in Human Behavior, 35*, 364–375. https://doi.org/10.1016/j.chb.2014.01.036

Twenge, J. M., & Campbell, W. K. (2019). Media use is linked to lower psychological well-being: Evidence from three datasets. *Psychiatric Quarterly, 90*(2), 311–331. https://doi.org/10.1007/s11126-019-09630-7

Waytz, A., & Gray, K. (2018). Does online technology make us more or less sociable? A preliminary review and call for research. *Perspectives on Psychological Science, 13*(4), 473–491. https://doi.org/10.1177/1745691617746509

Yeomans, M., Shah, A., Mullainathan, S., & Kleinberg, J. (2019). Making sense of recommendations. *Journal of Behavioral Decision Making, 32*(4), 403–414. https://doi.org/10.1002/bdm.2118

THE PSYCHOLOGY OF TECHNOLOGY

Introduction: Welcome to the Psychology of Technology

Sandra C. Matz

Our lives are inseparably interwoven with technology: We maintain our personal relationships via online social networks, order our groceries online, converse with chat bots, navigate with the help of our smartphones, work remotely via video conferencing, and more. The pervasiveness of technology in our daily routines raises several interesting questions for psychologists that can broadly be categorized as studying technology as (a) a psychological phenomenon or (b) a psychological tool. Focusing on technology as a psychological phenomenon, psychologists have been interested in how new technologies shape the human experience. How does technology change the way we work, communicate, or socialize? How do our interactions with technology affect our well-being and mental health? And how can new technological advances, such as robots and virtual reality, enrich the human experience? Focusing on technology as a tool, psychologists have gradually expanded their

https://doi.org/10.1037/0000290-001
The Psychology of Technology: Social Science Research in the Age of Big Data, S. C. Matz (Editor)
Copyright © 2022 by the American Psychological Association. All rights reserved.

analytical toolbox to integrate methodologies that allow them to analyze the large quantity of behavioral records we generate with every step we take in the digital world (e.g., social media profiles, credit card spending, smartphone logs).

Over the past decade, psychologists from all over the world have come together to study the psychology of technology in an interdisciplinary collaboration with researchers from the computer sciences, economics, engineering, sociology, political science, and many other disciplines. Although the field is still relatively young, the community is growing rapidly. There are more and more conferences and journals popping up each year that are dedicated to the topic and highlight the increasing demand for research in this area.

I have been part of this community for a number of years now. Trained as a personality psychologist, I got intrigued by the world of Big Data and computational psychology early in my graduate studies at the University of Cambridge. I was fascinated by the ways in which digital footprints could provide us with windows into people's behaviors, preferences, experiences, motivations, concerns, habits, and inner mental lives; or, in a nutshell, a window into their psychology. Today, I am a professor of management at Columbia Business School in New York. Together with my colleagues and students, I am conducting research at the intersection of psychology, computer science, and business. We explore the relationships between observable behaviors, personal dispositions, and inner mental states, with the goal of designing personalized interventions that can help individuals make better decisions and lead healthier and happier lives.

What intrigues me most about the psychology of technology is how incredibly rich and diverse this field is, both in terms of the questions asked and methodologies used to answer them. This, of course, makes the psychology of technology hard to pin down and define. If you bought a book on personality psychology, it would be pretty clear what you're going to get for your money. Sure, books might differ slightly in their focus on the particular frameworks they discuss, their coverage of the history of personality psychology, or the extent to which they highlight new ways of assessing psychological dispositions. But overall, you would probably

INTRODUCTION

have a good sense of what the book contained before opening it. This isn't necessarily the case for the book you are holding in your hands right now. There is no core curriculum of the psychology of technology, no agreed-upon set of fundamental topics and findings that simply have to appear in a volume like this one. At first glance, the topics discussed in this book might appear scattered. A chapter on human–robot interactions, a chapter on the psychology of mobility, and a chapter on the effects of social media on well-being? Yes! That's not a fluke or poor editorial decision making (at least as far as I'm concerned, but then I might be biased, of course). The breadth in topics is exactly what the psychology of technology is all about! It's a new field that is flourishing *because* it is diverse and brings together people and topics from all across academia. And it's what makes being part of this community so incredibly interesting and inspiring for me. It is not uncommon that I go through a full day of conference sessions learning about topics that I have never thought of before and don't have the slightest expertise in. Every single time I attend a conference or read a paper on the topic, I learn something new.

This book is my attempt to capture the richness and diversity that the psychology of technology provides for anyone who is interested in what technology has to offer psychologists in academia and industry. Leading experts in the field offer you a glimpse into their worlds and their unique take on the psychology of technology. The book is a curated selection of topics that have garnered attention among psychologists over the last couple of years, and that I believe are at the heart of the psychology of technology. There are many more chapters that could have been written (and maybe will in the future), which makes this volume one possible snapshot of what is happening in the field. Instead of being a comprehensive collection of research on the psychology of technology—which is a near-impossible task—this book is meant as an inspiration for young scholars in the field and a foundation to move the field forward. Think of it as an eclectic mix of stimulating ideas. It is quite likely that you will find some chapters more interesting than others. But my hope is that these pages will provide you with a sense of what is possible and what questions you might be able to ask and answer in your own careers.

So what exactly is this book going to teach you? We start our wondrous journey through the psychology of technology where I started mine: In the world of Big Data. Chapters 1 through 3 introduce you to different ways in which Big Data can be leveraged as a research tool to better understand the human psyche. Chapter 1 discusses how the digital traces we leave as we interact with technology and digital devices (e.g., Facebook Likes, status updates, profile pictures) can reveal an awful lot about who we are when it comes to our intimate psychological characteristics. Zooming in on language as a window into people's inner mental lives, Chapter 2 explores how new approaches to text analysis allow us to understand psychological phenomena across a whole range of psychological disciplines (e.g., personality, social, or organizational psychology). Zooming back out, Chapter 3 dives into the "behind the scenes" of Chapters 1 and 2 to provide readers with a practical, step-by-step guide for how to successfully handle and analyze Big Data.

Chapters 4 through 10 pivot to studying technology as a psychological phenomenon. Keeping with the notion of how our psychological experiences are inseparably interwoven with everyday technologies, Chapter 4 explores how our smartphones and other wearable devices not only affect our experience of the world around us but can also act as passive tracking tools of our daily habits and routines. Moving to the next level of immersive technologies, Chapter 5 explores the opportunities provided by virtual reality (VR), a tool that has the power to enhance the way we discover and experience the world around us. The chapter explores how psychologists can use VR to study important psychological phenomena in realistic settings and design interventions to change attitudes and behaviors through immersive experiences.

Shifting the focus to a topic that is widely debated in academic circles, as well as in the popular media, Chapters 6 and 7 explore the impact of social media on individuals and society. Building on a recent meta-analysis, Chapter 6 provides a nuanced take on how the use of social media is influencing people's mental health and well-being. Chapter 7 reviews the science of how the unique characteristics of the social media environment are changing the ways in which people express their moral emotions.

INTRODUCTION

Exploring the importance of new technologies in the workplace and organizational decision making, Chapters 8 through 10 offer insights into how Big Data and the introduction of machines and algorithms to the work force are reshaping our everyday work experiences. Chapter 8 discusses the regulatory environment around Big Data in organizational contexts and reviews applications of Big Data across a range of important human resource functions (e.g., selection or performance management). Chapter 9 focuses on a particular aspect of the future of work: the interaction between humans and robots. Building on recent empirical work of the authors, the chapter outlines challenges as well as opportunities associated with the introduction of robots into human teams and discusses how engineers and computer scientists stand to gain from taking psychological principles into account when designing human–machine interfaces. Building on the topic of human–computer interaction, Chapter 10 explores the "theory of machine" as a theoretical framework for psychologists to examine and document laypeople's theories about algorithmic decision making (e.g., how do people expect humans and algorithms to differ in the ways that they make decisions?).

Finally, Chapters 11 and 12 provide a higher level view of the psychology of technology. Chapter 11 discusses the ethics of Big Data in the context of privacy. It highlights how the availability of personal data poses severe challenges to current best practices for privacy protection and suggests a number of possible solutions for both practitioners and scientists to overcome these challenges. The afterword in Chapter 12 offers a bold outlook on the future of humanity in the context of technological advances. It makes predictions about how existing and emerging technologies will fundamentally change the human experience when it comes to the way we think about ourselves, the way we interact with others, and even the biological foundations underlying who we are.

1

What Our Data Reveal About Our Minds: Predicting Psychological Characteristics From Digital Footprints

Poruz Khambatta

The advent of digital technologies has had many noteworthy effects on humanity. One important development is an unprecedented increase in digital records of human behavior. As we use our computers, smartphones, and other digital devices to look up information on search engines, send emails, texts, or chat messages, read news articles, watch videos, listen to music, order food, find dates, manage our finances, and curate our online presence, we leave behind a trail of digital footprints. With the use of computational tools, these digital footprints can reveal things about our minds, such as who we are, how we feel, and what we believe. Such technologies may even understand us better than the people who know us best can and reveal things about our future that haven't happened yet. While this research is not without its challenges, ultimately, such approaches can further psychological science and pave the way for more effective psychological interventions.

https://doi.org/10.1037/0000290-002
The Psychology of Technology: Social Science Research in the Age of Big Data, S. C. Matz (Editor)
Copyright © 2022 by the American Psychological Association. All rights reserved.

In this chapter, we explore this new phenomenon and its many implications. We begin by noting the dramatic increase in digital records of human behavior in the current era and examining how machine learning approaches can be applied to such data sets to unearth new truths about the human mind. We then survey the emerging body of research in this area, examining digital footprints drawn from social media, smartphones, bank transactions, and more. The powerful advantages of this approach, such as its ability to capture human behavior in naturalistic contexts and its unparalleled precision, are examined. However, the challenges facing researchers who incorporate these tools into their work is also considered, including messy or inaccurate data, reliability and validity concerns, and issues related to privacy. Finally, we explore future directions. These include the ability to advance psychological theory and create personalized interventions that harness these powerful new tools to improve people's lives.

DIGITAL FOOTPRINTS

In 1949, a team of social scientists set out to conduct a unique investigation. For 14 hours, eight observers followed around an ordinary boy from rural Kansas, painstakingly documenting everything he said and did verbatim. Their full report was eventually compiled into a 435-page tome titled, *One Boy's Day: A Specimen Record of Behavior* (Barker & Wright, 1951), with the hope that this highly detailed record could be of use to social scientists. Today, we no longer need a team of eight meticulous observers to get a thorough sense of a person's day. Much of this information is already automatically recorded during our interactions with digital technologies.

Every day, nearly 1.79 billion people open Facebook (Statista, 2020a). Some post delicious cooking recipes to share with their friends. Others seek advice for the best exercise bike or board game. Many simply scroll through and soak up the information curated for them by the platform. All of these nearly 2 billion people have their own stories, unique trajectories through life that are interwoven with the stories of others to form the vibrantly diverse tapestry of humanity. Yet despite their many geographic and socioeconomic differences, one thing these users all have in common is that they are leaving behind digital footprints, electronic records of their activities. This is not only true for users of Facebook but those of most social media platforms. Teenagers liking a friend's post on Instagram, young adults sharing a video on TikTok, senior citizens forwarding an article on WhatsApp, and academics retweeting an intriguing idea on Twitter are all leaving behind digital records of their behavior.

Moreover, the record keeping does not end on social media. In this "age of engineering," we are relying on technology to help us perform an unprecedented number of tasks (Brown & Duguid, 2017; McFarland et al., 2016). As a result, individuals buying a coffee with their credit card, conjuring up directions on their smartphone, swiping for a romantic partner on a dating app, uploading a new profile picture on LinkedIn, or listening to a song on Spotify are all leaving digital traces as well.

The amount of data being generated by humanity every day is staggering (Marr, 2018). Every second, 456,000 tweets are sent on Twitter, and 46,740 photos are posted on Instagram. In the same amount of time, 510,000 comments are posted and 293,000 statuses are updated on Facebook. Meanwhile, 156 million emails and 16 million text messages are being sent. By 2025, it is estimated that there will be 175 zettabytes of data (Reinsel et al., 2018). How large is that exactly? That is 175 trillion (175,000,000,000,000) gigabytes. That is so much information that if you were to take all of this data and store it on DVDs, you could place those DVDs side-by-side and wrap all the way around the circumference of the earth. In fact, you would have enough DVDs left over to do it again. And again. This is so much data that your DVDs could be stacked around the circumference of the earth 222 times—quite a collection.

Many people may not realize just how much data is being captured and stored about them. Fewer still may recognize what all of this data can reveal.

WHAT YOUR LIKES SAY ABOUT YOU

In 2013, by default, Facebook Likes were publicly available. The whole world could bear witness to the fact that you liked curly fries, Barack Obama, Harley Davidson, camping, or Hello Kitty. But in April of that year, three researchers in the United Kingdom published a paper (Kosinski et al., 2013). Two weeks later, Facebook disabled public access to the Likes of over a billion users. What incited such a dramatic change? What did that paper show?

It turns out your Likes can say a lot about you. They can reveal our age ($r = .75$), our gender ($AUC = .93$)[1], our ethnicity ($AUC = .95$), our personality ($r = .29$ to $.43$), our intelligence ($r = .39$), our satisfaction with life ($r = .17$), our religious views ($AUC = .82$), our sexual orientation ($AUC = .75-.88$), our political identity ($AUC = .85$), whether we use addictive substances ($AUC = .65-.73$), and even whether our parents were separated when we were 21 ($AUC = .60$). How could you get all of that from Hello Kitty?

Well, it turns out that not everyone likes Hello Kitty. As with many things, a particular kind of person is more likely to like Hello Kitty on Facebook than others. For example, those who like Hello Kitty are more likely to be female. Those who like Barack Obama are more likely to be Democrats. Those who like science are more likely to be intelligent. And, of course, those who drink alcohol are more likely to endorse, "Dear Liver Thanks You're A Champ." This may seem straightforward so far. Nevertheless, there are some patterns between Facebook Likes and psychological traits that are not as easy to presuppose.

One of the most predictive Likes for high intelligence was curly fries. One of the most predictive Likes for low intelligence was Harley Davidson (which an English Harley Davidson dealer enthusiastically contested on the local news shortly after the paper was published[2]). One of the best predictors for a high satisfaction with life was "Proud to be Christian," which is consistent with research showing that religious individuals tend to be happier than those who are less religious (Myers, 2000). However, another predictor of high life satisfaction was even more perplexing: swimming. In

---

[1] AUC refers to area under the curve, a measure often used in machine learning to evaluate the prediction accuracy of a dichotomous classifier. For example, if you were trying to classify Democrats versus Republicans, and you selected one random person from each group, the AUC of a model represents the probability that it could accurately identify which person is the Democrat. Values range from 0 to 1 with higher numbers representing higher accuracy. Chance accuracy would be .50.

[2] https://www.youtube.com/watch?v=I4ftlpgfd8s

order to understand these mysterious phenomena, it is necessary to take a look under the hood at how digital records of our behavior can reveal our psychological dispositions.

HOW MACHINE LEARNING CAN PREDICT PSYCHOLOGICAL TRAITS FROM DIGITAL FOOTPRINTS

In the early days of artificial intelligence (AI), programmers set out to translate the world's information into a system of if–then statements computers could understand in order to build an artificially intelligent agent (Haenlein & Kaplan, 2019). This proved quite challenging. Imagine having to teach a computer what a "Z" looks like, so it could help you sort mail by automatically reading addresses. There is so much variation in handwriting, penmanship, and ink that inputting each and every possible rule for what counts as a "Z" manually would be rather intractable. A better approach would be to provide the computer with many different examples and have it teach itself what a "Z" looks like. This technique, in which computers automatically learn from experience, is known as machine learning (Jordan & Mitchell, 2015). Today, the U.S. Post Office uses machine learning to sort handwritten mail with high accuracy and efficiency (Mitchell, 2006).

The main metric used to evaluate a machine learning algorithm is out-of-sample accuracy (Tsamardinos et al., 2018). In other words, a machine learning model worth its salt should be able to accurately predict an outcome in a new sample of data that it hasn't seen before. A post office AI that performs very well on the handwritten addresses it was trained on but is inaccurate for new handwritten addresses would not be very useful. For this reason, many steps are needed to ensure that test error (the errors made on data that the computer was not trained on) is minimized. If these steps are not taken, the predictions the computer makes are unlikely to generalize to new data—a phenomenon termed overfitting. This is when a model fits the training data exceptionally well, too well in fact, and as a result performs poorly on new data.

Various machine learning algorithms can be used to predict psychological traits. They range in complexity from simpler approaches, such as linear or logistic regression (e.g., Kosinski et al., 2013), to more complex operations, such as random forest models (e.g., Gladstone et al., 2019) or deep learning neural networks (e.g., Y. Wang & Kosinski, 2017). A more comprehensive exploration of the methodologies involved in conducting such research and example code can be found in Chapter 14 of this volume.

Using machine learning, researchers were able to train computers to predict people's psychological traits from their Facebook Likes (Kosinski et al., 2013). A group of 58,000 volunteers had shared their Facebook Likes, demographic profiles, and the results of several psychometric tests with the researchers. By providing a machine learning algorithm with lots of examples of Facebook Likes and corresponding traits, the researchers were able to have the computer teach itself how to predict a trait from Facebook Likes. Sometimes, the links researchers found (e.g., between curly fries and intelligence) were links no researcher would have ever predicted or hypothesized. The computer had learned something new about the world that those who trained it had not known themselves.

While any single Like might not tell you all that much about a person, by combining multiple Likes, a fuller picture may begin to emerge. The average individual in the study had dozens of Likes (the median number of Likes was 68 per person). Some individuals had over 300. As the authors found, the more Likes the computer was given from a particular person, the more accurate its predictions became, though there were diminishing returns to more data.

As the authors argued, such predictive models could be of benefit to both individual users and scientists. For instance, if companies knew more about you, they could tailor their products, services, and advertisements to better fit your unique needs. A more outgoing individual might be particularly interested in visiting a new nightclub in town, while a less outgoing person might be more drawn to a less crowded venue. A liberal might be particularly inclined to learn more about a Democratic candidate in the upcoming primary, while a conservative might be more interested in familiarizing themself with the Republican contenders.

In addition to helping users, these technologies could enable new forms of psychometric assessment to help scientists better measure people's psychological traits. As discussed later in this chapter, such assessments may hold many advantages over the primarily questionnaire-based approach most commonly used in psychology.

However, the authors also cautioned that, as with many emerging technologies, such techniques could be used for malevolent purposes, such as to reveal things about people they would prefer not to divulge. To that end, they urged for greater transparency from organizations and for companies to imbue people with greater control over their personal information. In response, Facebook decided to no longer make Likes public by default.

A lot has happened since 2013. However, this highly cited paper by Kosinski et al. (2013) laid the foundation for many more papers to follow, which have extended and complemented these ideas in innovative new ways.

WHAT SMARTPHONE SENSORS, BANK TRANSACTIONS, STATUS UPDATES, AND MORE REVEAL ABOUT OUR MINDS

Much of the public discourse on the prediction of psychological traits from digital footprints focuses on social media. However, it is important to note that many other sources of data can also provide clues to our psychology.

Over 3 billion people around the world own a smartphone, and this number is constantly increasing (Statista, 2020b). Though tiny, these mobile computers are quite the companion. Smartphones are densely packed with sensors, including an accelerometer that detects motion, a microphone that picks up on ambient sounds and speech, a GPS sensor that can pinpoint our location, a Bluetooth radio that can detect nearby devices, and more (Harari, Müller, et al., 2017). These devices are also capable of storing this data as well as app usage patterns and relaying such information via the Internet to smartphone manufacturers, apps, or websites. While this information may help companies provide a better customer experience (or in more problematic cases, profit from user data without necessarily enhancing the experience), they can provide psychologists with an unprecedented window into human behavior.

Smartphone data can shed light on a person's behaviors, such as their physical activity, social interaction, or digital media consumption (Harari, Müller, et al., 2017, 2020). In addition, they can reveal where a person is and the features of this person's physical environment (Peak et al., 2010). And most important for our purposes, using the machine learning approaches described above, these data can also provide insight into psychological traits, such as an individual's personality (Stachl et al., 2020).

What can your smartphone reveal about you? Personality psychologists have identified five personality dimensions to characterize patterns of human thought, feelings, and behavior (John et al., 2008). Using data collected from smartphones, Stachl and colleagues (2020) found that random forest models could be trained to predict overarching personality traits—Openness ($r = .29$), Conscientiousness ($r = .31$), and Extraversion ($r = .37$)—and even more precise subdimensions, known as facets, such as openness to ideas ($r = .24$), ambition ($r = .26$), assertiveness ($r = .29$), and self-consciousness ($r = .32$). Moreover, the authors uncovered the behavioral signature for each of these Big Five personality traits. For example, individuals higher on openness had longer text messages. Individuals who were more conscientious were more likely to check the weather forecast at night. Extraverted folks called more people.

More unexpectedly, some of the biggest predictors of high openness were increased usage of sports news apps at night and more variation in how long the phone rang before the user answered it. Much like it did for curly fries or Harley Davidson, the machine learning algorithm identified patterns humans would likely not have anticipated. It seems our little digital companions may know more about us than they let on, and such insights could greatly advance psychological science. For those interested in learning more about research conducted using smartphone sensors, more detailed treatments of this topic can be found in Chapters 2 and 3.

Of course, our digital footprints extend far beyond the confines of our smartphones. Even our purchases, if made via credit card or online payments, can leave behind electronic records. Users of a British money management app who bravely volunteered to donate their financial transaction data to science helped researchers uncover the relationship between how we spend our money and our psychological traits (Gladstone et al., 2019).

On a 5-point scale, from (1) *not at all like me* to (5) *very much like me*, how much does the statement "I am good at resisting temptation" reflect your typical behavior? This was just one of the questions participants in this aforementioned study were asked. The researchers were interested in whether those who rated themselves higher on self-control would make different spending decisions than those who had a harder time controlling their impulses. In addition, the authors were also curious as to whether people who rate themselves as more materialistic on a self-reported scale have a different pattern of actual spending than those who report less of a hankering for luxury items. And finally, the authors also collected self-reported personality scores to see whether and how our personality manifests in our purchases. If all of these traits could be successfully predicted from purchases, that would mean that those who scored higher on each trait made different spending decisions than those who scored lower. On the other hand, if these traits had no relationship with our spending, then it would not be possible to tell who possessed more of a certain trait by simply examining what people bought, that is, the machine learning predictions would be no better than chance.

Digital records of spending decisions were categorized and clustered and then analyzed with random forest models to generate out-of-sample predictions for personality traits, materialism, and self-control. The authors found that all psychological traits could be predicted from spending behavior, though some, such as materialism ($r = .33$), were predicted more accurately than others, such as Openness ($r = .12$). This predictability suggests that there is a latent relationship between our psychological traits (as captured by self-reports) and our purchasing decisions (as captured by real financial transactions).

Examining the particular pattern of purchases associated with each trait yielded interesting insights. People high on Openness spent more money on flights and children's clothes and less on home do-it-yourself/repairs and pets. Conscientious individuals spent more on beauty treatments and personal savings and less on lunch or snacks. Those who said they were high on self-control spent more on investments, mortgage payments, and gym equipment, while those who felt they had less will power spent more of their money on snacks.

While some of these associations may seem straightforward, other results may come as more of a surprise. Medical expenses on dentists and eye specialists were one of the most strongly correlated predictors of a particular personality trait. Which would you suppose it is? Perhaps, it is most strongly associated with conscientious individuals, who are sure to get their routine check-up, or less conscientious people who have been eating too much candy or rubbing their eyes when they shouldn't. Alternatively, perhaps, it is neurotic individuals, who worry more about their health. In fact, it is none of these possibilities. It turns out that extraversion was the trait that was most strongly correlated with these medical expenses. Similarly perplexingly, buying toiletries was strongly correlated with agreeableness. Examples like this suggest that there may be some aspects to personality that we have yet to fully understand. As developed later in this chapter, such methods may not just find links that no scientist could have predicted but ultimately use those links to advance our scientific understanding of what it really means to be an extravert or an agreeable person. Thus, it seems our pattern of spending decisions can reveal the kind of people that we are. This is scientifically interesting, but it also raises important privacy concerns which will be addressed later in this chapter.

While spending data is typically not public, our posts on social media platforms, such as Twitter, often are. Although people may feel comfortable sharing this information with others, what users may not know is that the language they use can reveal things about them that they may feel less comfortable publicly disclosing.

For example, researchers have found that Facebook status updates can reveal whether a person suffers from depression, even 3 months before the depression was documented (Eichstaedt et al., 2018). While such information could help clinicians get people the treatment they need sooner, improving mental health outcomes and reducing human suffering, it also reveals the tremendous power of such technologies.

In the study, patients in the Emergency Department who consented to share their data with researchers provided Facebook status updates and electronic medical records. To predict depression, the authors used status update length, frequency of posting, temporal posting patterns, demographics, and the contents of posts. To analyze the language used in posts, they used LDA (Latent Dirichlet Allocation), a form of topic modeling common in natural language processing (Blei et al., 2001). LDA takes the words people write and automatically extracts higher order categories.

Feeding these language features (a feature is analogous to a predictor) into a model, the researchers were able to predict whether or not a person was depressed with an AUC of .69. In other words, if you randomly selected a participant who was depressed and another who was not, the computer could predict who the depressed person was 69% of the time.

What were the biggest clues researchers could use to tell if a person was depressed? Words associated with depressed mood (e.g., "tears," "cry," "pain"), loneliness (e.g., "miss"), and hostility (e.g., "hate," "ugh") were all strong indicators of depression. In addition, people who would ultimately be diagnosed as depressed were more likely to use first-person singular pronouns (e.g., "I," "me," "my"). This is consistent with a recent meta-analysis showing elevated use of first-person singular pronouns are indicative of depression (Edwards & Holtzman, 2017). They also found that depressed individuals were more likely to use language suggestive of rumination (e.g., "mind," "a lot") as well as words associated with anxiety (e.g., "scared," "worry"). Finally, the research revealed that depressed people talked more about other health-related issues, using words such as "head," "tired," "sick," "surgery," which is also consistent with past work (Simon et al., 1999). These clues' consistency with prior findings provides convergent evidence that they are valid signals of depression. Moreover, the innovative methodology used in this study could help clinicians more quickly and easily identify those who need help. Overall, this approach confirmed certain findings seen in previous studies, while providing new opportunities that had not previously been within reach.

In addition to depression, our language use can also reveal our personality. For instance, researchers have found that Facebook status updates can be used to predict personality traits (Park et al., 2015). Indeed, these automated language-based predictions converged with self-reports, informant-reports, and external validity criteria (e.g., number of friends or political attitudes), lending further credence to their veracity. Moreover, they demonstrated test-retest reliability over 6-month intervals. The authors suggested that this may enable a new way to detect personality (rather than relying on self-report questionnaires), a topic we return to later in this chapter.

Beyond the studies discussed above, researchers have also found that personality traits can be predicted from Twitter profiles (Quercia et al., 2011) and blog posts (Yarkoni, 2010), that mouse cursor movements can reveal whether a user is experiencing negative emotion (Hibbeln et al., 2017), and that how long an individual spends on a screen can shed light on their level of neuroticism (Mark et al., 2016). More recently, scientists have discovered that even images can illuminate something about our psychology. Researchers have found that the types of pictures a person likes on Flickr can reveal this individual's personality (Segalin, Perina, et al., 2017). People's own images (e.g., profile pictures) can be analyzed to identify their personality as well (Ferwerda et al., 2015; Segalin, Celli, et al., 2017). In fact, my own work and those of others shows that even a facial image itself can be used to predict psychological dispositions (Khambatta et al., 2018).

Finally, while most of the previously discussed studies involve recent digital footprints, it may also be possible to use machine learning to uncover insights from footprints left behind long ago. For instance, by analyzing two centuries of writings culled from Google Books and *The New York Times*, researchers found that linguistic positivity bias, the tendency of people to use more positive than negative words, has been decreasing over time and varies depending on the historical context (Iliev et al., 2016). This finding challenged what had been the dominant scientific view at the time—that linguistic positivity bias was a human universal unaffected by environmental influences. Using a similar approach, researchers have also found that, over the last 2 centuries, Western societies have increased their use of causal language (Iliev & Axelrod, 2016). This is theorized to be due to the heightened influence of education, science, and technology.

While certain digital footprints may provide greater insight into the minds of those who leave them than others (Gladstone et al., 2019), together, these findings have greatly augmented our understanding of human psychology. In the remainder of this chapter, the particular advantages, challenges, and future opportunities of these approaches are further explored.

ADVANTAGES OF AUTOMATED PSYCHOLOGICAL ASSESSMENTS FROM DIGITAL FOOTPRINTS

The advent of digital footprints opens many new vistas for the behavioral sciences. These new data sources provide the ability to capture naturalistic human behavior and enable a tremendous increase in power, accuracy, and representativeness.

Capturing Real Behavior in Naturalistic Contexts

In their quest to uncover the nature of our reality, scientists rely on technology to help them see farther than their natural human facilities permit. In many instances in science, it was a technological breakthrough that enabled striking new discoveries about the world. While Galileo did not invent the telescope, he innovated on its design to greatly increase its magnification (Wilde, 1995). This technological advancement enabled him to become the first person in history to bear witness to numerous astronomical phenomena. Perhaps most importantly, Galileo's observations provided strong evidence for the Copernican model, which posits that the earth revolves around the sun, thereby challenging the long-held geocentric view that the earth was the center of the universe (Gingerich, 1982).

In psychology, the new technological approaches discussed in this chapter may also yield powerful new insights. Traditionally, with the exception of behaviorism (Graham, 2019), psychology has relied heavily on questionnaires to capture a person's inner mental life (Paulhus & Vazire, 2007). While this approach undoubtedly has many strengths, it also carries significant limitations. For instance, people might not have introspective access to many of their mental traits (Nisbett & Wilson, 1977) or may actively be trying to self-present in a socially desirable way (Krosnick, 1999). For these reasons and others, many psychologists have argued for a greater focus on observing people's actual behavior in natural contexts (Baumeister et al., 2007; Boyd et al., 2020; Nave et al., 2018). However, this is typically easier said than done.

Most psychologists cannot follow around each participant with a team of eight observers as they did for *One Boy's Day* (Barker & Wright,

1951). Thus, the availability of digital devices and platforms that constantly record human behavior could serve as a great boon to psychologists (Kosinski et al., 2015; Yarkoni, 2012). By unobtrusively capturing real behavior in people's daily lives, we can get a better sense for what people are really like.

One notable illustration of the promise of this approach lies in the measurement of personality. The Big Five (the dominant personality conceptualization in modern psychology) were generated lexically, by clustering adjectives people use to describe personality traits (Goldberg, 1990). The thinking was that if a personality trait exists, then human language has likely evolved to capture it. Thus, by analyzing the words people use to describe themselves and others (e.g., wise, rebellious, warm) and how often individuals who rate themselves as high or low on one adjective also rate themselves as high or low on another, psychologists could identify higher-order patterns in traits. This has proved to be a very successful enterprise, as the personality traits derived from this approach can predict many important life outcomes, including happiness, health, spirituality, political views, the quality of one's relationships, career choice, job performance, criminal behavior, community involvement, and more (Ozer & Benet-Martínez, 2006).

Nevertheless, is everything relevant to a person's personality reflected in adjectives? Are there aspects of people's thoughts, feelings, and behavior that we do not quite have a word for? If we supplemented the linguistic lexicon with a lexicon of digital footprints, we might uncover previously unidentified aspects of personality and arrive at a better way of capturing traits (Bleidorn et al., 2017). These findings could greatly advance personality science.

As one such example, researchers found that by using digital footprints to predict personality, they could detect effects that traditional questionnaires could not (Youyou et al., 2017). While research on homophily would suggest that a person should tend to have a similar personality as their friends or spouse, empirical research using personality questionnaires has not borne out strong evidence for this. However, this may be due to the reference-group effect, the tendency of individuals to assess their personality by comparing themselves to salient social referents. In other words, if Melanie, a PhD student, is asked to rate how conscientious she is, she might compare herself with her diligent peers, such as Josh, and conclude that she is not very conscientious. However, as PhD students, both Melanie and Josh are likely to be more conscientious than the average person in the general population. Thus, because Melanie's social comparison set may not be representative of the general population, she might underestimate how conscientious she really is. As a result, if a researcher used a personality questionnaire to measure conscientious, the results would wrongly suggest that Melanie is not very similar to her peers. However, by using digital footprints (both Likes-based and language-based personality models), researchers were able to overcome the reference-group effect and show that both friends and romantic partners are indeed quite similar, more so than traditional questionnaires would suggest.

In addition, unlike questionnaires, which can be long and tedious to complete, psychometric assessments based on digital footprints can be collected with the touch of a button. Participants could agree to share data with researchers and in exchange receive instantaneous feedback on their predicted personality or other psychological traits. In addition, while psychological scales often measure in-the-moment experiences, by examining digital footprints, psychologists could measure variables longitudinally across large swaths of time. For instance, by sampling all Twitter accounts created between February 2008 and April 2009 and automatically detecting positive or negative affect in their tweets, researchers found that people from around the world follow similar temporal patterns in mood (Golder & Macy, 2011). For example, people tend to awaken in positive spirits, but negative affect increases as the day wanes on. Moreover, as the days get longer, baseline positive affect is higher, while when the days shrink, so do people's spirits.

Increased Power, Accuracy, and Representativeness

The "Big" in Big Data can provide many benefits. A common practice in psychology is for researchers to target a sample size that provides them with at least 80% power to detect a true effect (Adjerid & Kelley, 2018). While this may be an improvement over the underpowered studies of the past, it still means that 20% of the time, researchers may not detect an effect even though it exists. However, with a sample size in the thousands or even millions, researchers can be more assured that they are sufficiently powered to detect even small effects. Moreover, if an effect does not exist, with a very large sample size, researchers can be more confident it is a null effect (e.g., Kosinski, 2017).

Machine learning approaches can also enable highly accurate predictions. For instance, after training a model on data collected with smartphones, researchers were able to predict a student's GPA with an average accuracy of +/0.179 of the reported grade on a college transcript (R. Wang et al., 2015). Additionally, across the board, in an ever-increasing number of domains, machine learning models generate predictions that are more accurate than those generated by humans (Grove et al., 2000).

Computers may even be starting to understand us better than those who know us best. As one illustration, researchers have found that computer predictions based on Facebook Likes can predict people's personalities more accurately than ratings made by their friends (Youyou et al., 2015). The authors also found that these automated judgments demonstrated higher inter-rater agreement and better external validity than their human counterparts. For instance, computer-generated predictions better correlated with life outcomes, including substance use, number of friends, or physical health, than did personality ratings from friends. In addition, the more Likes the computer had for a given individual, the more accurate its predictions became. With enough Likes, it could predict people's personality traits even more accurately than a spouse could. As this work suggests, such technologies could enable psychologists to pinpoint thoughts, feelings, or behaviors with much higher precision than was previously possible.

Another important challenge psychologists typically face is that many studies rely on samples drawn from Western, educated, industrialized, rich, and democratic (WEIRD) populations that are not representative of humanity as a whole (Henrich et al., 2010). As smartphones and other digital devices further penetrate global markets, by analyzing digital footprints, researchers may be able to broaden their reach and access populations typically overlooked by traditional studies (Gosling et al., 2010).

CHALLENGES OF AUTOMATED PSYCHOLOGICAL ASSESSMENTS FROM DIGITAL FOOTPRINTS

Despite their many advantages, there are also, of course, challenges that arise from using these new tools. These include problems caused by erroneous data, reliability and validity issues, and privacy concerns.

Dealing With Messy, Inaccurate Data

As the adage goes, "On the internet, nobody knows you're a dog" (Fleishman, 2000). Against the backdrop of catfishing and misinformation campaigns, it may seem unwise for researchers to believe that people are who they say they are in the digital realm. For this reason, we may question the veracity of the digital footprints we encounter, and by extension, any psychological predictions based on them.

However, it is important to note that self-reports may also be subject to biases arising from factors such as intentional misrepresentation and social desirability (Kosinski et al., 2015). Moreover, multiple lines of converging evidence suggest that people do generally provide valid information on their digital profiles. For one thing, the high accuracies reported in previous studies could not be this high if people were entirely misrepresenting their identities online. In addition, research has shown that people's online profiles reflect their actual psychological traits rather than self-idealizations (Back et al., 2010) and that self-reports match profile information in the vast majority of cases (Kosinski et al., 2015). For these reasons, it does not appear that digital footprints are subject to significantly more misrepresentation than traditional methods. In certain instances and contexts, it may actually be harder for a participant to misrepresent themselves on their digital footprints (e.g., 6 months of Twitter posts) than in an anonymous online survey.

Nonetheless, unlike a 7-point survey, the world of digital footprints is a messy place. Researchers often have to make judgment calls, such as whether to include emojis as predictors in language models or what purchases count as toiletries. Also, as the inputs to models are less constrained than those in traditional questionnaires and subject to change at the sole discretion of the engineers of a platform, this can make a scientist's job significantly harder and increase researcher degrees of freedom. It may also affect the reliability of the model, as discussed in greater depth below.

Ensuring Reliability and Validity Over Time and Across Cultures

Liking curly fries on Facebook was once indicative of intelligence. However, after this finding became popularized in the media, it is plausible that many more individuals began liking curly fries merely to give the impression that they were intelligent, and this cue lost its diagnostic value. Even if this were not the case, would curly fries always be a valid cue to intelligence, even in 20 years? And would this be true for all cultures around the world?

While certain links between digital behavior and psychological traits may be well-grounded in past theory and empirical work (e.g., the link between extraversion and calling more people on the phone), other relationships uncovered by machine learning models may not be easily explained by past theorizing and empirical research (e.g., extraverts spending more at the eye doctor). Such unexpected relationships may be a boon for psychologists because they can teach us something new about how psychological traits manifest themselves. But they can also pose significant challenges for ensuring the generalizability of these unusual cues across different contexts.

Moreover, even at the same point in time, in the same geographic region, some digital footprints might indicate different things for different populations. For instance, if an 80-year-old woman likes Taylor Swift on Facebook, this might suggest that she scores high on the personality trait of Openness. However, if a 20-year-old indicates the same preference, it may be more normative and not as suggestive of particularly high openness (Bleidorn et al., 2017).

If a model did not take phenomena like this into account, it would perform differently for one group of people than another (e.g., overestimating or underestimating Openness), potentially leading to socially undesirable outcomes (Raghavan et al., 2020). For instance, a recommendation system on a job website may show women ads for lower paying jobs on average than the ads it shows to men (Kleinberg et al., 2017). Biases can result from a poorly trained model, but they often also stem from the data itself. A machine learning model may just be teaching itself the human biases already present in a data set (Caliskan et al., 2017), reinforcing existing social inequities. As women on average earn less money than men do (Blau & Kahn, 2000), rather than treating men and women equally, the algorithm may be learning to match women with lower paying jobs. In psychology, as the ground truth is often operationalized in terms of human ratings of psychological scales (e.g., informant ratings of personality), if raters are biased in their judgments, these biases may seep into the data and affect computational predictions (Tay et al., 2020). For example, if people who are more attractive are rated by others as having better character (Dion et al., 1972), even if this is not the case, an algorithm might also learn these erroneous relationships because it is training itself on biased human judgments.

Complicating matters further, digital platforms were not designed to maximize scientific value but to generate profits for companies and provide benefit to users (McFarland et al., 2016). To meet these goals, platforms themselves are often changing and may even be conducting their own experiments, showing certain people one type of content or feature while not doing so for others. This could influence the effectiveness of psychological instruments based on digital footprints. If a platform changes a particular feature for a particular person, this could lead the individual to alter their behavior, which would affect predictions based on this individual's digital footprints. For instance, if Google autocompletes a search query differently for one group of people than another to maximize ad revenue, these groups could have differences in their searches that are not purely the result of their own proclivities. Thus, using these differences to make inferences about people's psychology could yield erroneous findings. A platform-wide change can also be problematic. As one illustration, when Facebook status updates were first introduced, they always began with "[Person's name] is . . .," while today, they are unconstrained (Kosinski et al., 2015). This means that a language model trained on text from the early days of status updates may not work as well on newer data. If scientists did not keep themselves informed about such updates, they might mistake platform-induced changes for changes in psychological phenomena.

What can be done to address issues of reliability, validity, and consistency across contexts? It is important to note that these are not just problems facing computational social scientists. Data scientists in general struggle with patterns detected in certain data sets that cannot be found in other data sets or patterns that once existed, which have since disappeared. This phenomenon, termed concept drift, has already inspired solutions in computer science, such as combining the predictions of many different models (i.e., ensemble methods) and retraining individual models when they show signs of concept drift (Lu et al., 2019). Psychologists may benefit from adapting these solutions into their approaches.

Additionally, the field of psychology—and especially its subdiscipline of psychometrics, which is focused on topics of assessment and measurement—has provided many tools for such situations. For example, researchers could measure the test-retest reliability of their analytical approaches to ensure that they are indeed capturing patterns that persist across time (Tay et al., 2020). Additionally, researchers could engage in replication efforts to verify that their predictive models work on different samples. For instance, Iliev et al. (2016) examined both Google Books and *The New York Times* to see if the same pattern emerged regarding context-specific linguistic positivity bias.

It is important to note that many problems facing the prediction of psychological phenomena from digital footprints are often analogous to the problems scholars face when using traditional methods. Reliability and validity issues are present in traditional psychology research (John & Benet-Martínez, 2000), and cultural psychology has often noted that findings do not always translate across cultures (e.g., Markus & Kitayama, 1998). Nonetheless, it is important for psychologists to be mindful of these problems when creating and interpreting research based on digital footprints.

Privacy

There may have been a time when no one on the internet knew you were a dog. Today, however, digital platforms hold so much information on users that they would likely find ways to deduce this so that they could sell you more doggy treats. In other words, companies have a financial incentive to understand you and your needs so that they can sell you more things that match your preferences.

As one famous case example, the large American retailer Target found that they could predict whether a woman was pregnant, and even her expected delivery date, based on just her purchases (Hill, 2012). As with the other examples we've explored, some predictors were obvious (e.g., prenatal vitamins), while others were less intuitive (e.g., unscented lotion). This information allowed Target to send personalized coupons to these women for items like a crib. However, they also led many women to feel they were being spied on, and in at least one instance, caused someone else in the household who had not been told about the pregnancy to discover it. In response, Target reformed its practices by mixing pregnancy-related coupons with others to try to help pregnant women find the products they needed without making them uncomfortable.

In many ways, people have less privacy now than ever before. What they buy, where they go, what they do and more are continuously tracked by companies at scale and used to show them advertising. Targeted advertising may not always be a bad thing. People may appreciate seeing ads that are relevant to their unique psychological profile. However, as with any technology, there are likely going to be malevolent actors that will abuse the capabilities of psychological profiling to exploit weaknesses in people's character and encourage them to do things that are not in their best interest (Matz et al., 2017, 2020). Given the power of psychological profiling, several issues related to privacy are worth considering.

The distinction between public and private information becomes more nebulous in the digital realm. What the approaches above reveal is that if a model is well-trained, it may be able to predict the psychological traits of people it has never encountered before using the patterns it learned from those in its training data. In other words, if a model is trained to predict life satisfaction from tweets, it can analyze a new person's public tweets to infer this individual's life satisfaction, even though this person may not have shared their life satisfaction score. Similarly, if Facebook Likes can reveal intelligence, a person may not have to fill out an IQ scale in order for an algorithm to make an educated guess as to this individual's intellect. This is problematic because even if someone publicly shares a tweet or Facebook Like, they might not realize what additional information such sharing can reveal. Matters become even more complicated when considering that digital data often does not only pertain to one person but also to someone's friends. For instance, a photograph a person volunteers to share with researchers might contain other people who have not provided researchers with consent.

The issue of privacy involves multiple stakeholders—academics, participants/users, and technology companies. As the field evolves, it will become increasingly important for a consensus to emerge on what forms of data collection are appropriate and under what circumstances. For researchers, traditional notions of informed consent may not be well-suited to a world in which people's data can reveal more than they imagine and put others' privacy at risk. The public may certainly benefit from more open academic discussions about the power, promise, and perils of psychological profiling (Matz et al., 2020). However, in addition to discussion, action is also required.

As noted earlier, many of our spending decisions are now being tracked, and it is important for consumers to be aware that some financial companies may actually sell digital records of their purchases to other companies without their customers' awareness. The same is true of data collected on the web and from smartphones. If you do not want companies selling your data, it may be possible to opt out of this by changing an app's settings or contacting the company providing the service. For instance, many financial institutions allow users to opt out of having their purchase decisions shared with third parties. In addition, certain state and local governments, such as that of California, have passed legislation to protect consumers. In the coming years, better national consumer protections in the United States will hopefully give individuals more knowledge and agency over how their data is used. Privacy regulations, such as the European Union's General Data Protection Regulation, can help maintain people's privacy, while still allowing these technologies to benefit individuals and society.

For those interested in learning more about privacy and ethics, a more comprehensive discussion of these topics can be found in Chapter 17.

FUTURE DIRECTIONS

As these new tools are only in their infancy, their future prospects are rather boundless. In many ways, it is hard to fully foresee what may lie in store in years to come. Nevertheless, here we outline a few likely future directions that hold great promise.

Advancing Psychological Theory

Scientific theories have two overarching goals: prediction and explanation. As psychologists, we would like to predict what a person is going to do as a function of their personality or situation and also be able to explain why this occurs. These goals are not necessarily incompatible. However, the types of Big Data analytics discussed in this chapter reveal why they sometimes may be. Computational approaches often yield very accurate predictions, but due to the complexity of the algorithms used, it is often difficult to explain how these models arrived at their predictions. For example, while the random forest models that predict personality from spending behavior (Gladstone et al., 2019) or smartphone data (Stachl et al., 2020) are quite accurate, they also comprise a vast set of decision trees that would be very hard for humans to decipher and interpret. On the other hand, simpler models may be easier to comprehend, but they may not well capture the complex relationships within the data, leading to less accurate predictions. For instance, researchers found that nonlinear random forest models generally performed better than linear elastic net models when it came to both prediction accuracy and the number of criteria they could predict, suggesting that there may have been nonlinear relationships in the data that the simpler linear models failed to capture (Stachl et al., 2020).

Ultimately, there is no guarantee that a model that predicts human behavior with high accuracy will be simple enough for people to understand (Yarkoni & Westfall, 2017). Therefore, the best way to predict whether individuals will suffer from depression in the next 3 months may involve the use of a model that is so complex that even those who created it cannot fully understand how it works. While getting people the treatment they need is extremely important, there are certainly dangers to this approach. For one thing, algorithms may be biased against certain groups without anyone realizing it (Hajian et al., 2016). More broadly, if there is any mistake in the data or peculiarity in the prediction process, a complex model may not allow researchers to detect this. For instance, a convolutional neural network can be designed to diagnose pneumonia from X-rays. However, without anybody realizing it, the model could simply learn to detect whether the word "portable" is printed on the X-ray (rather than focusing on the patient) because it turns out patients who have pictures taken on a portable scanner have a different rate of pneumonia than other patients (Zech et al., 2018). While undoubtedly clever, this does not seem like a good idea. If the hospital stopped using film with the word "portable" on it, the model's predictions would suddenly be very different without doctors realizing that anything was wrong. A model that was more transparent about the cues it was using would avoid such problems. For these reasons and more, a big push in machine learning research is being made for more interpretable models (also referred to as "explainable AI"), and early efforts appear promising (Rudin, 2019).

If this endeavor is ultimately successful, this could greatly benefit psychological science. It may soon be possible to create computational theories of human behavior that are both highly accurate and relatively easy to understand. While traditional theories often require additional assumptions before they can be translated into concrete predictions (Higgins, 2004), models derived from machine learning can provide precise predictions whenever they are shown new data. This is because machine learning models have a level of mathematical formalization that is typically not found in traditional psychological theories, barring exceptions such as prospect theory (Kahneman & Tversky, 1979). If these precise models eventually become easier to interpret, they could greatly complement, and in certain cases, even replace traditional theories.

In the interim, we can use some existing approaches to gain insight into the psychological mechanisms that a black-box machine learning model is learning on its own. For example, we can examine which variables contribute most to accurate predictions (Sheetal et al., 2020; Stachl et al., 2020) or which variables are most strongly associated with an outcome (Gladstone et al., 2019; Kosinski et al., 2013). Combining machine learning models with such approaches can advance psychological theory by helping us both predict and explain human behavior.

Personalized Psychological Interventions

In addition to advancing our scientific understanding of a phenomenon, machine learning models can also help us more precisely tailor psychological interventions to each person's unique temperament. Over 50 years ago, Gordon Paul (1967), a pioneer in the field of clinical psychology, wrote that "the question towards which all outcome research should ultimately be directed is the following: What treatment, by whom, is most effective for this individual with that specific problem, and under which set of circumstances?" (p. 111). Indeed, even in the domain of physical health, the promise of personalized medicine is to "base medical decisions on individual patient characteristics, including molecular and behavioral biomarkers, rather than on population averages" (Fröhlich et al., 2018, p. 1). For social and personality psychologists, an analogous goal would be using digital footprints to predict which psychological intervention will be most effective for a particular person in a particular situation.

A small body of psychological research has already demonstrated the potential power of this approach. For instance, researchers have used Facebook Likes to infer personality and tailor persuasive appeals based on these personality inferences (e.g., appealing to those high in Openness by framing a crossword puzzle as an opportunity to exercise their creativity and imagination while appealing to those low in Openness by framing a crossword puzzle as an all-time favorite hobby that has challenged players for generations; Matz et al., 2017). Results revealed that when persuasive appeals were matched to a person's predicted personality traits, this generated up to 40% more clicks and up to 50% more purchases than mismatched or unpersonalized appeals. These numbers are only expected to increase as predictions become more and more accurate, and companies start to tailor not just the initial advertisement but the entire consumer process. While psychological targeting is often focused on helping marketers increase the efficiency of their campaigns and ultimately boost profits, these same approaches can be used by psychologists to improve people's well-being.

In addition, researchers have found that spending money on products that match one's personality can increase positive affect (Matz et al., 2016). Thus, by inferring personality traits from digital footprints, companies could not only increase their marketing efficiency and sales but also contribute to the happiness and satisfaction of their customers.

A similar approach could be applied to other psychologically relevant fields. For instance, the most prominent cause of death and disease in the world is people's behaviors (Glanz & Bishop, 2010). A large number of physical health issues (e.g., diabetes, heart disease, certain cancers) are caused or exacerbated by maladaptive behaviors (e.g., overeating, insufficient exercise, high stress, low social interaction; Barnes & Yaffe, 2011; Marx, 2002; Yusuf et al., 2020). If there was a way to persuade people to adopt healthier behaviors (e.g., stop smoking, exercise a little more), such interventions could greatly ameliorate these negative health consequences. Supporting the potential value of personalized interventions across different contexts, a meta-analysis has found that not only can persuasive appeals be effective, but they are even more effective when they are tailored (Noar et al., 2007). Thus, by automatically inferring people's demographics or psychological traits based on their digital footprints and using this information to give people the most helpful message for them, researchers and health professionals could potentially help people live longer and healthier lives.

Finally, people themselves may be wondering what their tweets say about them or whether they are likely to suffer from depression. In future work, researchers could consider providing people with automated feedback on their psychological traits, as predicted by digital footprints, to increase self-insight and enable people to intervene in their own psychology. Prior work has shown that people often enjoy learning more about themselves by examining what their digital footprints reveal about their behavior (Harari, Wang, et al., 2017). Thus, such tools would likely be well received.

However, special care is warranted when the predictions concern sensitive topics, such as mental health. Indeed, research suggests that receiving feedback could alter one's mindset and future behavior (Turnwald et al., 2019). This could lead to positive outcomes, such as if learning that they are at risk for depression encourages individuals to receive treatment sooner. However, researchers should be aware that negative outcomes may also occur and take special precautions to prevent such results. Lastly, the dissemination of psychological knowledge may alter the very behaviors on which that knowledge was derived (Gergen, 1973). Therefore, it might be necessary for researchers to retrain their models, if learning what a computer has predicted for you affects the accuracy of that very prediction.

CONCLUSION

What we do says a lot about who we are. The clues we leave behind when checking our email in the middle of the night or buying a bran muffin the next day provide glimpses into the nature of our preferences, penchants, and proclivities. These days, these clues are increasingly captured by digital platforms. With the use of advanced computational tools, our digital footprints can be automatically translated into psychological indicators, enabling others to understand us with an unprecedented level of ease and precision. This can allow scientists to gain deeper insight into the intricacies of the human mind and help them develop interventions that improve the lives of many. While these technologies do carry notable risks that warrant both discussion and action, they also hold great promise for enabling a world in which we can all be better understood.

REFERENCES

Adjerid, I., & Kelley, K. (2018). Big data in psychology: A framework for research advancement. *American Psychologist, 73*(7), 899–917. https://doi.org/10.1037/amp0000190

Back, M. D., Stopfer, J. M., Vazire, S., Gaddis, S., Schmukle, S. C., Egloff, B., & Gosling, S. D. (2010). Facebook profiles reflect actual personality, not self-idealization. *Psychological Science, 21*(3), 372–374. https://doi.org/10.1177/0956797609360756

Barker, R. G., & Wright, H. F. (1951). *One boy's day; A specimen record of behavior.* Harper.

Barnes, D. E., & Yaffe, K. (2011). The projected effect of risk factor reduction on Alzheimer's disease prevalence. *The Lancet Neurology*, *10*(9), 819–828. https://doi.org/10.1016/S1474-4422(11)70072-2

Baumeister, R. F., Vohs, K. D., & Funder, D. C. (2007). Psychology as the science of self-reports and finger movements: Whatever happened to actual behavior? *Perspectives on Psychological Science*, *2*(4), 396–403. https://doi.org/10.1111/j.1745-6916.2007.00051.x

Blau, F. D., & Kahn, L. M. (2000). Gender differences in pay. *The Journal of Economic Perspectives*, *14*(4), 75–99. https://doi.org/10.1257/jep.14.4.75

Blei, D. M., Ng, A. Y., & Jordan, M. T. (2001). Latent dirichlet allocation. *Journal of Machine Learning Research*, *3*(1), 993–1022. https://doi.org/10.5555/944919.944937

Bleidorn, W., Hopwood, C. J., & Wright, A. G. (2017). Using big data to advance personality theory. *Current Opinion in Behavioral Sciences*, *18*, 79–82. https://doi.org/10.1016/j.cobeha.2017.08.004

Boyd, R. L., Pasca, P., & Lanning, K. (2020). The personality panorama: Conceptualizing personality through big behavioural data. *European Journal of Personality*, *34*(5), 599–612. https://doi.org/10.1002/per.2254

Brown, J. S., & Duguid, P. (2017). *The social life of information*. Harvard Business Review Press.

Caliskan, A., Bryson, J. J., & Narayanan, A. (2017). Semantics derived automatically from language corpora contain human-like biases. *Science*, *356*(6334), 183–186. https://doi.org/10.1126/science.aal4230

Dion, K., Berscheid, E., & Walster, E. (1972). What is beautiful is good. *Journal of Personality and Social Psychology*, *24*(3), 285–290. https://doi.org/10.1037/h0033731

Edwards, T., & Holtzman, N. S. (2017). A meta-analysis of correlations between depression and first person singular pronoun use. *Journal of Research in Personality*, *68*, 63–68. https://doi.org/10.1016/j.jrp.2017.02.005

Eichstaedt, J. C., Smith, R. J., Merchant, R. M., Ungar, L. H., Crutchley, P., Preotiuc-Pietro, D., Asch, D. A., & Schwartz, H. A. (2018). Facebook language predicts depression in medical records. *Proceedings of the National Academy of Sciences*, *115*(44), 11203–11208. https://doi.org/10.1073/pnas.1802331115

Ferwerda, B., Schedl, M., & Tkalcic, M. (2015). Predicting personality traits with Instagram pictures. In *Proceedings of the 3rd Workshop on Emotions and Personality in Personalized Systems 2015* (pp. 7–10). Association for Computing Machinery. https://doi.org/10.1145/2809643.2809644

Fleishman, G. (2000, December 14). Cartoon captures spirit of the Internet. *The New York Times*. https://www.nytimes.com/2000/12/14/technology/cartoon-captures-spirit-of-the-internet.html

Fröhlich, H., Balling, R., Beerenwinkel, N., Kohlbacher, O., Kumar, S., Lengauer, T., Maathuis, M. H., Moreau, Y., Murphy, S. A., Przytycka, T. M., Rebhan, M., Röst, H., Schuppert, A., Schwab, M., Spang, R., Stekhoven, D., Sun, J., Weber, A., Ziemek, D., & Zupan, B. (2018). From hype to reality: Data science enabling personalized medicine. *BMC Medicine*, *16*(1), 150. https://doi.org/10.1186/s12916-018-1122-7

Gergen, K. J. (1973). Social psychology as history. *Journal of Personality and Social Psychology*, *26*(2), 309–320. https://doi.org/10.1037/h0034436

Gingerich, O. (1982). The Galileo affair. *Scientific American*, *247*(2), 132–143. https://doi.org/10.1038/scientificamerican0882-132

Gladstone, J. J., Matz, S. C., & Lemaire, A. (2019). Can psychological traits be inferred from spending? Evidence from transaction data. *Psychological Science*, *30*(7), 1087–1096. https://doi.org/10.1177/0956797619849435

Glanz, K., & Bishop, D. B. (2010). The role of behavioral science theory in development and implementation of public health interventions. *Annual Review of Public Health*, *31*(1), 399–418. https://doi.org/10.1146/annurev.publhealth.012809.103604

Goldberg, L. R. (1990). An alternative "description of personality": The Big-Five factor structure. *Journal of Personality and Social Psychology*, *59*(6), 1216–1229. https://doi.org/10.1037/0022-3514.59.6.1216

Golder, S. A., & Macy, M. W. (2011). Diurnal and seasonal mood vary with work, sleep, and daylength across diverse cultures. *Science*, *333*(6051), 1878–1881. https://doi.org/10.1126/science.1202775

Gosling, S. D., Sandy, C. J., John, O. P., & Potter, J. (2010). Wired but not WEIRD: The promise of the Internet in reaching more diverse samples. *Behavioral and Brain Sciences*, *33*(2–3), 94–95. https://doi.org/10.1017/S0140525X10000300

Graham, G. (2019). Behaviorism. In E. N. Zalta (Ed.), *The Stanford encyclopedia of philosophy*. Metaphysics Research Lab, Stanford University.

Grove, W. M., Zald, D. H., Lebow, B. S., Snitz, B. E., & Nelson, C. (2000). Clinical versus mechanical prediction: A meta-analysis. *Psychological Assessment*, *12*(1), 19–30. https://doi.org/10.1037/1040-3590.12.1.19

Haenlein, M., & Kaplan, A. (2019). A brief history of artificial intelligence: On the past, present, and future of artificial intelligence. *California Management Review*, *61*(4), 5–14. https://doi.org/10.1177/0008125619864925

Hajian, S., Bonchi, F., & Castillo, C. (2016). Algorithmic bias: From discrimination discovery to fairness-aware data mining. In *Proceedings of the 22nd ACM SIGKDD International Conference on Knowledge Discovery and Data Mining* (pp. 2125–2126). Association for Computing Machinery. https://doi.org/10.1145/2939672.2945386

Harari, G. M., Müller, S. R., Aung, M. S., & Rentfrow, P. J. (2017). Smartphone sensing methods for studying behavior in everyday life. *Current Opinion in Behavioral Sciences, 18*, 83–90. https://doi.org/10.1016/j.cobeha.2017.07.018

Harari, G. M., Müller, S. R., Stachl, C., Wang, R., Wang, W., Bühner, M., Rentfrow, P. J., Campbell, A. T., & Gosling, S. D. (2020). Sensing sociability: Individual differences in young adults' conversation, calling, texting, and app use behaviors in daily life. *Journal of Personality and Social Psychology, 119*(1), 204–228. https://doi.org/10.1037/pspp0000245

Harari, G. M., Wang, W., Müller, S. R., Wang, R., & Campbell, A. T. (2017). Participants' compliance and experiences with self-tracking using a smartphone sensing app. In *Proceedings of the 2017 ACM International Joint Conference on Pervasive and Ubiquitous Computing and Proceedings of the 2017 ACM International Symposium on Wearable Computers* (pp. 57–60). Association for Computing Machinery.

Henrich, J., Heine, S. J., & Norenzayan, A. (2010). The weirdest people in the world? *Behavioral and Brain Sciences, 33*(2–3), 61–83. https://doi.org/10.1017/S0140525X0999152X

Hibbeln, M., Jenkins, J. L., Schneider, C., Valacich, J. S., & Weinmann, M. (2017). How is your user feeling? *Management Information Systems Quarterly, 41*(1), 1–21. https://doi.org/10.25300/MISQ/2017/41.1.01

Higgins, E. T. (2004). Making a theory useful: Lessons handed down. *Personality and Social Psychology Review, 8*(2), 138–145. https://doi.org/10.1207/s15327957pspr0802_7

Hill, K. (2012). How Target figured out a teen girl was pregnant before her father did. *Forbes.* https://www.forbes.com/sites/kashmirhill/2012/02/16/how-target-figured-out-a-teen-girl-was-pregnant-before-her-father-did/?sh=31870d826668

Iliev, R., & Axelrod, R. (2016). Does causality matter more now? Increase in the proportion of causal language in English texts. *Psychological Science, 27*(5), 635–643. https://doi.org/10.1177/0956797616630540

Iliev, R., Hoover, J., Dehghani, M., & Axelrod, R. (2016). Linguistic positivity in historical texts reflects dynamic environmental and psychological factors. *Proceedings of the National Academy of Sciences, 113*(49), E7871–E7879. https://doi.org/10.1073/pnas.1612058113

John, O. P., & Benet-Martínez, V. (2000). Measurement: Reliability, construct validation, and scale construction. In H. T. Reis & C. M. Judd (Eds.), *Handbook of research methods in social and personality psychology* (pp. 339–369). Cambridge University Press.

John, O. P., Naumann, L. P., & Soto, C. J. (2008). Paradigm shift to the integrative Big Five trait taxonomy: History, measurement, and conceptual issues.

In O. P. John, R. W. Robins, & L. A. Pervin (Eds.), *Handbook of personality: Theory and research* (pp. 114–158). Guilford Press.

Jordan, M. I., & Mitchell, T. M. (2015). Machine learning: Trends, perspectives, and prospects. *Science, 349*(6245), 255–260. https://doi.org/10.1126/science.aaa8415

Kahneman, D., & Tversky, A. (1979). Prospect theory: An analysis of decision under risk. *Econometrica, 47*(2), 263–292. https://doi.org/10.2307/1914185

Khambatta, P., Carr, E. W., Schroeder, J., & Waytz, A. (2018). Digital impressions: Psychological mechanisms and societal implications. In S. Taneja (Ed.), *Academy of Management Proceedings, 2018*(1), 12754. https://doi.org/10.5465/AMBPP.2018.12754symposium

Kleinberg, J., Mullainathan, S., & Raghavan, M. (2017). Inherent trade-offs in the fair determination of risk scores. *Leibniz international proceedings in informatics. LIPIcs, 67*, 1–23. Schloss Dagstuhl—Leibniz-Zentrum fuer Informatik. https://doi.org/10.4230/LIPIcs.ITCS.2017.43

Kosinski, M. (2017). Facial width-to-height ratio does not predict self-reported behavioral tendencies. *Psychological Science, 28*(11), 1675–1682. https://doi.org/10.1177/0956797617716929

Kosinski, M., Matz, S. C., Gosling, S. D., Popov, V., & Stillwell, D. (2015). Facebook as a research tool for the social sciences: Opportunities, challenges, ethical considerations, and practical guidelines. *American Psychologist, 70*(6), 543–556. https://doi.org/10.1037/a0039210

Kosinski, M., Stillwell, D., & Graepel, T. (2013). Private traits and attributes are predictable from digital records of human behavior. *Proceedings of the National Academy of Sciences, 110*(15), 5802–5805. https://doi.org/10.1073/pnas.1218772110

Krosnick, J. A. (1999). Survey research. *Annual Review of Psychology, 50*(1), 537–567. https://doi.org/10.1146/annurev.psych.50.1.537

Lu, J., Liu, A., Dong, F., Gu, F., Gama, J., & Zhang, G. (2019). Learning under concept drift: A review. *IEEE Transactions on Knowledge and Data Engineering, 31*(12), 2346–2363. https://doi.org/10.1109/TKDE.2018.2876857

Mark, G., Iqbal, S. T., Czerwinski, M., Johns, P., & Sano, A. (2016). Neurotics can't focus: An in situ study of online multitasking in the workplace. In *Proceedings of the 2016 CHI Conference on Human Factors in Computing Systems* (pp. 1739–1744). Association for Computing Machinery. https://doi.org/10.1145/2858036.2858202

Markus, H. R., & Kitayama, S. (1998). The cultural psychology of personality. *Journal of Cross-Cultural Psychology, 29*(1), 63–87. https://doi.org/10.1177/0022022198291004

Marr, B. (2018). How much data do we create every day? The mind-blowing stats everyone should read. *Forbes.* https://www.forbes.com/sites/bernardmarr/

2018/05/21/how-much-data-do-we-create-every-day-the-mind-blowing-stats-everyone-should-read/#4b3ad3a660ba

Marx, J. (2002). Unraveling the causes of diabetes. *Science, 296*(5568), 686–689. https://doi.org/10.1126/science.296.5568.686

Matz, S. C., Appel, R. E., & Kosinski, M. (2020). Privacy in the age of psychological targeting. *Current Opinion in Psychology, 31*(2), 116–121. https://doi.org/10.1016/j.copsyc.2019.08.010

Matz, S. C., Gladstone, J. J., & Stillwell, D. (2016). Money buys happiness when spending fits our personality. *Psychological Science, 27*(5), 715–725. https://doi.org/10.1177/0956797616635200

Matz, S. C., Kosinski, M., Nave, G., & Stillwell, D. J. (2017). Psychological targeting as an effective approach to digital mass persuasion. *Proceedings of the National Academy of Sciences, 114*(48), 12714–12719. https://doi.org/10.1073/pnas.1710966114

McFarland, D. A., Lewis, K., & Goldberg, A. (2016). Sociology in the era of Big Data: The ascent of forensic social science. *The American Sociologist, 47*(1), 12–35. https://doi.org/10.1007/s12108-015-9291-8

Mitchell, T. M. (2006). *The discipline of machine learning* (CMU ML-06 108).

Myers, D. G. (2000). The funds, friends, and faith of happy people. *American Psychologist, 55*(1), 56–67. https://doi.org/10.1037/0003-066X.55.1.56

Nave, C. S., Feeney, M. G., & Furr, R. M. (2018). Behavioral observation in the study of personality and individual differences. In V. Zeigler-Hill & T. K. Shackelford (Eds.), *The SAGE handbook of personality and individual differences* (pp. 317–340). Sage Publications Ltd. https://doi.org/10.4135/9781526451163.n15

Nisbett, R. E., & Wilson, T. D. (1977). Telling more than we can know: Verbal reports on mental processes. *Psychological Review, 84*(3), 231–259. https://doi.org/10.1037/0033-295X.84.3.231

Noar, S. M., Benac, C. N., & Harris, M. S. (2007). Does tailoring matter? Meta-analytic review of tailored print health behavior change interventions. *Psychological Bulletin, 133*(4), 673–693. https://doi.org/10.1037/0033-2909.133.4.673

Ozer, D. J., & Benet-Martínez, V. (2006). Personality and the prediction of consequential outcomes. *Annual Review of Psychology, 57*(1), 401–421. https://doi.org/10.1146/annurev.psych.57.102904.190127

Park, G., Schwartz, H. A., Eichstaedt, J. C., Kern, M. L., Kosinski, M., Stillwell, D. J., Ungar, L. H., & Seligman, M. E. P. (2015). Automatic personality assessment through social media language. *Journal of Personality and Social Psychology, 108*(6), 934–952. https://doi.org/10.1037/pspp0000020

Paul, G. L. (1967). Strategy of outcome research in psychotherapy. *Journal of Consulting Psychology, 31*(2), 109–118. https://doi.org/10.1037/h0024436

Paulhus, D. L., & Vazire, S. (2007). The self-report method. In R. W. Robins, R. C. Fraley, & R. F. Krueger (Eds.), *Handbook of research methods in personality psychology* (pp. 224–239). Guilford Press.

Peak, J., Kim, J., & Govindan, R. (2010). Energy-efficient rate-adaptive GPS-based positioning for smartphones categories and subject descriptors. In *MobiSys '10: Proceedings of the 8th International Conference on Mobile Systems, Applications, and Services* (pp. 299–314). Association for Computing Machinery. https://doi.org/10.1145/1814433.1814463

Quercia, D., Kosinski, M., Stillwell, D., & Crowcroft, J. (2011). Our Twitter profiles, our selves: Predicting personality with Twitter. In *Proceedings of the 2011 IEEE International Conference on Privacy, Security, Risk, and Trust, and IEEE International Conference on Social Computing* (pp. 180–185). IEEE. https://doi.org/10.1109/PASSAT/SocialCom.2011.26

Raghavan, M., Barocas, S., Kleinberg, J., & Levy, K. (2020). Mitigating bias in algorithmic hiring: Evaluating claims and practices. In *FAT* 2020: Proceedings of the 2020 Conference on Fairness, Accountability, and Transparency* (pp. 469–481). Association for Computing Machinery. https://doi.org/10.1145/3351095.3372828

Reinsel, D., Gantz, J., & Rydning, J. (2018). *The digitization of the world: From edge to core* (IDC White Paper #US44413318). International Data Corporation (IDC). https://resources.moredirect.com/white-papers/idc-report-the-digitization-of-the-world-from-edge-to-core

Rudin, C. (2019). Stop explaining black box machine learning models for high stakes decisions and use interpretable models instead. *Nature Machine Intelligence, 1*(5), 206–215. https://doi.org/10.1038/s42256-019-0048-x

Segalin, C., Celli, F., Polonio, L., Kosinski, M., Stillwell, D., Sebe, N., Cristani, M., & Lepri, B. (2017). What your Facebook profile picture reveals about your personality. In *MM '17: Proceedings of the 2017 ACM International Conference on Multimedia* (pp. 460–468). Association for Computing Machinery. https://doi.org/10.1145/3123266.3123331

Segalin, C., Perina, A., Cristani, M., & Vinciarelli, A. (2017). The pictures we like are our image: Continuous mapping of favorite pictures into self-assessed and attributed personality traits. *IEEE Transactions on Affective Computing, 8*(2), 268–285. https://doi.org/10.1109/TAFFC.2016.2516994

Sheetal, A., Feng, Z., & Savani, K. (2020). Using machine learning to generate novel hypotheses: Increasing optimism about COVID-19 makes people less willing to justify unethical behaviors. *Psychological Science, 31*(10), 1222–1235. https://doi.org/10.1177/0956797620959594

Simon, G. E., VonKorff, M., Piccinelli, M., Fullerton, C., & Ormel, J. (1999). An international study of the relation between somatic symptoms and depression.

The New England Journal of Medicine, 341(18), 1329–1335. https://doi.org/10.1056/NEJM199910283411801

Stachl, C., Au, Q., Schoedel, R., Gosling, S. D., Harari, G. M., Buschek, D., Völkel, S. T., Schuwerk, T., Oldemeier, M., Ullmann, T., Hussmann, H., Bischl, B., & Bühner, M. (2020). Predicting personality from patterns of behavior collected with smartphones. *Proceedings of the National Academy of Sciences, 117*(30), 17680–17687. https://doi.org/10.1073/pnas.1920484117

Statista. (2020a). *Number of daily active Facebook users worldwide as of 2nd quarter 2020.* https://www.statista.com/statistics/346167/facebook-global-dau/

Statista. (2020b). *Number of smartphone users worldwide from 2016 to 2021.* https://www.statista.com/statistics/330695/number-of-smartphone-users-worldwide/

Tay, L., Woo, S. E., Hickman, L., & Saef, R. M. (2020). Psychometric and validity issues in machine learning approaches to personality assessment: A focus on social media text mining. *European Journal of Personality, 34*(5), 826–844. https://doi.org/10.1002/per.2290

Tsamardinos, I., Greasidou, E., & Borboudakis, G. (2018). Bootstrapping the out-of-sample predictions for efficient and accurate cross-validation. *Machine Learning, 107*(12), 1895–1922. https://doi.org/10.1007/s10994-018-5714-4

Turnwald, B. P., Goyer, J. P., Boles, D. Z., Silder, A., Delp, S. L., & Crum, A. J. (2019). Learning one's genetic risk changes physiology independent of actual genetic risk. *Nature Human Behaviour, 3*(1), 48–56. https://doi.org/10.1038/s41562-018-0483-4

Wang, R., Harari, G. M., Hao, P., Zhou, X., & Campbell, A. T. (2015). SmartGPA: How smartphones can assess and predict academic performance of college students. In *UbiComp '15: Proceedings of the 2015 ACM International Joint Conference on Pervasive and Ubiquitous Computing* (pp. 295–306). Association for Computing Machinery. https://doi.org/10.1145/2750858.2804251

Wang, Y., & Kosinski, M. (2017). Deep neural networks are more accurate than humans at detecting sexual orientation from facial images. *Journal of Personality and Social Psychology, 114*(2), 246–257. https://doi.org/10.1037/pspa0000098

Wilde, M. (1995). *Galileo's telescope.* The Galileo Project. http://galileo.rice.edu/bio/narrative_6.html

Yarkoni, T. (2010). Personality in 100,000 words: A large-scale analysis of personality and word use among bloggers. *Journal of Research in Personality, 44*(3), 363–373. https://doi.org/10.1016/j.jrp.2010.04.001

Yarkoni, T. (2012). Psychoinformatics: New horizons at the interface of the psychological and computing sciences. *Current Directions in Psychological Science, 21*(6), 391–397. https://doi.org/10.1177/0963721412457362

Yarkoni, T., & Westfall, J. (2017). Choosing prediction over explanation in psychology: Lessons from machine learning. *Perspectives on Psychological Science, 12*(6), 1100–1122. https://doi.org/10.1177/1745691617693393

Youyou, W., Kosinski, M., & Stillwell, D. (2015). Computer-based personality judgments are more accurate than those made by humans. *Proceedings of the National Academy of Sciences, 112*(4), 1036–1040. https://doi.org/10.1073/pnas.1418680112

Youyou, W., Stillwell, D., Schwartz, H. A., & Kosinski, M. (2017). Birds of a feather do flock together: Behavior-based personality-assessment method reveals personality similarity among couples and friends. *Psychological Science, 28*(3), 276–284. https://doi.org/10.1177/0956797616678187

Yusuf, S., Joseph, P., Rangarajan, S., Islam, S., Mente, A., Hystad, P., Brauer, M., Kutty, V. R., Gupta, R., Wielgosz, A., AlHabib, K. F., Dans, A., Lopez-Jaramillo, P., Avezum, A., Lanas, F., Oguz, A., Kruger, I. M., Diaz, R., Yusoff, K., . . . Dagenais, G. (2020). Modifiable risk factors, cardiovascular disease, and mortality in 155 722 individuals from 21 high-income, middle-income, and low-income countries (PURE): A prospective cohort study. *The Lancet, 395*(10226), 795–808. https://doi.org/10.1016/S0140-6736(19)32008-2

Zech, J. R., Badgeley, M. A., Liu, M., Costa, A. B., Titano, J. J., & Oermann, E. K. (2018). Variable generalization performance of a deep learning model to detect pneumonia in chest radiographs: A cross-sectional study. *PLOS Medicine, 15*(11), e1002683. https://doi.org/10.1371/journal.pmed.1002683

2

Saying More Than We Know: How Language Provides a Window Into the Human Psyche

M. Asher Lawson and Sandra C. Matz

Language is part of our human DNA. We are born prewired with the ability to acquire highly complex language skills, or as the American linguist Noam Chomsky put it, humans appear to have a "language instinct" (Chomsky, 1986). Language—both spoken and written—allows us to effectively communicate our thoughts, feelings, hopes, concerns, or intentions to other members of our species. It fosters cooperation and collaboration in a social world, and ultimately allows us to create a shared reality with others about the world we inhabit (Lau et al., 2001). Although we do not all speak the same language, the ability to express ourselves through words is a "human universal" that exists in every known culture around the world.

However, language is not just universal, it is also pervasive in our everyday lives. The average person spends the majority of their day (50%–80%) engaged in some kind of communication (Klemmer &

Snyder, 1972). We might greet a stranger in the subway, engage in a deep conversation with one of our close friends, write an email to a colleague, leave a voice message for our children, note down our thoughts in a diary, or post an update on Twitter. Given its pervasiveness in our daily lives, language captures important insights about who we are and how we relate to others (Pennebaker & Graybeal, 2001). As we show in this chapter, the words that people use can predict their intimate psychological traits (Park et al., 2015; Schwartz et al., 2013), the success of their romantic relationships (Ireland et al., 2011), the extent to which they are a good fit for the organizations they work in (Srivastava et al., 2018), and their likelihood of suffering from mental health problems (Pennebaker & King, 1999).

Although language has been an integral part of people's social life for millennia, the scientific inquiry into language as a window into people's psychology has only recently become popular among psychologists. A search for "natural language processing" in the category "psychology" on the Web of Science (https://clarivate.com/webofsciencegroup/solutions/web-of-science/) shows that there were only 146 articles mentioning the term in 2000. Over the years, this number has risen continuously, with 845 articles mentioning the term in 2019 (+479% from 2000). The main driver for this development is the ease with which both spoken and written language can be captured, stored, and processed in today's digital world (Boyd & Pennebaker, 2015). While studying human behavior and language in daily life was a painstakingly time-consuming and expensive exercise only 20 years ago, our intimate relationship with technology today makes it possible to capture language and its relationship to important psychological outcomes inexpensively, unobtrusively, and at scale (Boyd, 2017; Boyd & Pennebaker, 2015). In fact, many of the activities outlined in the previous section are mediated by digital devices that can keep track of every step we take and word we say. Whether it is the emails we send, the tweets we post, or our work meetings over Zoom that get recorded straight to the cloud, there is a growing "digital paper trail" of our activities that captures our use of language.

Social media platforms—such as Facebook, Twitter, and Reddit—are among the most commonly used data sources when it comes to studying the natural language of millions of people around the world. These services create a stage for users to communicate who they are, what they do, who they interact with, what they dream about, how they feel about the current state of the world, and more. In contrast to traditional psychological research, these expressions of one's inner mental state are unsolicited by scientists and represent an ecologically valid reflection of people's daily experiences (Boyd & Pennebaker, 2015). As such, they hold tremendous value for psychologists to better understand and predict human behavior.

In this chapter, we discuss the potential of using text analyses in psychological research. We first survey different analytical strategies for analyzing text, ranging all the way from counting words to capturing meaning through vectors in multidimensional space. We subsequently provide an overview of how text analyses have been used in different psychological subdisciplines—personality, health and clinical, social, organizational, consumer, developmental, and cognitive psychology—to both generate and empirically investigate psychological theories.

APPROACHES TO ANALYZING TEXT

There are many different approaches to analyzing text, and the range continues to expand. In conjunction with the revolution of data availability, another revolution occurred in the power and sophistication of the techniques that are available to gain psychological insights from text data. Early text analysis required the use of human "coders"—in academia often research assistants—people tasked with reading through text data and noting down particular features (e.g., the valence of the writing), or the occurrence of specific words (e.g., all words expressing doubt). This technique is still used today when it comes to analyzing smaller data sets because manual coding has the advantage of being able to harness human judgment to detect nuance in meaning and atypical expressions of common phenomena.

One classic study on the topic of rumor (Allport & Postman, 1947) provided participants with a drawing containing a White man and a Black man, where the White man was holding a razor. Allport and Postman asked participants to retell what was happening in the drawing in a chain of repeated tellings (sometimes called a broken telephone). In these chains of six to seven retellings, at some point in over half of the chains the razor shifted from being held by the White man to the Black man. Whilst analyzing this text might be relatively simple for a human coder—if the participants communicated clearly, it should be easy to tell who was holding the razor—it could be difficult to ascertain computationally. Suppose some participants referred to the people in the drawing as the person on the left and the person on the right, or the person wearing a hat versus the person wearing a sweatshirt. To write an automated script that could effectively capture all of these possibilities would require planning many different contingencies. Herein lies the strength of human coders.

However, there are many instances that require analyzing text data on a much greater scale than is feasible with the help of human coders. Modern data sources, such as the stream of Twitter users' tweets, produce such large quantities of data at such velocity (hundreds of millions of tweets a day) that attempting to employ human coders to analyze their sentiment or the presence of abusive content is simply not feasible. In addition to challenges related to scale, the use of human coders is also associated with the potential for individual biases and inconsistencies. The most common way to address this second possible challenge—computing interrater reliabilities—ensures some degree of consistency among the human coders employed (Gisev et al., 2013; Tinsley & Weiss, 1975). Yet, this measure is unable to detect whether the coders are *all* systematically biased. These two critical shortcomings of human coders—the labor requirement and potential for inconsistency and bias—are addressed by the use of computer-aided textual analysis (CATA). The use of computers to automate text analysis processes greatly expanded the range and scale of insights researchers could gain from text. As computational power and algorithmic knowledge continue to expand, so too does the depth and utility of these insights.

In order to use text analysis to study human psychology, we must consider the process of language generation and word choice. If we aim to learn something about individual-level variation in experienced uncertainty from people's use of qualifiers (e.g., "might," "may," "could"), for example, we must believe that individuals choose to use these words (either consciously or subconsciously) to reflect their inner mental life. This inner mental life can be a reflection of both individual dispositions (e.g., personality) and the situational context a person is in. Think of a presidential address for example. Part of the speech will be a reflection of the president's character, whether written directly by the president or by a speech writer. Yet, the speech is also a reflection of the situation in that it reflects the seriousness and authority associated with the office, as well as the particular context the speech was written for. The same reasoning can be applied to written language—certain documents are likely to offer the author greater degrees of freedom to express themselves. Such documents may display stronger relationships with the author's psychology than other documents. Considering to what extent a person's language is likely to be a reflection of their own dispositions or their situational constraints is important for deciding what insight one can draw from a particular text source. Regardless of what methodology is implemented to analyze a specific text source, researchers have to ask themselves whether the source makes it likely for people to consciously or subconsciously express themselves. In the following sections, we introduce three popular approaches to text analysis—frequency-based methods, topic modeling, and word embeddings—that each offer unique opportunities and challenges (for a recent review, see Dehghani & Boyd, in press).

Frequency-Based Methods

The original and most common application of automated text analysis uses computers to count how often certain words occur in a given text. Whereas looking for all the instances of psychologically meaningful words (e.g., "worry") might take a human coder many hours, a computer can iterate through a large corpus of text in minutes or even seconds. This type of analysis can be traced back to Father Roberto Busa, a Roman Catholic priest who handwrote 10,000 index cards that contained uses of the word "in" or words related to "in" from the Latin work of Thomas Aquinas (Busa, 1980) to learn about the concept of presence. The analytical strategy employed in frequency-based approaches is relatively simple: Count the number of occurrences of specific words and then normalize by how many words there are in the section of text to obtain the relative word frequencies. Despite its simplicity, this approach is extremely powerful and versatile. For example, in 1963, the two statisticians Frederick Mosteller and David Wallace used frequency-based text analysis to estimate who wrote the Federalist papers that had been published anonymously (Mosteller & Wallace, 1963).

One of the dominant approaches to frequency-based text analysis is to count specific words that indicate the *content* of the text being analyzed. For example, when looking to distinguish between company press releases that focus on financial performance versus employee well-being, one could compare the relative frequency of occurrence of the words "profit" and "happy." Following this strategy, one could analyze publicly available press releases to quantify differences across organizations in their relative focus on performance versus well-being. These differences might be indicative of different organizational values and cultures, job satisfaction levels among employees, and future responses to threats to profitability.

One particular form of frequency-based content analysis that has gained a lot of attention in computer science and the social sciences is rule-based sentiment analysis (e.g., Hutto & Gilbert, 2014). In its most basic form, this kind of sentiment analysis involves counting the relative occurrence of positive (e.g., "good") and negative (e.g., "bad") words, and inferring from these occurrences whether a section of text had a positive or negative tone. This technique has wide-ranging and important applications—for example, a company launching a new product could scrape all of the tweets that use a hashtag related to this product and perform sentiment analysis on these tweets. This would give the company a measure of how well the product was received by its customers in real time, which could help them to estimate consumer demand, and plan their production schedule.

However, rule-based sentiment analysis can only be as sophisticated as the rules you apply. Consider all of the negations people use in natural language (e.g., "this was not good"). Only searching for simple features such as the counts of single words (e.g., "good") will miss these negations, and add noise to the data. As the example shows, such frequency-based text analysis will inevitably lead to errors in the interpretation of text that arise from a lack of context. Yet, as one of the founders of frequency-based text analysis, James Pennebaker (2013), noted, "We have now run enough studies to determine that statistically, it is usually correct" (p. 9). In other words, on aggregate, frequency-based text analysis is generally quite accurate, even though there is a probability the method misinterprets individual words. There are also more advanced forms of sentiment analysis that have emerged to address some of these weaknesses, both through superior feature crafting (e.g., multiword phrases, inductively learning features from the data), and by using more sophisticated classification algorithms (e.g., naive Bayes, k-nearest neighbors, support-vector machines, neural network logistic regression). There are many excellent reviews in the area (see Feldman, 2013; Soleymani et al., 2017; Zhang et al., 2018).

While a large proportion of the literature is focused on counting the occurrence of content words, research has found that the seemingly unimportant function words that make up much of our language also offer psychological insights. In his book *The Secret Life of Pronouns: What Our Words Say About Us*, James Pennebaker (2013) discussed how people's use of particles (e.g., pronouns, conjunctions, auxiliary verbs) can reveal insights regarding emotional state, social identity, and cognitive styles (see also Pennebaker et al., 2003, and the section Applications of Text Analysis in Psychological Research in this chapter for more examples and details). While function words alone do not convey content, they can nevertheless reveal a lot about the person using them. The frequency with which people use first-person pronouns such as "I," "me" or "myself," for example, has been related to an elevated risk of suffering from depression (Rude et al., 2004). Hence, although function words (or style words) appear unassuming and are often overlooked by readers, they hold great value for studying people's linguistic styles as a window into their inner mental life.

Notably, these function words are often removed or downsampled in more complex algorithms, such as latent Dirichlet allocation (LDA) and word2vvec, because they appear too frequently to distinguish between instances (Blei et al., 2003; Mikolov, Sutskever, et al., 2013). Furthermore, when inductively generating insights from the data, nuances in the use of more interpretable words can be swamped by these commonplace aspects of speech. Yet, when combined with strong linguistic and disciplinary knowledge, differences in the use of these common words can be highly informative. Campbell and Pennebaker (2003) analyzed the relationship between respondents' writing style and the frequency of physician visits in the ensuing periods. They found that flexibility in the use of common words (especially pronouns) when processing trauma was related with positive health outcomes. Similarly, people's choices of pronouns—e.g., choosing whether to reference the self ("I") or the group ("we")—can inform us about the strength of their identification with a group, their relative self or other focus, and more.

The most commonly used software for frequency-based text analysis with regard to both content and function word is Linguistic Inquiry and Word Count (LIWC, pronounced "Luke"; Pennebaker, Booth, et al., 2015; Pennebaker, Boyd, et al., 2015). This software allows users to extract different aspects of language through a user-friendly interface that does not require any coding or programming skills. LIWC is based on *dictionaries*, sets of words that are associated with specific language categories reflective of both the content of a text (e.g., words related to work, money, and religion) and its linguistic style (e.g., authenticity, present focus, and emotionality). The software is widely used among social scientists and at the time of writing has been cited nearly 6,000 times on Google Scholar.

Limitations of Frequency-Based Analysis

Frequency-based analysis has the potential to offer intriguing insights into human psychology. However, the method is entirely dependent on a user specifying *what* to search for and therefore requires high levels of domain expertise. Which words should be included in a dictionary measuring the psychological concept of loneliness? When using existing software such as LIWC, much of this work has already been done (e.g., the latest version of LIWC2015 contains over 100 dictionaries), and the dictionaries' relations to the psychological constructs they aim to capture have been validated in various samples. However, many of the idiosyncratic psychological constructs researchers are interested in might not be represented in existing software, leaving the task of creating valid dictionaries to the researchers themselves.

The process of selecting words for a particular dictionary involves a number of choices and some degree of arbitrariness. Herein lies a major limitation of frequency-based approaches. Researchers have to specify relationships between words and psychological constructs a priori, based on their domain expertise and own understanding of the construct, which can lead to systematic biases. There are many potential pitfalls. For example, researchers might include words that might be theoretically associated with a given construct but that are not indicative of a certain trait when used in people's everyday language (e.g., the word "demagogue" is conceptually related to authoritarianism, but it is unlikely that a person who uses the word would have a higher likelihood of subscribing to such an ideology). Moreover, researchers are likely to miss out on important aspects of language that could be highly indicative of certain psychological constructs but have not yet been identified in the literature. Errors of omission are likely to occur in dictionary construction. If these dictionaries are uncritically adopted by other researchers, these errors can propagate to create streams of spurious or misinterpreted results.

A second major weakness of frequency-based analysis is its neglect of the semantic structure of sentences. Here it is necessary to distinguish between words and tokens. A token is a nonunique instance of a word; the sentence "I think I am learning" has five tokens, but only four words. Frequency-based analysis views text as a matrix (known as document-term matrix), where the strings of tokens that form language are fully summarized by two columns—one containing the unique words, and one for the frequency of each word's occurrence. By viewing documents in this way, many other aspects of language are lost. While there are some additional insights that can be derived by looking at the combination of words in a sentence (e.g., the average complexity of the language can be calculated using the Flesch–Kincaid Grade Level [Kincaid et al., 1975]), other aspects of meaning cannot be recovered when using frequency-based approaches. It is impossible, for example, to detect sarcasm, irony, or even negations, because the context of the words is lost in the analytical process.

Topic Modeling

While frequency-based analyses based on domain-knowledge and pre-defined dictionaries are easy to implement, it is oftentimes advantageous to make use of the full complexity of text by letting algorithms learn patterns in the text inductively in an unstructured way. Over the past decade, topic models have emerged as the method of choice when using a data-driven approach to understanding text. A topic model is a statistical model that discovers semantic structure in collections of text documents. There are many different types of topic models, including probabilistic latent semantic analysis (PLSA; Hofmann, 1999) and hierarchical latent tree analysis (HLTA; Liu et al., 2014), but the most commonly used topic model is LDA, and that will be our primary focus here (Blei et al., 2003). LDA is not strictly limited to the analysis of text documents, but given the nature of this chapter, our description of LDA focuses on the context of text analyses. For other uses, see Kosinski et al. (2016).

Suppose you wanted to analyze 1,000 news stories published by *The Washington Post*. Each document contains many different words, each with varying frequency. The process of LDA posits that each document and each word has some probability of belonging to a set of k underlying topics or latent dimensions. LDA is a clustering algorithm that will estimate these probabilities to summarize rich text data in terms of the distributions of the documents and words across the k clusters. There are small technical differences between different approaches to topic modeling (e.g., variational expectation–maximization vs. collapsed Gibbs sampling), but the main outputs are similar in nature (Blei et al., 2003; Griffiths & Steyvers, 2004). First, there is a matrix reflecting the probability of each document having been generated by a particular topic (the theta matrix). The theta matrix can be used to express the documents (in this case, our 1,000 news stories) in terms of a topic distribution (i.e., what proportion of each document is associated with each topic). In addition, there is a matrix reflecting the probabilities that particular words belong to particular topics (depending on the model, there may be a gamma matrix, or both a gamma matrix and a phi matrix). One way to interpret an LDA model's topics is by recovering the most probable words from each topic's gamma or phi matrix, akin to interpreting the dimensions of factor analyses in questionnaire development based on the items that are associated with each dimension. Looking at these words will give you a sense of what content is associated with a given topic. Additionally, it may be advantageous to look at the most probable *unique* words associated with each topic. To do this, you might look at the 50 most probable words for each topic, and then remove any duplicates between the lists. Once these lists have been generated, researchers can ask either expert judges or study participants to adjudicate what word or phrase best summarizes each topic's content (e.g., the topic "sports" if the most probable words are "training," "baseball," "basketball," "league," & "team").

LDA has the advantage that each document can be summarized as a distribution over all of the topics. Rather than having to classify a document as belonging specifically to a particular topic, each document is expressed as a composite of all topics. This offers the possibility of more nuanced comparisons between documents via their topic distributions. Suppose you wanted to compare two of *The Washington Post* stories mentioned earlier, having estimated an LDA model with 10 topics ($k = 10$). By comparing the two documents' topic distributions over these 10 topics, you can observe both similarities and differences. Perhaps both documents have a similar distribution over Topics 1 to 8, but differ in the relative weights of Topics 9 and 10. Being able to compare documents in terms of topics helps to provide greater granularity regarding the differences between documents than other clustering algorithms (e.g., k-means) that only allow each document to belong to one cluster.

LDA is frequently used in predictive modeling: The ability to summarize large corpora of text in terms of topic distributions is an effective way of preprocessing text data to form a smaller and more parsimonious number features that can be used to train machine learning algorithms. Given the complexity of text, the raw document-term matrices (with every unique word being represented as a column) can be enormous, including hundreds of thousands of columns. In many cases, researchers will not have large enough samples to allow for estimating that many coefficients in their predictive models, or they may simply not have enough computing power to train their models using large input matrices. Reducing the dimensionality of text using LDA—or similar dimensionality reduction methods such as singular value decomposition (SVD; Wall et al., 2003)—can often be helpful in turning raw text into manageable data sets for prediction. For example, one might express a corpus of 1 million tweets with an LDA model with 100 topics. Suppose 10,000 of these tweets are tagged as malicious, and the other 990,000 as safe. An analyst could then train an algorithm to predict whether the tweet was malicious using each tweet's distribution over the 100 topics as the predictor variables and the malicious tags as the target variable.

For researchers in the social sciences, the greatest challenge associated with topic modeling is often the interpreting of topics. While computer scientists are predominantly concerned with prediction accuracy, psychologist often aim to understand the underlying dynamics of the relationships they observe. It is nice to know that Dimension 8 is highly predictive of whether a tweet is malicious or not, but what does Dimension 8 mean? Whilst we outlined an approach for interpreting topics, the available methods are far from perfect. Topics can seem incoherent (e.g., there is no clear theme) or overlapping (e.g., within the domain of health care, it may not be possible to make fine-grained distinctions). Further, when looking to yield behavioral insights from text data using LDA, the interpretability of topics is often in direct contention with metrics of model fit or predictive validity.

Topic modeling using LDA has clear advantages over frequency-based approaches because it does not rely on the a priori identification of dictionaries based on domain expertise. This reduces researcher degrees of freedom, and the need for deep disciplinary knowledge to specify features. However, topic modeling can also be inferior to frequency-based approaches when it comes to interpretation. The latent topics generated by data-driven models may be difficult or impossible to interpret in a distinct way (or the process of interpretation may impose researchers' biases on the analytical process). Moreover, as with frequency-based analysis, LDA fails to account for the semantic structure and meaning of sentences.

Word-Embedding-Based Models

The methods of computerized text analysis discussed so far analyze text as a table of words with their associated frequencies of occurrence. This precludes researchers from gaining deeper insights into the meaning of words or documents that are associated with the colocation of words, rather than just their presence or absence. Recent advances in natural language processing algorithms provide computational methods for analyzing words as they are embedded in sentences, capturing nuanced variation in semantic meaning. Word-embedding approaches are among the most commonly used, with the algorithm word2vvec being widely used by researchers across numerous disciplines (Mikolov, Chen, et al., 2013; Mikolov, Sutskever, et al., 2013).

Word2vvec is a method of analyzing large corpora of text using neural networks. What does that mean? And how does it work? The quotation by the linguist John Rupert Firth, "You shall know a word by the company it keeps" (1957, p. 11) reflects the basic intuition behind word2vvec. In the continuous bag-of-words (CBOW) model architecture, the word2vvec algorithm scans through the corpus of text, trying to predict a focal word using the words around it. Suppose your corpus of text was just one sentence: "A dog is going to the park in order to play with another dog, as well as so its owner can get a breath of fresh air." You can think of word2vvec as running a window across the text, considering only a portion at a time. The algorithm would try to predict that it should select the word "dog" by looking at the surrounding words in the two contexts in which the word dog occurs (e.g., "a [dog] is" and "another [dog] as"). Seeing a word occur in many different contexts enables the algorithm to "learn" the word's meaning. There is also a skip-gram model architecture, where the model uses one focal word to predict the surrounding context words (Y. Goldberg & Levy, 2014). The skip-gram architecture takes longer to run but is more accurate for infrequent words (Mikolov, Sutskever, et al., 2013). The output of the word2vvec algorithm is a dictionary of high dimensional vectors (the number of dimensions is defined by the researcher), with each vector relating to a specific word in the dictionary. These are called *word vectors* or *word embeddings*.

Suppose you chose to estimate a word2vvec model with 100-dimensional vectors using a corpus of 100 million tokens. For every word that occurred in your text corpus, you would recover a word vector with 100 dimensions. There are of course some caveats to this—researchers typically exclude words that occur less than a minimum number of times, as it would be difficult for the algorithm to learn the meanings of these words based on their occurrence in a limited number of contexts. The algorithm estimates these high-dimensional vectors by optimizing their performance in predicting focal words from their context. In the process of doing so, the word vectors gradually encode the words' semantic meanings across many different contexts. For those familiar with factor analysis or other dimensionality reduction methods, word2vvec can be thought of as summarizing each word's semantic meaning in terms of variation along hundreds of latent dimensions (100 in the example above). At face value, this output sounds no more interpretable than the topic distributions of an LDA model. A vector with hundreds of dimensions is clearly not easy to interpret by eye. It is true that interpreting single dimensions of word vectors is almost impossible (as opposed to factor loadings in the analogy used). The beauty of word vectors comes from comparing their relationships with each other. For example, consider training a word2vvec model using all novels in the 20th century, and pulling out the word vectors for the words "man," "woman," and "king." You could compute the following combination of the vectors: King + Man + Woman, and it would return a vector that was near-identical to the vector for the word "queen." What the model does is simply combine what it means to be a woman with the aspects of what it means to be a king that are not associated with men, to infer that the combination of the two must be "queen." This relational property of word vectors, whereby they encode the relationships between different words, can be used to reveal a wide range of insights regarding language use.

For example, studying the similarity between the word vectors associated with gendered words (e.g., "he" or "she") and words that describe traits (e.g., agentic words such as "competent" or "dominant") makes it possible to study stereotypes encoded in language (Caliskan et al., 2017;

DeFranza et al., 2020; Garg et al., 2018). As we describe in more detail in the applications section later in this chapter, men have traditionally been more associated with leadership qualities than women. By comparing the similarity of the vectors "she" and "competent" with the similarity of "he" and "competent," one can capture the extent to which people's language use reflects these gender stereotypes. An example of the kind of insights word2vvec can discover was the highly publicized case where a word2vvec model trained using historical text data identified that "man is to computer programmer as woman is to homemaker" (Bolukbasi et al., 2016).

While word embeddings can also be used as a dimensionality reduction technique for prediction contexts (similar to topic modeling), what makes them unique is that they allow researchers to study nuances in the semantic meaning of words in a way that has previously been impossible. Suppose you were a researcher interested in studying how people understand the concept of motivation. Using word vectors, you could look at the cosine similarity between words relevant to motivation (e.g., "motivation," "motivated") and other related concepts (e.g., "effort" vs. "passion"). If the cosine similarity between "motivation" and "passion" was much greater than the cosine similarity between "motivation" and "effort," you might conclude that people's understanding of motivation is more about people's passion for and enjoyment of that task, rather than the act of expending effort on it. Whilst the individual dimensions of word vectors are hard to interpret, one can garner unique insights by looking at words' colocation with related concepts. When combined with the disciplinary knowledge necessary to identify which words to analyze and what to contrast them against, the output of word2vvec becomes a very powerful tool for social science research.

A notable limitation of methods such as word2vvec is that they require large amounts of data to train. This requirement can limit the contexts in which word2vvec models can be implemented and the level of granularity at which they can be applied. For example, a researcher might be interested in estimating a word2vvec model using a company's publicly available text in 1-year periods. Yet, even a full year of company data might not be enough to recover reliable estimates of these word vectors. This means that it is often necessary to collapse across units (e.g., firm) or time periods to ensure the reliability of estimates. One widespread technique that helps to partially mitigate this concern is the use of transfer learning, where a word2vvec model is first pretrained on a related data set (e.g., Google books, or a corpus of business newspaper articles) and subsequently fine-tuned with the data of interest. This means that the word vectors will reliably encode semantic meaning (due to the large corpus of text used to pretrain them), but can also capture the nuances of the specific texts on which they were fine-tuned. This process allows research to aim for more granular levels of analyses with fewer data points per unit. But pretraining is not a silver bullet. If there is only a small amount of text available for the updating of the pretrained model, subtle differences across observations risk being swamped by the pretrained vectors (which serve as a strong prior).

In addition to data requirements, word vectors also create several challenges when it comes to interpreting the semantic meaning of words. Some words, for example, can only be understood as part of a phrase (e.g., idioms such as "piece of cake"), and other words have multiple meanings in different contexts (e.g., "the rabbit bounded" vs. "bounded rationality").

In order to address that some words hold multiple meanings, researchers at Google produced a new language representation model: BERT (Devlin et al., 2019). This model uses bidirectional context—both the words before and the words after the focal word—to effectively infer the meaning of a word from its context. word2vvec is a context-free model (i.e., the orders of the words preceding and following the focal word are not used) that estimates a single semantic meaning for a focal word from the surrounding words (in the CBOW implementation). In contrast, BERT can distinguish between different meanings of a focal word, based on the words that surround it. While we have mainly focused on a single algorithm—Google's word2vvec—BERT is just one example of an important advancement that will doubtlessly influence social science research in the years to come. Others important approaches include Stanford's GloVe vectors (Pennington et al., 2014) and Facebook's fastText models (Bojanowski et al., 2017).

Overall, word embedding models such as word2vvec are among the most exciting methodological advancements for social science researchers when it comes to text analysis. The ability to focus on differences in the semantic *meaning* of words in their contexts offers some of the richest insights into the human psyche through language. Yet, as with any other methodology, there are limitations associated with word embeddings. As noted, the heavy data requirements and complexity of considering semantic meaning necessitates expertise in working with data and computational methods in order to effectively deploy these models.

APPLICATIONS OF TEXT ANALYSIS IN PSYCHOLOGICAL RESEARCH

Text analysis has been used in essentially every major subdiscipline of psychology, and its popularity has rapidly increased over the past decade. In the following section, we provide a taster of what this research looks like. Given the vast amount of research that is out there, we cannot provide a systematic and comprehensive literature review. Instead, we will present a selection of intriguing research findings in each subdiscipline to highlight the potential of the text analysis approach outlined in the previous sections for generating novel ideas and testing existing theories in real-life settings of high ecological validity.

Language in Personality Psychology

Personality psychology has traditionally relied on self-report questionnaires to measure a person's relatively stable psychological traits. Such questionnaires typically ask respondents to indicate their agreement with statements like "I am the life of the party" and then, depending on their response, infer whether the person is more or less extraverted than other individuals in the given population. Although self-report questionnaires are a convenient way to access people's self-views, they are not without drawbacks. There is a vast literature showing how self-reports are susceptible to (un)intentional response biases that can distort the validity of assessments obtained from questionnaires (Podsakoff et al., 2003). For example, individuals might intentionally present themselves in a more positive light (i.e., social desirability bias; Krumpal, 2013), the order in which questions are presented might influence their responses (i.e., order effects), or people might simply misremember instances of their behavior based on how accessible and salient they are in a given moment (i.e., recency effect). In addition to response biases, self-reports also suffer from often being difficult to scale (Kosinski et al., 2015). While collecting data from a few hundred subjects in the lab is feasible, it is much more difficult to collect data on thousands of people from all around the world.

Recent years have seen the rise of automated personality assessments from digital footprints (Azucar et al., 2018). That is, instead of requiring participants to complete self-report questionnaires, their personalities can be predicted from the digital traces they leave online, including their Facebook Likes (Kosinski et al., 2013; Youyou et al., 2015), spending records (Gladstone et al., 2019), or social media pictures (Segalin et al., 2017). Among the many digital traces that could be used to make such predictions, text is the most widely used source to date (Boyd & Pennebaker, 2017; Kern et al., 2014). Specifically, research has shown that personality traits can be predicted from personal blogs (Yarkoni, 2010), tweets (Qiu et al., 2012; Quercia et al., 2011), Facebook status updates (Park et al., 2015; Schwartz et al., 2013), and Reddit posts (Gjurković & Šnajder, 2018). By extracting word frequencies, n-grams, LIWC categories, latent topics, or word-embeddings from the individual's language and relating it to their self-reported personality profiles using supervised machine learning, these studies have shown that a person's personality can be predicted with high levels of accuracy.

For example, Park et al. (2015) analyzed the Facebook status updates of 66,732 Facebook users to predict their self-reported Big Five personality traits. They observed correlations of $r = .38$ to $.46$ between the self-reported and predicted scores, which is similar to the scale reliabilities of short personality questionnaires (e.g., the Ten Item Personality Inventory; Gosling et al., 2003). Lending credibility and theoretical depth to these predictions, the univariate relationships between words and personality traits have been shown to exhibit high face validity. For example, extraverted users are more likely to post updates containing the words "party," "amazing," and "exciting," whereas introverted users are more likely to talk about "computers," "internet," or "reading." Importantly, the research by Park et al. (2015) also highlights that automated predictions of personality from digital footprints conform to established psychometric standards such as high retest reliability and internal consistency, as well as predictive validity. For example, they showed that the predicted personality scores were as good at predicting external criteria, such as a person's life satisfaction, their political ideology, or IQ, as the self-reported scores.

Interestingly, the use of language in the context of personality assessment takes us back to the origin of current personality models. In the late 1930s, the famous psychologist Gordon Allport was the first to theorize that important personal traits should be encoded in the language we use to describe ourselves and others (i.e., the lexical hypothesis; Allport, 1937). Together with Henry Odbert, Allport surveyed the English dictionary and extracted a total of 17,953 adjectives that could be used in the character description of people (Allport & Odbert, 1936). Building on the lexical approach, the famous Big Five model of personality (also known as the five-factor model) was developed using factor-analytical methods to reduce the space of character words to a number of meaningful dimensions that could explain a large amount of the original variance in how people were describing themselves and others (L. R. Goldberg, 1992).

Linking current approaches to text analyses back to the origins of personality traits highlights an opportunity for advancing personality psychology in the current age of Big Data. Whereas the majority of studies to date have focused on replicating self-report responses with the help of automated predictions (i.e., using the self-reported scores as the gold standard and then linking them to the words we use), there have been recent attempts to use language in a more bottom-up, unsupervised way. For example, Kulkarni et al. (2018) used data from 49,139 Facebook users to identify novel dimensions of personality by applying a purely data-driven clustering approach. Instead of using language to predict the self-reported Big Five personality traits of participants, the authors let the "data speak for themselves" to determine how certain words and phrases would cluster together naturally, and to infer novel latent traits. The results are behavior-based linguistic traits (BLTs) that represent personality dimensions in language space. Supporting the feasibility and validity of such a bottom-up approach, the BLTs showed similar psychometric properties as traditional self-report measures, and correlated moderately with self-reported Big Five scores.

Although the majority of work in this space has focused on predicting the Big Five personality traits, language can be used to predict psychological characteristics more broadly. For example, research has shown that the words people use can predict personal values (Boyd et al., 2015), political ideology, or sexual orientation (Kosinski et al., 2013). In addition to relatively stable psychological traits, language has also been related to more malleable psychological states such as mood or discrete emotions (Bollen et al., 2011). In fact, language-based analyses of both traits and states makes it possible to study their development over time, offering the potential to replace cumbersome and expensive longitudinal data collections and conducting within-person research at scale (Eichstaedt & Weidman, 2020). For example, researchers could use a person's tweets or Facebook status updates to track their mood over time and relate it to external events such as national elections, economic crises, or global pandemics such as COVID-19.

Language in Health and Clinical Psychology

Improving the physical and mental health of individuals constitutes one of the biggest global challenges of modern times, with the United Nations' Sustainable Development Goals aiming to "reduce by one third premature mortality from noncommunicable diseases through prevention and treatment and promote mental health and well-being" by 2030 (World Health Organization, 2020, p. 1). Even in the most developed countries, millions of people die from preventable or curable diseases every year. In the United States, cardiovascular disease kills one person every 36 seconds (655,000 annually; Centers for Disease Control and Prevention [CDC], n.d.-a), and mental-health-related suicide ends the life of one person every 11 minutes (48,000 annually; CDC, n.d.-b). Cardiovascular disease is the leading cause of death among Americans, and depression the second leading cause among younger adults (< 35 years; CDC, n.d.-a).

One of the most critical elements to improving physical and mental health is the ability to diagnose potential health issues as early as possible. Instead of trying to treat people who have had a heart attack or suffered from depression, the early detection of populations or individuals at risk could not just save billions of dollars but also thousands of lives. However, early diagnostics of physical and mental health issues are often difficult as they require the active monitoring of symptoms over time. Effective diagnostic tools need to be able to detect deviations from a person's regular functioning in real time, and to create a holistic impression of a person's health that goes beyond the moment when they decide to see a doctor.

Advances in predictive technologies and the availability of large amounts of user-generated data have introduced new diagnostic tools that can monitor people's physical and mental health unobtrusively over time. For example, mobile devices such as smartphones have been used to capture people's physical activity via the GPS sensors embedded in the devices to predict depression (Canzian & Musolesi, 2015), heart disease (Raihan et al., 2016), or schizophrenia (Fulford et al., 2021). Given that language is a powerful indicator of our inner mental lives, it is unsurprising that the words we use can also reveal a lot of information about our physical and mental health (Pennebaker & King, 1999; Pennebaker et al., 2003). An individual suffering from severe chronic pain, for example, might be more likely to complain about symptoms and less likely to rave about all the exciting outdoor adventures they have recently engaged in. Similarly, a person with clinical depression might be more likely to think about and express more negative sentiment about their current experiences than an individual without depression.

Testing this proposition empirically, Eichstaedt et al. (2015) showed that the language used in Twitter posts is predictive of heart disease rates on the level of U.S. counties. Analyzing over 148 million tweets across 1,347 U.S. counties, they showed that tweets alone predicted heart-disease related mortality rates with high accuracy (person correlation between actual and predicted mortality rates $r = 0.42$). Notably, Twitter language alone was found to be a better predictor of age-adjusted mortality from heart disease than a combination of previously identified sociodemographic, economic and health-related risk factors (e.g., gender, income, obesity). While language describing negative social relationships, indicating psychological disengagement, and expressing negative emotions (particularly anger) was related to higher levels of mortality, the expression of positive emotions and psychological engagement were related to lower levels of mortality and therefore better physical health. Given that language provides insights about people's inner mental states rather than their physical fitness, why does language provide such a powerful window into people's physical health? The authors argue that language captures important psychological precursors of heart disease including hostility and chronic stress, which can be passively observed in the language people use to communicate their experiences on social media. Hence, language provides insights into our physical world through our related mental experiences.

Focusing on mental health at the individual level, Schwartz et al. demonstrated that a person's Facebook posts can be used to predict depression in medical records (Eichstaedt et al., 2018). Analyzing the status updates of 683 patients over a 6-month period, they were able to accurately predict whether a person was suffering from depression in 72% of the cases (where chance is 50%). While 72% is certainly not perfect, it roughly matches the accuracy of short screening surveys that are currently used in mental health diagnostics. Similar to the prediction of heart disease, the relationships underlying these predictions have high face validity. For example, clinical levels of depression were associated with the expression of negative emotions (e.g., sadness), interpersonal challenges (e.g., loneliness, hostility), as well as cognitive processes that are known to be detrimental to mental health (e.g., rumination, a strong focus on the self).

In addition to providing a diagnostic tool that can predict whether an individual might be at an elevated risk to experience physical or mental health problems at a given point in time, text analysis also holds the promise of being able to track people's health continuously over time (e.g., Eichstaedt & Weidman, 2020). At the moment, most medical assessments are one-time checkups that are ad hoc and provide only a snapshot of the individual's health. In most cases, these checkups occur when individuals already suffer from health problems and consult with their physicians on how to alleviate them. The ability to track people over time by passively observing their language posted on social media, for example, makes it possible to get a much more holistic understanding of health trajectories as well as the contexts under which individuals might experience more or less symptoms. Instead of contrasting the language and physical health of an individual to those of the broader population, this approach allows researchers and clinicians to compare people's current language use and symptoms to those observed at other times, hence enabling them to conduct more fine-grained within-person analyses. Taken together, the ability to cheaply and unobtrusively measure the physical and mental health of individuals at scale holds great promise for revolutionizing the way in which we diagnose and treat people.

Language in Social Psychology

Although language is often an expression of our inner mental states, and therefore indicative of intraindividual characteristics, words can also reveal a lot about our relations to those around us. The way we speak with or about other people provides insights about how we feel about them, what attitudes we have towards them, as well as whether the relationship between us and these people is thriving or struggling to survive.

Up to this point we have predominantly focused on what language can reveal about the person producing the language. A lot of the work on language in the context of social psychology focuses on how the language of two or more people overlaps and aligns with regard to the topics they speak about or the style they use (Gonzales et al., 2010). In line with the proverb "birds of a feather flock together," research has shown that people are more likely to have friends who are similar in their linguistic styles (Kovacs & Kleinbaum, 2020; Youyou et al., 2017), and high levels of linguistic similarity have been shown to predict both the success of initial romantic interactions between strangers (e.g., in a dating context), as well as the stability of newly formed romantic couples over time (Ireland et al., 2011). For example, Ireland et al. (2011) analyzed the transcript of 40 speed dating encounters, finding that the more similar participants were in their linguistic styles, the more likely they were to show romantic interest in their counterpart. They also examined the instant messages of 86 couples to show that those with high levels of linguistic similarity had a significantly higher chance of still being together after 3 months. Importantly, the effect of similarity in language on social relationships is not a one-way street. Similarity in language not only increases the likelihood that we will enter or stay in a romantic relationship (i.e., language similarity serves as a selection mechanism), but the more time we spend with another person, the more similar we become in terms of our linguistic styles (language similarity also serves as an assimilation mechanism; Kovacs & Kleinbaum, 2020).

Another important topic that has garnered a lot of attention in the context of social psychology is the way in which groups use language to create a sense of shared reality with other ingroup members, and distinguish themselves from outgroup members. For example, research has shown that the language used by supporters of different political parties on social media is markedly different (Sterling et al., 2020) and that politicians differ greatly in the content and style they use with their constituents (making their message resonate more or less strongly with different parts of the population; Ahmadian et al., 2017). Sterling et al. (2020), for example, used text analysis to study the language of nearly 25,000 Twitter users who could be categorized as liberals and conservatives, as well as moderates versus extremists. Their findings show that conservatives are more likely to use language related to security, anger, power, and tradition, whereas liberals were more likely to tweet about topics related to benevolence. Similarly, extremists were more likely to tweet content associated with anger, anxiety, affiliation, or resistance to change, whereas moderates were more likely to use benevolent language. In addition, language has been shown to be a powerful barometer of voter sentiment and an effective predictor of voting behavior and election outcomes. For example, in a working paper, Ward et al. (2021) used over 1.5 billion tweets to show how negative emotions predicted support for populist candidates and movements in both the 2016 U.S. presidential election of Donald Trump as well as the 2016 Brexit referendum in the United Kingdom. Highlighting the unique role of emotions (e.g., anger, sadness, anxiety), the effect of language-based sentiment was predictive of populist support above and beyond traditional economic and sociocultural factors.

Language in Organizational Psychology

It has been estimated that 80% to 95% of all data generated by businesses are unstructured, with the majority of this unstructured data being collected in the form of text (Gandomi & Haider, 2015). Every day, billions of emails are sent, phone calls are made, internal reports are written, and external shareholder reports are released. On a macrolevel, this information can be used to analyze the sentiment in an organization, to understand the strategic direction a company is headed, and to track organizational culture. On a microlevel, the language produced in organizational contexts can provide insights into the well-being of employees, be used to navigate the talent management pipeline in the pursuit of proactively matching employees to the best jobs, and track group dynamics within and across different teams in the organization.

One of the most prominent research streams in organizational psychology is that of organizational fit (also known more broadly as person–environment fit; Kristof, 1996). Research on person–organization fit (P-O fit) suggests that compatibility between an organization and the employees within that organization is provided by a strong corporate culture and a shared set of values and beliefs among coworkers. The benefits of P-O fit to both employees and organizations are bountiful. For example, P-O fit has been related to higher levels of trust and commitment among employees as well as a stronger sense of community (O'Reilly et al., 1991). As a downstream consequence, individuals who experience higher levels of organizational fit are more likely to be satisfied with their employment, to display good citizenship behavior, and to stay with the organization for longer (Hoffman & Woehr, 2006). In a seminal paper, Srivastava et al. (2018) used the language from over 10 million work emails of a technology company as an indicator of organizational culture and as the foundation for measuring

P-O fit (Srivastava et al., 2018). Specifically, they investigated how different employees adapt to the organizational culture over time—measured as the fit in linguistic styles between an employee's email and those that employee received from other members of the organization—and how different enculturation trajectories relate to consequential outcomes such as performance and retention. There findings suggest that those employees who quickly adjust to the dominant organizational culture in the way they use language reduces the likelihood that they will have to leave the organization involuntarily. In contrast, a decrease in language-based culture fit foreshadowed a voluntary departure from the organization.

Shifting focus from the fit between employees and the wider organizational culture to the fit between employees and their jobs (e.g., accountant, salesman, or barista; person–job fit; Edwards, 1991), Kern et al. (2019) used the tweets from over 128,000 Twitter users across 3,513 occupations to develop a comprehensive psychological occupation map (Kern et al., 2019). That is, they used language from tweets to automatically predict the personality of users and then clustered the average personality profiles of occupations to identify groups of occupations with similar psychological profiles (i.e., personality and values; see the section Language in Personality Psychology in this chapter for more details on how language can be used to predict individual differences). Mapping this higher dimensional-space to a 2D representation, their map provides a compelling and easily accessible overview of the psychological underpinnings of different occupations that allows practitioners to actively recommend roles based on the fit between a person's psychological traits and those of the job. Knowing the personality of a job seeker, for example, one could recommend a number of occupations that are similar in their psychological orientation and that all cluster together.

Language can also be used to track changes in organizational culture in response to certain events. One of the major challenges organizations face today is the lack of women and people of color in positions of leadership (Catalyst, 2020). The underrepresentation of women has been explained, in large parts, by prevailing gender stereotypes that associate men—but not women—with agentic characteristics that are typically seen as conducive to successful leadership (e.g., competent, assertive, dominant;

Eagly & Karau, 2002; Heilman, 2001). In some of our own work, we have shown that these stereotypes transmitted in language can be mitigated when companies hire a woman CEO. Using word embeddings (word2vvec; see the section on Frequency-Based Methods for an explanation of the method) to analyze over 43,000 public shareholder documents of S&P 500 companies, we showed that the language of companies that had hired a female CEO shifted towards associating women more strongly with agentic terms than comparable companies that had not hired a woman CEO. We suggest that this happens because companies that hire a woman CEO are motivated and incentivized to signal the competency of their woman CEO, as well as the ability of women to lead more generally. This interpretation is supported by the fact that the shift in language was predominantly driven by the positive aspects of agency (e.g., competent) and not the negative ones (e.g., aggressive). Our findings highlight the power of language in revealing important dynamics within organizations that allow one to track the development of certain linguistic features over time and in response to important external events or internal interventions (Obschonka et al., 2017).

Language in Consumer Psychology

Language has been an integral part of consumer psychology for decades: Marketing is built on the persuasive power of words. Yet, the digitization of much of our shopping and consumption experiences has led to a new surge in research examining language as a tool to generate marketing insights. As with organizational data more generally, there is no shortage of consumer data: Customers write online reviews, talk about their favorite brands on social media, and call customer service hotlines (which can be transcribed to written text). Together, these data offer new opportunities to investigate consumer, firm, and market behavior at an unprecedented scale and level of granularity (see Berger et al., 2020, for an excellent review article) and to provide answers to both old and new questions: What words are most effective in persuading a consumer to buy a certain product? What makes a message go viral? What language do brands use to create a certain image among consumers? Can social media posts crowdsourced from consumers be used to predict market developments and trends?

A topic that has garnered an increasing amount of attention in the age of social media is why and when content goes viral. There are numerous examples of social media content going viral, ranging from playful games such as the ice bucket challenge, to sociopolitical movements like those seen in the Arab spring, to the dangerous spread of misinformation in the context of political elections. Although word-of-mouth has been an important way of communicating information throughout human history, the ability to broadcast to millions of people with as little as one click has fundamentally changed the speed and scale at which words can travel and influence the masses. It is therefore not surprising that recent years have seen a surge in research dedicated to the question of what drives virality. Of course, there are many aspects of online content that might explain its popularity and spread beyond language (e.g., the popularity of the person sharing content and the structure of the network). However, a large proportion of the existing literature in the context of consumer behavior has focused on identifying patterns in the language of content that spreads rapidly and goes viral. This research has shown, for example, that emotional content (in particular, high-arousal emotions such as awe, anger, or anxiety; Berger & Milkman, 2012, 2013; Guerini & Staiano, 2015; Kramer et al., 2014) and content that is highly moralized (Brady et al., 2019) is more likely to spread on social media, with the combination of moral and emotional content being the most likely to go viral (Brady et al., 2017; Valenzuela et al., 2017). Analyzing the moral and political discourse among Twitter users in over 500,000 posts, Brady et al. (2017) showed that for every moral–emotional word that is added to a tweet (e.g., fight, greed, destroy), the tweet experiences an increase in sharing of approximately 20%.

A specific case of virality is the spread of disinformation (also called "fake news"), which refers to information that is factually incorrect and disseminated with the intent to foster confusion, uncertainty, and distrust (e.g., Del Vicario et al., 2016). In a seminal paper, Vosoughi et al. (2018) analyzed about 126,000 stories on Twitter to show that misinformation spreads "significantly farther, faster, deeper and more broadly than the truth" (Vosoughi et al., 2018). On average, false stories were 70% more likely to be retweeted than true ones. Although the same effects were found across all categories of disinformation, they were particularly pronounced for political news. Diving deeper into the content of the tweets using text analyses—both topic modeling and a dictionary-based approach—the authors provide potential explanations for this effect. First, fake stories were more likely to be novel when compared with content to which users had previously been exposed. Given that novelty has previously been shown to attract attention and facilitate information sharing (Itti & Baldi, 2009), the novelty of fake news stories might increase their likelihood of going viral. Second, compared to true stories, fake stories were found to evoke user reactions characterized by high-arousal negative emotions (e.g., fear and disgust) in their replies, which are emotions that have previously been linked to higher levels of virality (Berger & Milkman, 2012).

A third domain in consumer psychology in which text analysis been used extensively is the study of online reviews. Some scholars have characterized online reviews as the new word of mouth (Chevalier & Mayzlin, 2006; Hu et al., 2006). While people have traditionally relied on their friends and local experts for product recommendations and feedback, online reviews today have become a central part of consumers' decision-making process. According to a 2017 report, 93% of customers say that online reviews have influenced their purchase decisions (Podium, 2017). How do reviews influence consumers' purchasing decisions? What makes for an influential review? As one would expect, reviews that are positively valenced are associated with an increase in conversion rates (i.e., number of consumers deciding to buy the product) while negative reviews are associated with a decrease (Ludwig et al., 2013). Yet, there exists an asymmetry in that negative reviews are more valued by consumers and more influential in impacting their decision-making (Charlett et al., 1995; Lee et al., 2008). In addition, the extent to which users perceive reviews as useful has been linked to content that is more compelling in its argumentation and that signals the expertise of the reviewer (Willemsen et al., 2011). However, it is not just the content of a review that predicts whether another user will follow the recommendation but also the extent to which the review is aligned in its linguistic style with the typical style observed in a given interest group (Ludwig et al., 2013).

FUTURE DIRECTIONS

The future of text analysis in psychological research depends on advances in its ability to detect fine-grained nuances in meaning, and the practicality of the data demands to do so. Pursuing the twin goals of deep insight with limited data requirements is a challenging task, as they are often diametrically opposed. Yet, to continue to rise in popularity, text analysis will have to be able to detect more subtle differences in meaning. For example, can models truly understand humor? Will they be able to interpret language in a way analogous to how people form representations of concepts? As it stands, there are many aspects of any given passage of text that are not well encoded by the available tools.

As we experience advances in the tools available to analyze language, advancements are also occurring in the technology used to record and produce language. Will the proliferation of speech-to-text transcription fundamentally change written language? More broadly, how do the different technologies (e.g., iPads, phones, virtual reality headsets) we use to generate and communicate language affect how we construct and share meaning? While such interfaces offer great potential for studying language, they also may affect how it continues to develop.

Finally, the strategic use of insights from language and new technologies by which language is communicated offer fruitful paths for interventions. Whilst AI chatbots have thus far largely served as novelties rather than practical solutions, they offer huge potential for solving societal issues. For example, Andrew Ng et al. (Woebot Health, 2021) developed a chat bot aimed at counseling people with mental health issues (https://woebothealth.com/board-members/). Combining insights from natural language processing to adapt interventions to people's preferences, as one would in a conversation, and technology to administer them could offer a lower-cost, scalable solution to many different issues in society. Given that cost makes services such as mental health care inaccessible to some of the world's most vulnerable people, such a revolution could have profound, positive implications.

CONCLUSION

As the digital revolution continues to produce an explosion in behavioral data, one of the most promising avenues for generating psychological insights is the use of natural language processing of text data. Within the domain of text analysis, there are numerous different methods that can yield such insights. Frequency-based analyses are easy to implement, but because they require researchers to formulate dictionaries a priori and do not take context into account, they are somewhat limited in their scope and potential for uncovering novel relationships. Topic modeling is useful for investigating the relationship between latent dimensions in language and consequential real-world outcomes, but the meaning of the topics can be difficult to interpret. Word embeddings offer the deepest insights into the semantic meaning of the language we use. These models have revolutionized both the ability of algorithms to predict target outcomes from text, and the ability of computers to understand the meaning of human language. However, the large data requirements of word embeddings can pose a considerable challenge to researchers and limit the granularity with which a phenomena can be studied. For example, it would be fascinating to estimate word vectors for each and every individual to understand how they encode semantic relationships in their speech, or how that meaning changes across their lifespan, but only very few individuals will have generated enough text to build such models reliably (this might change rapidly as speech to text translations becomes more reliable). Which method to choose depends on the specific research question, the availability of data, and the skill set of the researcher. However, we believe that the ability to use the tools of natural language processing is becoming increasingly important for social scientists if they want to keep up with their colleagues in the computer sciences and engineering. Psychologists have been at the forefront of studying human behavior. For us to remain central players in this field, we will have to adjust our tool kits to leverage the novel opportunities provided by Big Data.

REFERENCES

Ahmadian, S., Azarshahi, S., & Paulhus, D. L. (2017). Explaining Donald Trump via communication style: Grandiosity, informality, and dynamism. *Personality and Individual Differences, 107*(1), 49–53. https://doi.org/10.1016/j.paid.2016.11.018

Allport, G. W. (1937). *Personality: A psychological interpretation.* Holt.

Allport, G. W., & Odbert, H. S. (1936). Trait-names: A psycho-lexical study. *Psychological Monographs, 47*(1), i–171. https://doi.org/10.1037/h0093360

Allport, G. W., & Postman, L. (1947). *The psychology of rumor.* Henry Holt.

Azucar, D., Marengo, D., & Settanni, M. (2018). Predicting the Big 5 personality traits from digital footprints on social media: A meta-analysis. *Personality and Individual Differences, 124*(1), 150–159. https://doi.org/10.1016/j.paid.2017.12.018

Berger, J., Humphreys, A., Ludwig, S., Moe, W. W., Netzer, O., & Schweidel, D. A. (2020). Uniting the tribes: Using text for marketing insight. *Journal of Marketing, 84*(1), 1–25. https://doi.org/10.1177/0022242919873106

Berger, J., & Milkman, K. L. (2012). What makes online content viral? *Journal of Marketing Research, 49*(2), 192–205. https://doi.org/10.1509/jmr.10.0353

Berger, J., & Milkman, K. L. (2013). Emotion and virality: What makes online content go viral? *NIM Marketing Intelligence Review, 5*(1), 18–23. https://doi.org/10.2478/gfkmir-2014-0022

Blei, D. M., Ng, A. Y., & Jordan, M. I. (2003). Latent Dirichlet allocation. *Journal of Machine Learning Research, 3*(1), 993–1022. https://doi.org/10.5555/944919.944937

Bojanowski, P., Grave, E., Joulin, A., & Mikolov, T. (2017). Enriching word vectors with subword information. *Transactions of the Association for Computational Linguistics, 5*, 135–146. https://doi.org/10.1162/tacl_a_00051

Bollen, J., Mao, H., & Zeng, X. (2011). Twitter mood predicts the stock market. *Journal of Computational Science, 2*(1), 1–8. https://doi.org/10.1016/j.jocs.2010.12.007

Bolukbasi, T., Chang, K.-W., Zou, J. Y., Saligrama, V., & Kalai, A. T. (2016). Man is to computer programmer as woman is to homemaker? Debiasing word embeddings. In D. D. Lee, M. Sugiyama, U. V. Luxburg, I. Guyon, & R. Garnett (Eds.), *Advances in neural information processing systems, 29* (pp. 4356–4364). Curran Associates. http://papers.nips.cc/paper/6228-man-is-to-computer-programmer-as-woman-is-to-homemaker-debiasing-word-embeddings.pdf

Boyd, R. L. (2017). Psychological text analysis in the digital humanities. In S. Hai-Jew (Ed.), *Data analytics in digital humanities: Multimedia systems and applications* (pp. 161–189). Springer. https://doi.org/10.1007/978-3-319-54499-1_7

Boyd, R. L., & Pennebaker, J. W. (2015). A way with words: Using language for psychological science in the modern era. In C. V. Dimofte, C. P. Haugtvedt, & R. F. Yalch (Eds.), *Consumer psychology in a social media world* (pp. 222–236). Routledge/Taylor & Francis Group.

Boyd, R. L., & Pennebaker, J. W. (2017). Language-based personality: A new approach to personality in a digital world. *Current Opinion in Behavioral Sciences*, *18*, 63–68. https://doi.org/10.1016/j.cobeha.2017.07.017

Boyd, R. L., Wilson, S. R., Pennebaker, J. W., Kosinski, M., Stillwell, D. J., & Mihalcea, R. (2015). Values in words: Using language to evaluate and understand personal values. *Proceedings of the Ninth International AAAI Conference on Web and Social Media*, *9*(1), 31–40. https://ojs.aaai.org/index.php/ICWSM/article/view/14589

Brady, W. J., Wills, J. A., Burkart, D., Jost, J. T., & Van Bavel, J. J. (2019). An ideological asymmetry in the diffusion of moralized content on social media among political leaders. *Journal of Experimental Psychology: General*, *148*(10), 1802–1813. https://doi.org/10.1037/xge0000532

Brady, W. J., Wills, J. A., Jost, J. T., Tucker, J. A., & Van Bavel, J. J. (2017). Emotion shapes the diffusion of moralized content in social networks. *Proceedings of the National Academy of Sciences*, *114*(28), 7313–7318. https://doi.org/10.1073/pnas.1618923114

Busa, R. (1980). The annals of humanities computing: The Index Thomisticus. *Computers and the Humanities*, *14*(2), 83–90. https://doi.org/10.1007/BF02403798

Caliskan, A., Bryson, J. J., & Narayanan, A. (2017). Semantics derived automatically from language corpora contain human-like biases. *Science*, *356*(6334), 183–186. https://doi.org/10.1126/science.aal4230

Campbell, R. S., & Pennebaker, J. W. (2003). The secret life of pronouns: Flexibility in writing style and physical health. *Psychological Science*, *14*(1), 60–65. https://doi.org/10.1111/1467-9280.01419

Canzian, L., & Musolesi, M. (2015). Trajectories of depression: Unobtrusive monitoring of depressive states by means of smartphone mobility traces analysis. In *UbiComp '15: Proceedings of the 2015 ACM International Joint Conference on Pervasive and Ubiquitous Computing* (pp. 1293–1304). Association for Computing Machinery. https://doi.org/10.1145/2750858.2805845

Catalyst. (2020, August 11). *Quick take: Women in management*. https://www.catalyst.org/research/women-in-management/

Centers for Disease Control and Prevention. (n.d.-a). *Heart disease facts*. U.S. Department of Health & Human Services. Retrieved September 8, 2020, from https://www.cdc.gov/heartdisease/facts.htm

Centers for Disease Control and Prevention. (n.d.-b). *Facts about suicide*. U.S. Department of Health & Human Services. Retrieved May 22, 2021, from https://www.cdc.gov/violenceprevention/suicide/fastfact.html

Charlett, D., Garl, R., & Marr, N. (1995). How damaging is negative word of mouth? *Marketing Bulletin, 6*, 42–50. http://marketing-bulletin.massey.ac.nz/V6/MB_V6_N1_Charlett.pdf

Chevalier, J. A., & Mayzlin, D. (2006). The effect of word of mouth on sales: Online book reviews. *Journal of Marketing Research, 43*(3), 345–354. https://doi.org/10.1509/jmkr.43.3.345

Chomsky, N. (1986). *Knowledge of language: Its nature, origin, and use*. Praeger.

DeFranza, D., Mishra, H., & Mishra, A. (2020). How language shapes prejudice against women: An examination across 45 world languages. *Journal of Personality and Social Psychology, 119*(1), 7–22. https://doi.org/10.1037/pspa0000188

Dehghani, M., & Boyd, R. (in press). *The atlas of language analysis in psychology*. Guilford Press. https://philpapers.org/rec/DEHTAO-2

Del Vicario, M., Bessi, A., Zollo, F., Petroni, F., Scala, A., Caldarelli, G., Stanley, H. E., & Quattrociocchi, W. (2016). The spreading of misinformation online. *Proceedings of the National Academy of Sciences, 113*(3), 554–559. https://doi.org/10.1073/pnas.1517441113

Devlin, J., Chang, M.-W., Lee, K., & Toutanova, K. (2019). *BERT: Pre-training of deep bidirectional transformers for language understanding*. arXiv. http://arxiv.org/abs/1810.04805

Eagly, A. H., & Karau, S. J. (2002). Role congruity theory of prejudice toward female leaders. *Psychological Review, 109*(3), 573–598. https://doi.org/10.1037/0033-295X.109.3.573

Edwards, J. R. (1991). Person-job fit: A conceptual integration, literature review, and methodological critique. In C. L. Cooper & I. T. Robertson (Eds.), *International review of industrial and organizational psychology, 1991* (Vol. 6, pp. 283–357). John Wiley & Sons.

Eichstaedt, J. C., Schwartz, H. A., Kern, M. L., Park, G., Labarthe, D. R., Merchant, R. M., Jha, S., Agrawal, M., Dziurzynski, L. A., Sap, M., Weeg, C., Larson, E. E., Ungar, L. H., & Seligman, M. E. P. (2015). Psychological language on Twitter predicts county-level heart disease mortality. *Psychological Science, 26*(2), 159–169. https://doi.org/10.1177/0956797614557867

Eichstaedt, J. C., Smith, R. J., Merchant, R. M., Ungar, L. H., Crutchley, P., Preotiuc-Pietro, D., Asch, D. A., & Schwartz, H. A. (2018). Facebook language predicts depression in medical records. *Proceedings of the National Academy of Sciences, 115*(44), 11203–11208. https://doi.org/10.1073/pnas.1802331115

Eichstaedt, J. C., & Weidman, A. C. (2020). Tracking fluctuations in psychological states using social media language: A case study of weekly emotion. *European Journal of Personality, 34*(5), 845–858. https://doi.org/10.1002/per.2261

Feldman, R. (2013). Techniques and applications for sentiment analysis. *Communications of the ACM, 56*(4), 82–89. https://doi.org/10.1145/2436256.2436274

Firth, J. R. (1957). *A synopsis of linguistic theory 1930–1955*. Oxford University Press.

Fulford, D., Mote, J., Gonzalez, R., Abplanalp, S., Zhang, Y., Luckenbaugh, J., Onnela, J.-P., Busso, C., & Gard, D. E. (2021). Smartphone sensing of social interactions in people with and without schizophrenia. *Journal of Psychiatric Research, 137*, 613–620. https://doi.org/10.1016/j.jpsychires.2020.11.002

Gandomi, A., & Haider, M. (2015). Beyond the hype: Big data concepts, methods, and analytics. *International Journal of Information Management, 35*(2), 137–144. https://doi.org/10.1016/j.ijinfomgt.2014.10.007

Garg, N., Schiebinger, L., Jurafsky, D., & Zou, J. (2018). Word embeddings quantify 100 years of gender and ethnic stereotypes. *Proceedings of the National Academy of Sciences, 115*(16), E3635–E3644. https://doi.org/10.1073/pnas.1720347115

Gisev, N., Bell, J. S., & Chen, T. F. (2013). Interrater agreement and interrater reliability: Key concepts, approaches, and applications. *Research in Social and Administrative Pharmacy, 9*(3), 330–338. https://doi.org/10.1016/j.sapharm.2012.04.004

Gjurković, M., & Šnajder, J. (2018). Reddit: A gold mine for personality prediction. In M. Nissim, V. Patti, B. Plank, & C. Wagner (Eds.), *Proceedings of the Second Workshop on Computational Modeling of People's Opinions, Personality, and Emotions in Social Media* (pp. 87–97). Association for Computational Linguistics. https://doi.org/10.18653/v1/W18-1112

Gladstone, J. J., Matz, S. C., & Lemaire, A. (2019). Can psychological traits be inferred from spending? Evidence from transaction data. *Psychological Science, 30*(7), 1087–1096. https://doi.org/10.1177/0956797619849435

Goldberg, L. R. (1992). The development of markers for the Big-Five factor structure. *Psychological Assessment, 4*(1), 26–42. https://doi.org/10.1037/1040-3590.4.1.26

Goldberg, Y., & Levy, O. (2014). *Word2vvec explained: Deriving Mikolov et al.'s negative-sampling word-embedding method*. arXiv. http://arxiv.org/abs/1402.3722

Gonzales, A. L., Hancock, J. T., & Pennebaker, J. W. (2010). Language style matching as a predictor of social dynamics in small groups. *Communication Research, 37*(1), 3–19. https://doi.org/10.1177/0093650209351468

Gosling, S. D., Rentfrow, P. J., & Swann, W. B., Jr. (2003). A very brief measure of the Big-Five personality domains. *Journal of Research in Personality, 37*(6), 504–528. https://doi.org/10.1016/S0092-6566(03)00046-1

Griffiths, T. L., & Steyvers, M. (2004). Finding scientific topics. *Proceedings of the National Academy of Sciences, 101*(Suppl. 1), 5228–5235. https://doi.org/10.1073/pnas.0307752101

Guerini, M., & Staiano, J. (2015). Deep feelings: A massive cross-lingual study on the relation between emotions and virality. In *Proceedings of the 24th International Conference on World Wide Web* (pp. 299–305). Association for Computing Machinery. https://doi.org/10.1145/2740908.2743058

Heilman, M. E. (2001). Description and prescription: How gender stereotypes prevent women's ascent up the organizational ladder. *Journal of Social Issues, 57*(4), 657–674. https://doi.org/10.1111/0022-4537.00234

Hoffman, B. J., & Woehr, D. J. (2006). A quantitative review of the relationship between person–organization fit and behavioral outcomes. *Journal of Vocational Behavior, 68*(3), 389–399. https://doi.org/10.1016/j.jvb.2005.08.003

Hofmann, T. (1999). Probabilistic latent semantic indexing. In *SIGIR '99: Proceedings of the Twenty-Second Annual International SIGIR Conference* (pp. 50–57). Association for Computing Machinery. https://doi.org/10.1145/312624.312649

Hu, N., Pavlou, P. A., & Zhang, J. (2006). Can online reviews reveal a product's true quality? Empirical findings and analytical modeling of online word-of-mouth communication. In *Proceedings of the 7th ACM Conference on Electronic Commerce* (pp. 324–330). Association for Computing Machinery. https://doi.org/10.1145/1134707.1134743

Hutto, C., & Gilbert, E. (2014). VADER: A parsimonious rule-based model for sentiment analysis of social media text. *Proceedings of the International AAAI Conference on Web and Social Media, 8*(1), 216–225. https://ojs.aaai.org/index.php/ICWSM/article/view/14550

Ireland, M. E., Slatcher, R. B., Eastwick, P. W., Scissors, L. E., Finkel, E. J., & Pennebaker, J. W. (2011). Language style matching predicts relationship initiation and stability. *Psychological Science, 22*(1), 39–44. https://doi.org/10.1177/0956797610392928

Itti, L., & Baldi, P. (2009). Bayesian surprise attracts human attention. *Vision Research, 49*(10), 1295–1306. https://doi.org/10.1016/j.visres.2008.09.007

Kern, M. L., Eichstaedt, J. C., Schwartz, H. A., Dziurzynski, L., Ungar, L. H., Stillwell, D. J., Kosinski, M., Ramones, S. M., & Seligman, M. E. P. (2014). The online social self: An open vocabulary approach to personality. *Assessment, 21*(2), 158–169. https://doi.org/10.1177/1073191113514104

Kern, M. L., McCarthy, P. X., Chakrabarty, D., & Rizoiu, M.-A. (2019). Social media-predicted personality traits and values can help match people to their ideal jobs. *Proceedings of the National Academy of Sciences, 116*(52), 26459–26464. https://doi.org/10.1073/pnas.1917942116

Kincaid, J. P., Fishburne, J., Rogers, R. L., & Chissom, B. S. (1975). *Derivation of new readability formulas (automated readability index, fog count and Flesch reading ease formula) for Navy enlisted personnel* (#ADA006655) [Report]. Naval Technical Training Command Millington TN Research Branch. https://apps.dtic.mil/sti/citations/ADA006655

Klemmer, E. T., & Snyder, F. W. (1972). Measurement of time spent communicating. *Journal of Communication, 22*(2), 142–158. https://doi.org/10.1111/j.1460-2466.1972.tb00141.x

Kosinski, M., Matz, S. C., Gosling, S. D., Popov, V., & Stillwell, D. (2015). Facebook as a research tool for the social sciences: Opportunities, challenges, ethical considerations, and practical guidelines. *American Psychologist, 70*(6), 543–556. https://doi.org/10.1037/a0039210

Kosinski, M., Stillwell, D., & Graepel, T. (2013). Private traits and attributes are predictable from digital records of human behavior. *Proceedings of the National Academy of Sciences, 110*(15), 5802–5805. https://doi.org/10.1073/pnas.1218772110

Kosinski, M., Wang, Y., Lakkaraju, H., & Leskovec, J. (2016). Mining Big Data to extract patterns and predict real-life outcomes. *Psychological Methods, 21*(4), 493–506. https://doi.org/10.1037/met0000105

Kovacs, B., & Kleinbaum, A. M. (2020). Language-style similarity and social networks. *Psychological Science, 31*(2), 202–213. https://doi.org/10.1177/0956797619894557

Kramer, A. D. I., Guillory, J. E., & Hancock, J. T. (2014). Experimental evidence of massive-scale emotional contagion through social networks. *Proceedings of the National Academy of Sciences, 111*(24), 8788–8790. https://doi.org/10.1073/pnas.1320040111

Kristof, A. L. (1996). Person–organization fit: An integrative review of its conceptualizations, measurement, and implications. *Personnel Psychology, 49*(1), 1–49. https://doi.org/10.1111/j.1744-6570.1996.tb01790.x

Krumpal, I. (2013). Determinants of social desirability bias in sensitive surveys: A literature review. *Quality & Quantity: International Journal of Methodology, 47*(4), 2025–2047. https://doi.org/10.1007/s11135-011-9640-9

Kulkarni, V., Kern, M. L., Stillwell, D., Kosinski, M., Matz, S., Ungar, L., Skiena, S., & Schwartz, H. A. (2018). Latent human traits in the language of social media:

An open-vocabulary approach. *PLOS ONE, 13*(11), e0201703. https://doi.org/10.1371/journal.pone.0201703

Lau, I. Y.-M., Chiu, C., & Lee, S. (2001). Communication and shared reality: Implications for the psychological foundations of culture. *Social Cognition, 19*(3), 350–371. https://doi.org/10.1521/soco.19.3.350.21467

Lee, J., Park, D.-H., & Han, I. (2008). The effect of negative online consumer reviews on product attitude: An information processing view. *Electronic Commerce Research and Applications, 7*(3), 341–352. https://doi.org/10.1016/j.elerap.2007.05.004

Liu, T., Zhang, N. L., & Chen, P. (2014). Hierarchical latent tree analysis for topic detection. In T. Calders, F. Esposito, E. Hüllermeier, & R. Meo (Eds.), *Machine learning and knowledge discovery in databases* (pp. 256–272). Springer. https://doi.org/10.1007/978-3-662-44851-9_17

Ludwig, S., de Ruyter, K., Friedman, M., Brüggen, E. C., Wetzels, M., & Pfann, G. (2013). More than words: The influence of affective content and linguistic style matches in online reviews on conversion rates. *Journal of Marketing, 77*(1), 87–103. https://doi.org/10.1509/jm.11.0560

Mikolov, T., Chen, K., Corrado, G., & Dean, J. (2013). *Efficient estimation of word representations in vector space.* arXiv. http://arxiv.org/abs/1301.3781

Mikolov, T., Sutskever, I., Chen, K., Corrado, G. S., & Dean, J. (2013). Distributed representations of words and phrases and their compositionality. In C. J. C. Burges, L. Bottou, M. Welling, Z. Ghahramani, & K. Q. Weinberger (Eds.), *Advances in neural information processing systems* (Vol. 26, pp. 3111–3119). Curran Associates.

Mosteller, F., & Wallace, D. L. (1963). Inference in an authorship problem: A comparative study of discrimination methods applied to the authorship of the disputed Federalist papers. *Journal of the American Statistical Association, 58*(302), 275–309. https://doi.org/10.1080/01621459.1963.10500849

O'Reilly, C. A., Chatman, J., & Caldwell, D. F. (1991). People and organizational culture: A profile comparison approach to assessing person–organization fit. *Academy of Management Journal, 34*(3), 487–516. https://doi.org/10.5465/256404

Obschonka, M., Fisch, C., & Boyd, R. (2017). Using digital footprints in entrepreneurship research: A Twitter-based personality analysis of superstar entrepreneurs and managers. *Journal of Business Venturing Insights, 8*, 13–23. https://doi.org/10.1016/j.jbvi.2017.05.005

Park, G., Schwartz, H. A., Eichstaedt, J. C., Kern, M. L., Kosinski, M., Stillwell, D. J., Ungar, L. H., & Seligman, M. E. P. (2015). Automatic personality assessment through social media language. *Journal of Personality and Social Psychology, 108*(6), 934–952. https://doi.org/10.1037/pspp0000020

Pennebaker, J. W. (2013). *The secret life of pronouns: What our words say about us.* Bloomsbury Press.

Pennebaker, J. W., Booth, R. J., Boyd, R. L., & Francis, M. E. (2015). *Linguistic inquiry and word count: LIWC2015.* Pennebaker Conglomerates.

Pennebaker, J. W., Boyd, R. L., Jordan, K., & Blackburn, K. (2015). *The development and psychometric properties of LIWC2015* [Unpublished manuscript]. University of Texas at Austin. https://repositories.lib.utexas.edu/handle/2152/31333

Pennebaker, J. W., & Graybeal, A. (2001). Patterns of natural language use: Disclosure, personality, and social integration. *Current Directions in Psychological Science, 10*(3), 90–93. https://doi.org/10.1111/1467-8721.00123

Pennebaker, J. W., & King, L. A. (1999). Linguistic styles: Language use as an individual difference. *Journal of Personality and Social Psychology, 77*(6), 1296–1312. https://doi.org/10.1037/0022-3514.77.6.1296

Pennebaker, J. W., Mehl, M. R., & Niederhoffer, K. G. (2003). Psychological aspects of natural language use: Our words, our selves. *Annual Review of Psychology, 54*(1), 547–577. https://doi.org/10.1146/annurev.psych.54.101601.145041

Pennington, J., Socher, R., & Manning, C. (2014). GloVe: Global vectors for word representation. In A. Moschitti, B. Pang, & W. Daelemans (Eds.), *Proceedings of the 2014 Conference on Empirical Methods in Natural Language Processing (EMNLP)* (pp. 1532–1543). Association for Computational Linguistics. https://doi.org/10.3115/v1/D14-1162

Podium. (2017). *State of online reviews* [Annual report]. https://www.podium.com/resources/podium-state-of-online-reviews/

Podsakoff, P. M., MacKenzie, S. B., Lee, J.-Y., & Podsakoff, N. P. (2003). Common method biases in behavioral research: A critical review of the literature and recommended remedies. *Journal of Applied Psychology, 88*(5), 879–903. https://doi.org/10.1037/0021-9010.88.5.879

Qiu, L., Lin, H., Ramsay, J., & Yang, F. (2012). You are what you tweet: Personality expression and perception on Twitter. *Journal of Research in Personality, 46*(6), 710–718. https://doi.org/10.1016/j.jrp.2012.08.008

Quercia, D., Kosinski, M., Stillwell, D., & Crowcroft, J. (2011). Our Twitter profiles, our selves: Predicting personality with Twitter. In *Proceedings of the 2011 IEEE International Conference on Privacy, Security, Risk, and Trust, and IEEE international Conference on Social Computing* (pp. 180–185). IEEE. https://doi.org/10.1109/PASSAT/SocialCom.2011.26

Raihan, M., Mondal, S., More, A., Sagor, M. O. F., Sikder, G., Majumder, M. A., Manjur, M. A. A., & Ghosh, K. (2016). Smartphone based ischemic heart disease (heart attack) risk prediction using clinical data and data mining approaches, a prototype design. In *2016 19th International Conference on Computer and Information Technology (ICCIT)* (pp. 299–303). IEEE. https://doi.org/10.1109/ICCITECHN.2016.7860213

Rude, S., Gortner, E.-M., & Pennebaker, J. (2004). Language use of depressed and depression-vulnerable college students. *Cognition and Emotion, 18*(8), 1121–1133. https://doi.org/10.1080/02699930441000030

Schwartz, H. A., Eichstaedt, J. C., Kern, M. L., Dziurzynski, L., Ramones, S. M., Agrawal, M., Shah, A., Kosinski, M., Stillwell, D., Seligman, M. E. P., & Ungar, L. H. (2013). Personality, gender, and age in the language of social media: The open-vocabulary approach. *PLOS ONE, 8*(9), e73791. https://doi.org/10.1371/journal.pone.0073791

Segalin, C., Perina, A., Cristani, M., & Vinciarelli, A. (2017). The pictures we like are our image: Continuous mapping of favorite pictures into self-assessed and attributed personality traits. *IEEE Transactions on Affective Computing, 8*(2), 268–285. https://doi.org/10.1109/TAFFC.2016.2516994

Soleymani, M., Garcia, D., Jou, B., Schuller, B., Chang, S.-F., & Pantic, M. (2017). A survey of multimodal sentiment analysis. *Image and Vision Computing, 65*, 3–14. https://doi.org/10.1016/j.imavis.2017.08.003

Srivastava, S. B., Goldberg, A., Manian, V. G., & Potts, C. (2018). Enculturation trajectories: Language, cultural adaptation, and individual outcomes in organizations. *Management Science, 64*(3), 1348–1364. https://doi.org/10.1287/mnsc.2016.2671

Sterling, J., Jost, J. T., & Bonneau, R. (2020). Political psycholinguistics: A comprehensive analysis of the language habits of liberal and conservative social media users. *Journal of Personality and Social Psychology, 118*(4), 805–834. https://doi.org/10.1037/pspp0000275

Tinsley, H. E., & Weiss, D. J. (1975). Interrater reliability and agreement of subjective judgments. *Journal of Counseling Psychology, 22*(4), 358–376. https://doi.org/10.1037/h0076640

Valenzuela, S., Piña, M., & Ramírez, J. (2017). Behavioral effects of framing on social media users: How conflict, economic, human interest, and morality frames drive news sharing. *Journal of Communication, 67*(5), 803–826. https://doi.org/10.1111/jcom.12325

Vosoughi, S., Roy, D., & Aral, S. (2018). The spread of true and false news online. *Science, 359*(6380), 1146–1151. https://doi.org/10.1126/science.aap9559

Wall, M. E., Rechtsteiner, A., & Rocha, L. M. (2003). Singular value decomposition and principal component analysis. In D. P. Berrar, W. Dubitzky, & M. Granzow (Eds.), *A practical approach to microarray data analysis* (pp. 91–109). Springer. https://doi.org/10.1007/0-306-47815-3_5

Ward, G., Schwartz, H. A., Giorgi, S., Menger, J., & Matz, S. C. (2021). *Negative emotions and populist voting* [Working paper]. Sloan School of Management, MIT.

Willemsen, L. M., Neijens, P. C., Bronner, F., & de Ridder, J. A. (2011). "Highly recommended!" The content characteristics and perceived usefulness of online consumer reviews. *Journal of Computer-Mediated Communication, 17*(1), 19–38. https://doi.org/10.1111/j.1083-6101.2011.01551.x

Woebot Health. (2021). *Board members.* https://woebothealth.com/board-members/

World Health Organization. (2020). *Mental health included in the UN Sustainable Development Goals.* https://apps.who.int/iris/handle/10665/340847

Yarkoni, T. (2010). Personality in 100,000 words: A large-scale analysis of personality and word use among bloggers. *Journal of Research in Personality, 44*(3), 363–373. https://doi.org/10.1016/j.jrp.2010.04.001

Youyou, W., Kosinski, M., & Stillwell, D. (2015). Computer-based personality judgments are more accurate than those made by humans. *Proceedings of the National Academy of Sciences, 112*(4), 1036–1040. https://doi.org/10.1073/pnas.1418680112

Youyou, W., Stillwell, D., Schwartz, H. A., & Kosinski, M. (2017). Birds of a feather do flock together: Behavior-based personality-assessment method reveals personality similarity among couples and friends. *Psychological Science, 28*(3), 276–284. https://doi.org/10.1177/0956797616678187

Zhang, L., Wang, S., & Liu, B. (2018). Deep learning for sentiment analysis: A survey. *WIREs Data Mining and Knowledge Discovery, 8*(4), e1253. https://doi.org/10.1002/widm.1253

3

The Big Data Toolkit for Psychologists: Data Sources and Methodologies

Heinrich Peters, Zachariah Marrero, and Samuel D. Gosling

As human interactions have shifted to virtual spaces and as sensing systems have become more affordable, an increasing share of people's everyday lives can be captured in real time. The availability of such fine-grained behavioral data from billions of people has the potential to enable great leaps in our understanding of human behavior. However, such data also pose challenges to engineers and behavioral scientists alike, requiring a specialized set of tools and methodologies to generate psychologically relevant insights.

In particular, researchers may need to utilize machine learning techniques to extract information from unstructured or semistructured data, reduce high-dimensional data to a smaller number of variables, and efficiently deal with extremely large sample sizes. Such procedures can be computationally expensive, requiring researchers to balance computation time with processing power and memory capacity. Whereas modeling

https://doi.org/10.1037/0000290-004
The Psychology of Technology: Social Science Research in the Age of Big Data, S. C. Matz (Editor)
Copyright © 2022 by the American Psychological Association. All rights reserved.

procedures on small data sets will usually take mere moments to execute, applying modeling procedures to Big Data can take much longer with typical execution times spanning hours, days, or even weeks depending on the complexity of the problem and the resources available. Seemingly subtle decisions regarding preprocessing and analytic strategy can end up having a huge impact on the viability of executing analyses within a reasonable timeframe. Consequently, researchers must anticipate potential pitfalls regarding the interplay of their analytic strategy with memory and computational constraints.

Many researchers who are interested in using Big Data report having problems learning about new analytic methods or software, finding collaborators with the right skills and knowledge, and getting access to commercial or proprietary data for their research (Metzler et al., 2016). This chapter serves as a practical introduction for psychologists who want to use large data sets and data sets from nontraditional data sources in their research (i.e., data not generated in the lab or through conventional surveys). First, we discuss the concept of Big Data and review some of the theoretical challenges and opportunities that arise with the availability of ever larger amounts of data. Second, we discuss practical implications and best practices with respect to data collection, data storage, data processing, and data modeling for psychological research in the age of Big Data.

WHAT IS BIG DATA?

The term "Big Data" is discussed across many scientific disciplines, but there is still no single widely accepted definition of the concept. In fact, different disciplines define Big Data in vastly different ways. In the fields of computer science and engineering, Big Data is usually defined as data that are generated with such volume, variety, and velocity that special infrastructure and software is needed to store, process, and analyze it (see Borgman, 2015). Companies like Amazon, Google, and Facebook, for example, maintain data centers consisting of hundreds of thousands of servers needed to manage the quantities of data that are generated on their platforms. Another common definition states that data sets that do not fit in the memory of a single machine can be considered Big Data. Of course, such definitions depend on the tools and technologies that are available in a given field at a given point in time, rendering the concept of Big Data a moving target. Until recently psychologists have rarely had access to quantities of data that pose challenges to state-of-the-art hardware and software, and the term "Big Data" has commonly been used to describe data from nontraditional data sources, such as social network sites or smartphone apps. Even though the vast majority of such data sets would not be considered "big" in the fields of computer science or engineering, they typically (a) are much larger than the data sets psychologists have traditionally been generating through lab experiments and surveys, and (b) require and enable analytic techniques that have traditionally not been part of behavioral scientists' repertoires. In other words, when psychologists use the term Big Data, they often refer to data that is "big" relative to the standard of the field and that mimics "actual" Big Data in terms of how it is generated, processed, and analyzed.

A PSYCHOLOGICAL PERSPECTIVE ON BIG DATA: BIG N, BIG V, BIG T

Raymond B. Cattell (1953) famously argued that the methods used in psychological research should be informed by the structure of psychological data. Cattell suggested that data can be classified along three dimensions: persons, variables, and occasions. This taxonomy can be leveraged to describe the size and complexity of data sets, with sample size (Big N), number of variables (Big V),[1] and temporal structure (Big T) serving as the defining criteria of Big Data in psychological research (Adjerid & Kelley, 2018).

Big N refers to data from a large number of individuals. In the context of behavioral science, these data include sample sizes of hundreds of thousands or millions of individuals. Person-level data collected from social

---

[1] In computer science and statistics, the numbers of rows and columns in a matrix are often denoted as n and p, such that p represents the number of variables. We use this notation in a later section on computational complexity, but we will continue to use "Big V" when we refer to the taxonomy of Big Data introduced by Adjerid and Kelley (2018).

media profiles or through large-scale online surveys usually fall into this category. The high participation rate in digital platforms allows for the collection of diverse samples and gives researchers access to demographics that have been traditionally underrepresented in psychological research (Gosling et al., 2004; Kosinski et al., 2015). Furthermore, large samples allow for the investigation of small but meaningful effects (Matz et al., 2017) and the use of complex models while avoiding overfitting (Yarkoni & Westfall, 2017). The latter point is especially important in light of the tendency of underpowered studies to produce not only Type II but also Type I errors, which may have contributed to the so-called replication crisis (Open Science Collaboration, 2015).

The term *Big V* data refers to the number, diversity, and complexity of variables a data set contains. Big V data are prevalent in exploratory contexts where a multitude of potential predictors is considered; this is often the case when semistructured or unstructured data are used. Many methods in natural language processing (NLP), for example, rely on the analysis of word-frequency distributions within and across documents. Text data are typically represented in the form of document-term matrices, where the frequency of each individual word is treated as a variable. Hence, text data are typically very high dimensional. Similarly, image data are usually high dimensional because a variety of features can be extracted from images. Big V data can also be generated through feature engineering (i.e., creating large numbers of derived variables from lower dimensional raw data). Big V data pose special challenges to statistical procedures (e.g., overfitting, computational costs) and call for specialized techniques, including regularization and variable selection or dimensionality reduction.

Big T refers to data that contains large amounts of temporal information (i.e., a large number of measurement occasions per person). This property is native to studies involving log data or physiological measures (e.g., smartphone sensing or EEG studies), where data can be recorded through a device's sensing systems with near continuous granularity, which can result in hundreds of thousands of measurements per person. Such granularity of observations offers unprecedented opportunities to describe human behavior in great detail and to understand within-person psychological processes. Complex longitudinal data call for sophisticated aggregation methods in order to derive interpretable insights. Machine learning techniques can be used to cluster or classify events and activities based on their underlying temporal structure. Finally, Big T data may require the use of specialized statistical models for longitudinal data, such as time series regression, growth curve modeling, or time-to-event regression, and specialized preprocessing in order to account for time lags or to ensure stationarity.

Importantly, data can be "big" on one, two, or all three dimensions, and in practice the boundaries between Big V, and Big T may be fuzzy because a Big T data set can be turned into a Big V data set by extracting large numbers of variables that capture how each individual behaves over time. GPS data, for instance, usually consist of only three main variables (latitude–longitude pairs and time stamps),[2] with thousands of data points collected per individual; these longitudinal data can be aggregated in numerous ways to create meaningful person-level variables and enriched with additional information about the characteristics of the areas and places visited by each person (e.g., demographics, events, weather), resulting in an even richer set of variables.

THEORETICAL CONSIDERATIONS: EXPLANATION VERSUS PREDICTION

Psychology has traditionally been chiefly concerned with causal explanations of human behavior. Explanatory research uses statistical modeling to test causal hypotheses derived from theory (Shmueli, 2010). The predominant methods of explanatory research in the field of psychology revolve around controlled experiments and statistical inference, where conclusions about the broader population are derived from the analysis of in-sample relationships, and models are used to confirm predictions from theory. These procedures follow a linear process that relies on researcher

---

[2] GPS records can also contain other variables, such as altitude, bearing, and accuracy estimates.

expertise and statistical theory. First, researchers must select variables and forms of association (e.g., interactions, curvature) that are plausibly related to the dependent variable. Second, researchers select an appropriate data model (e.g., linear model), which is explicitly defined in statistical theory (usually frequentist), and estimate its associated model parameters from the sample data. Third, researchers apply statistical inference tests (i.e., significance tests) to the model parameters based on statistical theory and then confirm that the observed test score (e.g., T score, p value, F ratio) is valid by ensuring that the statistical assumptions required by the test have been satisfied. Fourth, if the assumptions are satisfied and the researcher has concluded that there is a low probability that the data could have been generated by chance, the test is interpreted as evidence that the model is generalizable. Alternatively, if the final model did not yield sufficient evidence, researchers are discouraged from repeating the analysis with different analytic decisions on the same data because doing so corrupts the statistical inference (e.g., Type I error inflation). Despite the widespread popularity of the explanatory approach among psychologists, the current replication crisis suggests that existing explanatory models in the field of psychology generalize poorly to new samples (Camerer et al., 2018; Yarkoni & Westfall, 2017). This may be explained in part by an overreliance on small sample sizes in experimental research and in part by questionable research practices and procedural overfitting (i.e., p hacking).

Predictive research, in contrast to explanatory research, is generally not concerned with causal theory and focuses squarely on the empirical generalizability of statistical models to new or future observations (Shmueli, 2010). For example, whereas the traditional explanatory procedure uses a number of assumptions to conclude that an estimated model is representative of any sample from the population, predictive models are specifically selected based on how well a model estimated in one sample accounts for variation in a different sample. To do this, predictive modeling relies heavily upon algorithmic procedures that are far less demanding on statistical theory (e.g., cross-validation, hyperparameter tuning, feature selection). As in explanatory modeling, researchers must first select the variables of interest, but the theoretical relationships between the variables are usually not a primary concern. Second, researchers have to prepare their data for analysis according to the algorithms they wish to apply. Third, researchers apply the algorithm(s) and evaluate the effectiveness using evaluation metrics that capture how well a model performs on new data. Fourth, a judgment is made regarding the quality of the prediction, usually by comparing the performance against a reference model (e.g., a model predicting the mean in the case of continuous data or the mode in the case of categorical data).

In the past, behavioral and social scientists have largely avoided this purely predictive, less theory-driven approach, but the rise of Big Data with its wealth of unstructured, associative, behavioral data has made predictive research more attractive. The attraction of predictive models is driven by three key factors. First, large real-world data sets often contain complex patterns and relationships that have little grounding in existing theory and are hard to cast in terms of simple hypotheses. Second, the traditional methodological toolkit of psychological scientists is quickly pushed to its limits when confronted with Big Data. Significance levels lose their meaning when applied to very large samples because even the tiniest effects become significant (Big N) and traditional statistical methods are ill equipped to deal with an abundance of predictors (Big V) or complex temporal structures with many measurement instances (Big T). Third, Big Data enables researchers to employ "data-hungry" machine learning techniques and to test the generalizability of their models "out-of-sample," which constitutes a key aspect of the predictive framework. At the same time, the methods that are most suitable to analyze large, complex data sets, such as shrinkage models (Lasso, Ridge, elastic net), ensemble models (random forests, gradient boosting), or "black-box" models like neural networks, are not well suited for explanatory research. Penalized linear models, for example, produce biased coefficient estimates, and random forests, gradient boosting, and neural networks capture complex nonlinearities and interactions, which are hard or impossible to infer from model parameters and would be too complex to interpret in light of existing theory.

Nonetheless, as Shmueli (2010) pointed out, the predictive approach still serves several vital scientific functions: First, by taking into account complex data and a multitude of potential relationships, predictive modeling can uncover new phenomena and causal mechanisms, thereby generating new hypotheses and starting points for theoretical advancement and explanatory approaches. Second, predictive modeling can be used to discover new measures and operationalizations; for example, personality scores extracted from social media data seem to predict a range of outcome variables better than questionnaire-based trait scores (Kulkarni et al., 2018). Third, predictive modeling can serve as a reality check for existing explanatory theories by assessing their predictive power and generalizability in real-world contexts. To leverage the benefits of predictive modeling, psychologists must expand their methodological toolkit. The next section serves as an introduction for psychologists who want to take advantage of nontraditional data sets and engage in predictive research. For an in-depth discussion of explanation and prediction in statistics and psychology, interested readers are directed to Shmueli (2010), Stachl et al. (2020), and Yarkoni and Westfall (2017).

PRACTICAL RECOMMENDATIONS AND TOOLS

This section introduces some of the practical aspects of working with large, complex data sets. Given the scope of the present chapter, we can only scratch the surface of the many issues relevant to data collection, storage, processing, and modeling. However, this section may serve as a starting point for researchers who are planning to work with Big Data.

Data Collection

Digital platforms and other companies collect an abundance of behavioral data that can be used for psychological research. In addition to collecting their own data, psychologists can gain access to interesting secondary data in multiple ways, including by querying data that is made available by companies through specific application programming interfaces (APIs), scraping unstructured data from the web, or by collaborating with companies to gain access to their user base and data. The present section briefly outlines each of these options.

Application Programming Interfaces (APIs)

An API is a computing interface that allows applications to communicate with each other. As such, APIs allow for the transfer of data between applications and enable the integration of third-party applications. For instance, if a user logs into an app with their Facebook or Google account, this process is handled in the background by a Facebook or Google API, which confirms the user's authenticity and gives the app access to some of the user's data in a standardized process. In the context of data collection for psychological research, APIs can be utilized to authenticate and track users across different platforms (given informed consent), and researchers can query the APIs of digital platforms to access data that is made available by the respective provider. Generally, APIs are the official infrastructure that a provider wants third parties to use. Hence, they should be the first option to consider when data have to be collected from social media platforms or other websites.

The most common type of APIs are representational state transfer (REST) APIs. REST is a standard that defines architectural constraints for web services to ensure interoperability. REST APIs can access and manipulate web resources as identified by their uniform resource locators (URLs) through the hypertext transfer protocol (HTTP). The most important types of HTTP commands enabled by REST APIs are: GET (fetch data), PUT (edit existing data), POST (add new data), and DELETE (delete data). For researchers interested in data collection, GET is the most important command. GET takes as input a uniform resource identifier (URI) consisting of a base URL and a list of query parameters, usually including an authentication key/token.

The response of a REST API call contains the requested data, typically formatted in XML or JSON. REST API calls are facilitated by libraries like *httr* in R and *requests* in Python, which allow users to efficiently manage query parameters and extract relevant data from the response. Besides REST APIs there are other APIs, such as the Facebook graph API, which allows third parties to access the network graph representing users, their relationships, and records of their personal data. The graph API can be queried just like a REST API by combining a base URL with a list of query parameters. Prominent research that made use of the Facebook API was published by Kosinski et al. (2013), who collected behavioral data along with self-report data from dozens of psychometric tests from more than 10 million Facebook users (Kosinski et al., 2015) who volunteered to donate their data. The same API was later abused by Cambridge Analytica in violation of user agreements to harvest Facebook profiles of millions of users and run targeted political ads (Isaak & Hanna, 2018; Schneble et al., 2018). In response to the increasing societal awareness of privacy concerns, businesses have become increasingly strict with regard to the data they allow third parties to access through their APIs, which poses challenges for researchers.

Web-Scraping

The term "web-scraping" describes a range of methods that automatically extract information from websites by transforming the unstructured source data of a web page into a structured format. Taking advantage of regularities in URL structure, web page structure, and content structure, it is often possible to scrape large amounts of data by looping over many web pages. Each web page has a source code written in markup language (e.g., HTML), which contains its information in a machine readable format. Web browsers, for instance, read the source code to allow them to display the content of a web page to the user. Web-scraping typically follows a five-step process to extract information from the source code of web pages: First, it is necessary to locate the URLs of the web pages that contain the relevant data. Second, the source code needs to be fetched from each URL. Third, the source code is parsed into logical units according to its markdown structure (typically HTML elements). Fourth, it is possible to search for the elements that contain the relevant data. And fifth, the often-unstructured data need to be reformatted into a structured format for further analysis. These steps can be automated using loops and other control structures in conjunction with powerful libraries, such as *requests* and *BeautifulSoup* in Python or *httr* and *rvest* in R.

Such libraries facilitate the process considerably but web-scraping has several important drawbacks: It is still relatively laborious and error prone, data quality is often less than optimal, and it relies critically on decisions made by third parties. In fact, many online platforms prohibit scraping and protect themselves against it, for example by imposing rate limits on IP addresses, or requiring login credentials. For most legitimate websites, researchers can find out to what extent they are allowed to scrape, by appending "/robots.txt" to the website's base URL. As a rule of thumb, whenever a platform has an API, it is advisable to use this official interface and use scraping only as a last resort.

Industry Collaborations

Large data sets can be collected through research collaborations with companies and other organizations. Many companies collect large amounts of data as a part of their day-to-day business operations. Such secondary data can be of great value for researchers. Alternatively, companies are sometimes willing to participate in the collection of primary data by leveraging their platforms and customer base. For example, relatively large data sets of psychological data have been collected through surveys in collaboration with the BBC (Rentfrow et al., 2015) and *Time* magazine (Ebert et al., 2019). High-profile industry collaborations promise high rewards, but there are usually high barriers that prevent companies from sharing proprietary data with researchers. These barriers include privacy and legal concerns resulting from the high level of public scrutiny to which large companies are subjected. However, companies sometimes make data available to the public in anonymized and aggregated form in the context of "data for good" initiatives. For example, Facebook made available aggregate mobility data from roughly 29 million users to enable researchers to study the adoption of social distancing during the COVID-19 crisis (Facebook Data for Good, 2020; see also Peters et al., 2020). But it is not only tech giants that accumulate data of interest to psychologists; startups often collect substantial amounts of data that could be used for psychological research. It is sometimes easier to collaborate with startups because they tend to impose lower bureaucratic hurdles and usually benefit more from researchers working with their data; for example, collaboration with researchers can lend credence to a company's product or provide management with valuable information about its user base.

Data Storage

Collecting large amounts of data requires dealing with issues of data storage. Smaller data sets that do not require frequent updates can easily be held as files on local drives but larger, more complex data require a principled approach to data storage in order to maintain data integrity and to ensure that the data can be easily accessed, updated, and shared; this can be achieved through traditional relational database management systems (RDBMSs) or distributed systems optimized for Big Data processing, such as Apache Hadoop or data warehouse solutions like Amazon Redshift or Google BigQuery.

The term RDBMS refers to databases that store data in tables or "relations" consisting of rows and columns. Tables represent real-world entities and their relationships with each other, such that insights can be generated by merging and filtering related tables. For example, one table could contain demographic data for users of an app, while other tables would contain time-stamped behavioral records, indicating every time a user has opened and closed the app or interacted with any feature within the app. A researcher can then calculate derived metrics like overall app-usage time per user, merge this derived table with the demographic data and filter the resulting table to obtain data only from individuals born in a specific geographic region. Every relation in an RDBMS is represented by a set of schemata: A *logical schema* is a conceptual model of the data, which describes data types and variable names, and so on. A *physical schema* describes the data layout (files, indices) and how the data are represented in secondary storage. A *virtual schema* (view) refers to derived tables, which are generated by way of merging and filtering to derive insights from data. Most popular RDBMSs use the Structured Query Language (SQL) for querying and maintenance. SQL is a data manipulation language in the sense that it allows users to query, insert, delete, and modify entries, but also a data definition language because it allows users to define relational schemata (i.e., create, alter, and delete tables and their attributes). RDBMSs are useful because they provide a highly reliable, standardized, and intuitive way to organize structured or semistructured data. However, they are quickly stretched to their limits once the volume of data exceeds the realm of gigabytes, especially in the case of unstructured data.

If truly massive amounts of data need to be stored and processed (i.e., data that are too large for a single server), distributed files systems are a viable option. A popular open-source file system is the Hadoop Distributed File System (HDFS), which organizes large files across multiple machines. Data stored in HDFS can be processed with MapReduce, a programming model that is optimized for parallel and distributed processing. A MapReduce-command is typically composed of a map function, which filters and sorts entries according to specified criteria, and a reduce function, which performs summary operations. Importantly, HDFS also enables queries in SQL-like syntax through systems like Apache Hive or Apache Spark SQL. Apache Spark additionally contains a library of parallelized, distributed data manipulation commands and distributed machine-learning algorithms optimized for Big Data applications. These libraries are crucial to make efficient use of distributed storage and computational resources. All of the Big Data systems mentioned above have programming interfaces for R and Python.

RDBMS as well as distributed file systems can be hosted either on private servers or in the cloud (e.g., Amazon Web Services, Google Cloud Platform, Microsoft Azure). Using cloud systems offers several advantages. First, cloud platforms provide access to preconfigured infrastructure and environments, so very little effort is required for setup and maintenance, compared to a private server. Second, storage is virtually unlimited and can easily be extended or reduced depending on the user's current needs. Third, if large amounts of data need to be processed and analyzed, this can be done within the cloud environment, using the computational resources of the cloud, without the need to transfer any data. Private servers, on the other hand, can in rare cases offer more fine-grained control, and may be required due to privacy protocols (e.g., university ethics guidelines may prevent researchers from storing sensitive data on cloud platforms). Another cloud-based alternative to traditional RDBMSs and distributed file systems are data warehousing solutions like Amazon Redshift or Google

BigQuery, which mimic the look and feel of traditional SQL query tools but are optimized for Big Data and provide an extensive array of additional analytic capabilities.

We have summarized a variety of options for data storage, ranging from local drives to traditional RDBMSs, open source distributed files systems, and commercial data warehousing solutions. The optimal solution for a given research project depends on the scale, access frequency, and update frequency of the data. Typically, large scale projects with high access and update frequency require more principled storage solutions (i.e., RDBMS, distributed file systems, or data warehousing). Finally, researchers do not have to become database engineers, but it is certainly advisable to learn the basics of SQL and how to set up and connect to remote databases or distributed file systems.

Data (Pre) Processing

Data preprocessing is the transformation of raw data into a format that is suitable for modeling. Working with complex data requires a flexible, but principled approach to preprocessing. Big Data, like any data that were not collected in a controlled lab setting, poses specific challenges to preprocessing:

- Raw data often contain a significant amount of low quality data and information that is irrelevant to the research question at hand.
- Information is usually distributed across many different data sources, which need to be filtered and merged.
- The data are often very sparse and contain a lot of missing or corrupted values.
- Psychologically relevant variables need to be derived from fine-grained behavioral data, and the level of granularity often needs to be reduced in order to answer psychologically relevant questions.

To address these challenges, researchers need to make important decisions with regard to filtering and merging data from different data sources, handling missingness and sparsity, as well as feature engineering.

Filtering and Merging

Big Data is often secondary data that were not originally collected for the purpose of psychological research and may include a lot of information that is irrelevant to the research question at hand. At the same time, the information is usually distributed across many different data sources that need to be combined to answer psychologically relevant questions. As a consequence, researchers need to filter the raw data to extract the relevant information and to merge data to bring together records of the same entities in the appropriate units of analysis. The removal of irrelevant information is a complex process because sophisticated analyses may be needed to classify the data. The merging of data may be challenging because there is not always a unique identifier that allows records to be matched.

Filtering and merging can, in principle, be performed at two stages in the data analysis process: (a) when querying the data from a database or (b) after reading the data into the data analysis environment. If it is already clear which records are required for an analysis, and there is also a unique identifier that allows related records to be matched, filtering and merging should be performed early on (i.e., when querying the database). This approach reduces analytic flexibility, but it is usually preferable when working with large data sets because it reduces the use of memory and computational resources. Performing the filtering and merging in the data analysis environment has the disadvantage of loading irrelevant data into memory only to be discarded later on, which can easily become a problem when large amounts of data are processed. Both R and Python can be used to query RDBMS or HDFS and other cloud systems.

However, it is not always possible to filter the required data at the query stage because doing so often requires a deep understanding of the data and may be based on complex exploratory analyses. For example, when analyzing data from social media platforms, researchers may be interested only in specific types of content, such as posts discussing a certain topic. Filtering content can require the use of classification algorithms as part of the preprocessing procedure. Similarly, it may be necessary to distinguish between activity generated by actual users and activity generated by bots, which would be considered noise in most psychological contexts. An estimated 15% of Twitter "users" are bots, and identifying bots may require sophisticated classification models (Varol et al., 2017).

In many applications a critical component of data quality is the amount of data that are available per person. Especially when raw data are transformed into derived metrics, the measurement quality of the derived metrics depends on the amount of raw data available for each person (e.g., Youyou et al., 2015). In the context of user data, the amount of data generated per user often follows a gamma distribution where many users generate very little data and an increasingly smaller number of users generate a lot of data. For example, many people install an app and never use it or open it just a few times, whereas a much smaller subset of people will use the app regularly over a long period of time. Consequently, derived measures will differ greatly from user to user in terms of quality. It is therefore important to implement sensible exclusion criteria as part of the preprocessing procedure. Setting a low threshold sacrifices measurement quality while preserving sample size. Setting a high threshold can increase measurement quality while reducing sample size but can bias the data in subtle ways, especially when the absence of data is not just the result of idiosyncrasies in data collection but is indicative of the absence of behavior. In GPS-tracking studies, for example, people's location trajectories are sometimes recorded though event-based sampling (i.e., a person's location is recorded only if the person moves from one location to the next); by implication, the people who show the least amount of mobility will produce the smallest number of data points, while the lack of mobility may be related to key variables (e.g., mental health outcomes; Müller et al., 2020). Setting a strict threshold on the amount of mobility data may therefore discard an important part of the distribution and decrease the chance of detecting meaningful relationships between mobility and relevant outcome variables.

Merging data can be a challenge, especially when there is no unique identifier across multiple data sources, which is often the case for secondary data collected from multiple sources. If researchers have control over data collection the identifiability of records should be a primary concern. For example, it is possible to track users across different platforms by using APIs for authentication or by recording device identifiers. If there is no unique identifier, data can still be merged on the basis of other identifying information such as names, addresses, and birth dates or other demographic data and GPS data. However, such data is rarely formatted identically across data sources. Nonetheless it is possible to merge such data based on fuzzy matching techniques, which take into account the similarity, rather than exact congruence, of one or more identifying variables. The packages *fuzzyjoin* in R and *fuzzywuzzy* in Python enable fuzzy merging, but they do not perform very well at scale. For truly large scale data sets in the range of millions of records, it is necessary to use more complex techniques, such as the fastlink algorithm (Enamorado et al., 2019), which is implemented in the *fastlink* package for R. The same algorithm is implemented in the PySpark package *splink*, which works well at even greater scale (100 million records or more). Fuzzy merging algorithms can also be used for the purpose of deduplicating (i.e., removing duplicated records from a data set).

Missing and Sparse Data

Missing data are a prevalent phenomenon in Big Data contexts. In many cases a substantial proportion of the data is missing, especially when data sets are generated by combining data from multiple sources. The issue of missing data deserves particular attention because missingness can have significant impact on statistical analyses and may be handled differently, depending on the researchers' choice to engage in explanatory or predictive research. Missing data can be categorized into three categories: (a) missing completely at random (MCAR), when missingness depends neither on observed variable values nor on unobserved variable values themselves; (b) missing at random (MAR), when missingness does not depend on the unobserved values themselves, but is explained by observed variable values; and (c) missing not at random (MNAR), when missingness depends on the unobserved values themselves (i.e., the data are neither MCAR nor MAR). The missingness mechanism is important because it can inform researchers' choices with respect to preprocessing and modeling.

Generally, missing data can be addressed in three different ways. Researchers may decide to (a) delete records affected by missingness, (b) impute missing values, or (c) use the fact that a datapoint is missing as additional information in a predictive model. First, deleting records with missing values is the simplest strategy, but it may reduce sample size and power; this is usually not a concern in Big N contexts, but it can be prohibitive in Big V contexts, in which almost all records may have some missing values. Additionally, deletion can lead to biased coefficient estimates in the absence of MCAR. Second, imputation of missing values can be performed by substituting missing values with measures of central tendency (e.g., mean) or through model-based and multiple imputation, where missing values are predicted from observed values. Model-based imputation can be computationally costly in Big Data contexts, but it leads to more realistic distributions compared to mean substitution. Imputation, especially mean substitution, increases sample size relative to the amount of information that is captured in a data set, which may bias error estimates and render imputation problematic in the context of explanatory modeling and statistical inference. Third, in certain cases missingness can be highly informative of the target variable (when the data are not MCAR). Therefore the performance of predictive models can sometimes be dramatically improved by including dummy variables indicating missingness. In financial fraud prediction, for example, failure to report certain metrics may be indicative of fraudulent behavior. Similarly, failure to respond to self-report surveys or record one's behaviors may be indicative of the very construct that is being measured, such as when tracking alcohol intake or monitoring depressive symptoms. Explicitly modeling missing data can thus be helpful in predictive research, while it is usually not expedient in explanatory research.

Related but distinct from the issue of missing data is the issue of sparse data. Especially when dealing with behavioral records, it is crucial for researchers to know whether behavior might have occurred but failed to be recorded (missing data), or whether behavior was correctly not recorded because it did not occur (sparse data). Behavioral data are typically captured as a list of time-stamped events, which are transformed into derived metrics. When such a table is merged with other data, it is very likely that some records cannot be matched because not every individual exhibited every type of behavior. This assumption is especially true in the context of Big Data, where the data may be composed of a wide variety of different behaviors and include a large number of individuals with idiosyncratic behavioral profiles. Hence, a combined table will contain a large number of missing values, which actually represent the absence of behavior rather than the failure to record behavior. Such data should never be treated as missing. Instead, researchers have to rely on their knowledge of the data-generating process to fill in the correct values. In most scenarios the absence of behavior will be coded as zero, resulting in data matrices consisting of zeros to a substantial degree. The same is true in many applications involving text analysis and natural language processing techniques, where document-term matrices are often sparse and consist mostly of zeros. Sparse data pose several unique challenges to preprocessing and modeling. For example, both R and Python support a special data format for sparse matrices, which is much more efficient in terms of memory and computation than are matrices containing explicit zeros, especially when it comes to larger data sets. Additionally, some of the routine preprocessing steps may have unintended consequences when applied to sparse feature matrices. For example, it is usually not advisable to normalize sparse data because it turns zero cells into nonzero cells, thereby creating a dense matrix, which can dramatically increase memory usage. Similarly, some implementations of principal components analysis center the input data before computing the singular value decomposition. In such cases it is recommended to use alternatives such as truncated singular value decomposition, also known as latent semantic analysis. Sparse target variables may also require specific statistical techniques, such as zero-inflated models.

Feature Engineering

Depending on the structure of the data, a significant proportion of effort has to be devoted to feature engineering. *Feature engineering* describes the process of transforming raw data into meaningful metrics, which can be used as predictors in a model. Psychological constructs have traditionally been measured through questionnaires such that variables can easily be generated by summing up individual item responses. This is not possible in the case of complex multivariate and longitudinal behavioral data. The most common feature engineering tasks in psychological research consist of (a) deriving psychologically relevant measures from very granular, semistructured, or unstructured data; (b) deriving person-level variables from complex longitudinal within-person data; and (c) deriving variables that capture the latent structure of high-dimensional data. The most useful techniques for feature engineering involve supervised learning, unsupervised learning, and the extraction of distribution parameters.

Supervised learning can be useful for extracting features from very granular or unstructured data. The term *supervised learning* refers to machine learning techniques that are used to predict values of a continuous (regression models), ordinal (ordinal regression), or discrete (multi) nominal (classification models) target variable. For example, classification models are useful for inferring behaviors and events from multimodal sensing data (e.g., sleeping patterns, modes of transportation, classes of physical activity). Similarly, object recognition algorithms can be used to extract features from image data (e.g., the presence of people, distinct facial expressions, objects that are relevant to the research question at hand). Supervised learning models can be trained from scratch for the specific purpose of feature engineering, but it is often more efficient to use pretrained models. For example, object recognition can be performed with YOLO (Redmon et al., 2016), and abstract features can be extracted from image data using the convolutional bases of pretrained convolutional neural network models available in the Keras library for R and Python (Chollet, 2015). Similarly, pretrained word-embedding models can be used to derive features from natural language (e.g., Pennington et al., 2014). When dealing with longitudinal data, it can be helpful to fit time series regressions at the within-person level and extract intercepts and slopes as between-person level predictors capturing idiosyncratic behavioral trends over time. For example, researchers could use the slopes of within-person time series regressions to capture whether individuals increased (positive slope) or decreased (negative slope) their use of certain features of an academic app over time. These variables can then be used to predict other outcomes, such as user satisfaction, attitudes toward features of the app, or academic performance measures.

Unsupervised learning is useful for identifying structure in complex data without explicitly modeling the relationships between predictors and a target variable. Cluster analysis can be useful to categorize data points based on similarity and dimensionality reduction techniques can help reduce the number of features by reducing the feature space to a smaller number of underlying dimensions. This process is especially useful when dealing with high dimensional data, such as text data. In the context of text data, other unsupervised techniques like topic modeling can be used for feature extraction, too. Topic modeling techniques, such as latent Dirichlet allocation (Blei et al., 2003), leverage latent structure in texts and identify themes that exist within a corpus of text based on the co-occurrence of words across and within documents.

The simplest and most common feature engineering technique for longitudinal behavioral data is the extraction of distribution parameters. When presented with a time series of behaviors or events, researchers may be interested in their frequency and variability over time. Distribution descriptives like measures of central tendency and dispersion, entropy or skewness, and kurtosis are easy to extract and can provide a relatively good description of behavioral variation over time.

Of course, the three feature extraction strategies outlined above can be used in conjunction. It is often helpful to identify behaviors or behavioral clusters through supervised or unsupervised learning and then calculate summary statistics to aggregate these features to the person level. In the analysis of GPS data, for example, unsupervised and supervised learning may be used to identify places where individuals tend to spend a lot of time (i.e., home, work, social places). Then, in a second step, researchers can analyze how people distribute their time across these different types of places to derive person-level individual difference measures that are related to person-level outcomes.

Finally, feature extraction is affected by researchers' decisions to engage in explanatory or predictive research. Explanatory research benefits from a low number of strong, highly interpretable features, whereas predictive research benefits from any feature that improves prediction performance. For this reason the feature engineering process in explanatory research is usually a top-down process, where researchers try to operationalize existing constructs. Feature engineering in predictive research, on the other hand, is often bottom-up, such that researchers first generate a multitude of atheoretical features and then try to optimize model performance through feature selection or shrinkage techniques.

MODELING

Explanatory modeling and predictive modeling both benefit from an abundance of data because large data sets provide statistical power, safeguard against overfitting, and enable researchers to test their models out of sample without sacrificing precious training data. That is, large samples can be divided into sufficiently large subsamples, some of which are used for model fitting, while others are used to assess model performance on data that has not been previously used for modeling (e.g., through cross-validation; see below). On the other hand, Big Data poses unique challenges to the scalability of statistical methods and may require substantial computational resources. Generally, it is important to strike a tradeoff between the benefits of using more data and the corresponding increase in computational requirements. Needless to say, it is not always necessary to use the maximum amount of data available to draw scientific conclusions with a sufficient degree of confidence. For most psychological research questions, the difference between hundreds of thousands of data points and hundreds of millions of data points would be unlikely to fundamentally change the nature of the results. However, adding orders of magnitude of data can quickly increase computational requirements to a level where specialized hardware and software is needed or where the problem cannot be solved within a reasonable timeframe. The following sections discuss how explanatory and predictive modeling approaches depend on the structure of Big Data and explain common strategies to deal with Big N, Big V, and Big T data.

Statistical Inference for Big Data

Statistical inference and the explanatory approach have their place in the world of Big Data, especially when Big Data is used to model causal mechanisms or to test theory-derived hypotheses, and when the assessment of uncertainty is a primary concern. In the field of psychology, the dominant approach to statistical inference is frequentist and almost exclusively based on general linear models, which include linear regression models, ANOVA, *t*-tests, and *F*-tests. Having access to large sample sizes has clear advantages in the explanatory framework. Most important, large sample sizes can help to prevent overfitting because larger samples represent the underlying population more accurately than smaller samples. Consequently, it is possible to fit more complex models without picking up on sample-specific patterns in the data, which do not exist in the broader population (Yarkoni & Westfall, 2017). Similarly, even if large data sets are not representative, they can enable researchers to draw representative subsamples or to focus on populations that are otherwise underrepresented in psychological research. For example, while Facebook users may on average be younger and better educated than the general population, the sheer size of data sets collected through Facebook provides access to sufficiently large numbers of individuals from underrepresented groups (Kosinski et al., 2015). However, the traditional statistical methods used by psychologists are not always well equipped to deal with large and complex data sets.

Traditional statistical analyses (i.e., general linear models) work reasonably well with Big N, and some of the statistical assumptions, such as the assumptions of normality of residuals or homoscedasticity, may even be relaxed if sample size is sufficiently large (Maas & Hox, 2004). However, sample size has obvious implications for significance testing and computation time. Specifically, large sample sizes make it easy to detect significant effects even when effect sizes are very small. Importantly, the types of real-world behavioral records that are characteristic of Big Data are usually noisy and multidetermined, which imposes bounds on the effect sizes that researchers can expect to find (Matz et al., 2017). With respect to computation, larger sample sizes entail longer execution times. For most of the common inference statistical methods, runtime is a linear or quadratic function of sample size. In the rare case in which execution time becomes prohibitive due to sample size, researchers can easily subsample their data set and run the analysis on a subsample.

Big V data poses several unique problems for statistical inference methods. First, traditional statistical methods cannot computationally handle a large number of variables. Linear regression, for example, is computationally infeasible when the number of features exceeds the number of observations (V > N) because there can be no unique coefficient estimate. Second, even when the number of variables does not exceed the number of observations (V < N) linear regression may not scale well computationally because the relationship between the number of variables and runtime tends to be polynomial. Third, models become prone to overfitting with a growing number of variables. Fourth, models with a large number of variables are prone to multicollinearity, which causes imprecise coefficient estimates and inflated p values. The last two points can, to some extent, be addressed by using regularized models (at the expense of biased coefficient estimates), dimensionality reduction (at the expense of interpretability), and variable selection (at the expense of potentially discarding relevant information).

Big T data require special attention to the various potential sources of temporal effects beyond the effects that a researcher may be interested in. Modeling temporal data is perhaps the most difficult modeling scenario because there are various types of overlapping temporal effects (e.g., day of week, month of year, holiday) that can easily distort interpretations. Moreover, there are additional complications with the transformation of time into a discrete measure. When an effect is small, distortions from suboptimal transformations can overwhelm the real signal by improperly weighing information. With respect to statistical analyses, techniques such as growth curve models, multilevel models and autoregressive models can either make a specific temporal trend the focal point in an analysis or alternatively account for some temporal effects while focusing on more general patterns. In other cases it may be sufficient to extract person-level metrics that capture the properties of the underlying time series and model relationships at the person level as in two stage regression (Gelman, 2005).

Predictive Modeling for Big Data

Predictive modeling is synonymous with supervised machine learning, where a target metric is predicted from a set of features. There are a variety of algorithms specifically designed for predictive modeling. These models range from regularized linear and logistic regression models (e.g., Tibshirani, 1996) to support vector machines (e.g., Cortes & Vapnik, 1995), random forests (Breiman, 2001), gradient boosting (Friedman, 2000), and neural networks (for an introduction, see Hastie et al., 2016). The algorithms used for predictive modeling are in some cases identical to the algorithms used for explanatory modeling, but the focus of the two approaches is fundamentally different. In contrast to explanatory modeling and statistical inference, predictive modeling does not place much emphasis on theoretical relationships represented by model coefficients, but instead emphasizes prediction performance. In place of significance tests and goodness-of-fit measures, predictive modeling requires the empirical assessment of generalizability by evaluating the model's prediction performance on new data; this is typically achieved through resampling and cross-validation (e.g., Stone, 1974).

Cross-validation in its most basic form describes the practice of randomly splitting a data set into two subsets, one of which is used to train a model (training set), while the other one is used to assess prediction performance on data that has not been previously used to fit the model (testing set). Subsequently, the roles of the two data sets are reversed, such that the subset that has been previously used for testing is now used for training, while the subset that has been previously used for training is now used for testing. K-fold cross validation generalizes this process by repeating it with k splits, such that the model is trained on k-1 subsets and tested on the remaining subset in each iteration. This is repeated k times until each data point in the original data set has been used for training and testing. Model performance is then assessed as the average prediction performance over all k testing sets.

There are many variations to the cross validation approach. Leave-one-out cross-validation maximizes the size of the training set by performing n splits and repeatedly testing the model on a testing set with a sample size of one (e.g., Lachenbruch & Mickey, 1968). Monte Carlo cross validation does not use mutually exclusive splits but draws a random subsample as a training set in each iteration; this has the advantage that the number of iterations becomes independent of the proportion of the data that is used for training and testing in each iteration (Xu & Liang, 2001). Finally, there is nested cross-validation, which performs one or more additional splits within each training set. The inner-loop cross-validation cycle (looping through splits within the training set) is used for hyperparameter tuning (i.e., finding model configurations with high predictive power) while the outer loop (looping through train-test splits) is used to assess generalized model performance of the best models on testing data.

Hyperparameters are parameter values that can be specified by researchers to influence how a model learns (i.e., to control overfitting and learning speed). Hyperparameter tuning is an important step in predictive modeling because the hyperparameter configuration of an algorithm can have great impact on predictive performance. In the case of Lasso regression (Tibshirani, 1996), for instance, there is an additional regularization parameter (typically denoted as alpha or lambda), which determines how heavily coefficient values are penalized in the optimization process. In contrast to OLS regression, the Lasso algorithm minimizes not only the residual sum of squares but also the sum of the regression coefficients weighted by alpha. This ensures that the coefficients of unimportant predictors are set to zero, effectively removing these predictors from the model. By choosing the value of alpha, researchers can determine how heavily coefficients are penalized. Choosing a high alpha value will emphasize penalization relative to the impact of squared residuals. Choosing a low alpha value will emphasize squared residuals. Because the level of penalization can greatly affect generalization performance, it is advisable to search a range of alpha values and pick the one that leads to the best results on new data. Other models, like random forests, gradient boosting, and neural networks, have a much higher number of hyperparameters that can be tuned. In these cases, hyperparameter configurations constitute a multidimensional search space.

Multiple techniques have been suggested to optimize hyperparameter search, ranging from very simple options like grid search or random search (Bergstra & Bengio, 2012) to more sophisticated approaches like Hyperband (Li et al., 2016) or Bayesian optimization (Hutter et al., 2011), where additional machine learning models are used to optimize the parameters of the focal model.

The predictive approach and the cross-validation framework profit from Big N data because it enables researchers to use hold-out sets for hyperparameter tuning and performance assessment without sacrificing training data. K-fold cross-validation can considerably increase computational requirements because each model needs to be fitted multiple (k) times. In a Big N context, however, it is easy to generate very large training sets and very large testing sets, which allows for a precise assessment of generalization performance without the need for a high number of cross-validation iterations. In addition to the fact that Big N data facilitates hyperparameter search and the assessment of generalization performance, some of the most popular supervised learning algorithms are data hungry and require large training sets (e.g., neural networks).

Predictive modeling is usually not concerned with the interpretation of coefficients, so researchers can engage in feature engineering to produce large feature spaces and maximize prediction performance. At the same time, dimensionality reduction, regularization, and variable selection techniques are effective tools to tackle overfitting, which is a common problem with respect to Big V data. Dimensionality reduction techniques, like matrix factorization, UMAP (McInnes et al., 2018), or autoencoder models (Rumelhart et al., 1986), can reduce the feature space to a smaller number of dimensions while retaining most of the information captured in the original feature space. A potential disadvantage of these techniques is the reduced interpretability of the new features. Regularized linear models like Lasso, Ridge, or elastic net regressions have a specific hyperparameter (alpha) that controls how heavily model coefficients are penalized. In heavily regularized models (high alpha), only the features with the highest predictive power will have nonzero coefficients, while features that have a negligible impact on model fit will be filtered out. This reduces model complexity and therefore counteracts overfitting.

Model complexity can also be reduced through model-based and model-agnostic feature selection techniques. Model-based feature selection works with feature importance metrics derived from a fitted model, such as the magnitude of coefficients in linear models or impurity decrease in random forest models. A common technique is recursive model-based selection, in which a model is repeatedly fitted, while the feature with the lowest importance score is discarded in each iteration. Model-agnostic feature selection techniques, on the other hand, do not rely on feature-importance metrics derived from the model itself, but they observe how prediction performance reacts to changes in the feature space. Permutation importance, for instance, permutes the values of a feature to eliminate all associations between feature values and target values. The subsequent reduction in prediction performance quantifies the importance of the focal feature. Based on permutation importance, features can be selected in a forward (select most important features first) or backward (eliminate least important features first) manner. Feature selection algorithms usually rely on fitting the model repeatedly so their computational costs can be prohibitive in Big Data contexts.

As mentioned earlier, inference statistical models can be used for predictive modeling in conjunction with cross validation, but can machine learning methods be used for explanation rather than prediction? Research concerned with model interpretability and model explainability has made tremendous progress in recent years and produced many methodological innovations enabling researchers to better understand the predictions made by black-box models. However, being able to explain predictions of a model is not the same as being able to explain the underlying data generating process or causal mechanisms. Feature importance scores derived from machine learning models, for example, are fundamentally different from coefficient estimates in statistical models. Feature importance scores, such as permutation importances, are generally nonlinear. In the case of kernel support vector machines (SVMs), random forest models, gradient boosting models, and neural networks, feature importance scores capture polynomial relationships and complex interactions. It is therefore difficult to infer directionality or interpret effects in isolation from the effects of other features. Similarly, in the case of penalized regression models, the coefficient estimates are biased because the objective function minimizes the combined sum of squared residuals and coefficient values. Coefficients therefore depend on the regularization parameter set by the researcher, and in the presence of correlated features the algorithm arbitrarily removes some of them, rendering the remaining coefficients uninterpretable.

Nonetheless, as pointed out earlier, predictive modeling can support explanatory research in at least three major ways (Shmueli, 2010). First, it can uncover new phenomena and causal mechanisms by taking into account complex data and a multitude of potential relationships. Second, predictive modeling can be used to discover new measures and operationalizations through bottom-up feature engineering. Dimensionality reduction, regularization, and variable selection techniques can be used to find meaningful predictive relationships in rich feature spaces, which can then be integrated into existing theory or serve as starting points for theoretical advancement. In practice that means that researchers would use simple explanatory models to better understand the findings uncovered by complex predictive models. Third, predictive modeling can serve as a reality check for explanatory models. In practice this means that theoretical findings produced by explanatory modeling can be subsequently assessed with regard to their prediction performance using the cross-validation approach. We recommend that researchers routinely validate the generalizability of inference statistical findings by means of cross validation. Finally, recent research suggests that algorithms traditionally associated with the predictive approach can be used for causal inference and the estimation of treatment effects (e.g., Farrell et al., 2021; Shi et al., 2019; Steinkraus, 2019).

RESOURCE CONSIDERATIONS

As noted earlier, the prospect of running a computational procedure in an acceptable amount of time depends on the difficulty of the computational problem and the computational resources available. Computational resources, in turn, depend on the hardware setup, most importantly memory capacity and processing power. It is therefore helpful for researchers to assess how preprocessing and modeling procedures scale with respect to computational requirements.

Computational Complexity

In the field of computer science, computational complexity theory is chiefly concerned with the classification of algorithms according to their resource requirements. Psychologists certainly do not need to become experts in this highly theoretical area of research, but developing a basic intuition of computational complexity can help avoid certain pitfalls when working with large data sets.

In general, the computational requirements of an algorithm grow with the size of the input data. The relationship between input size and computational requirements can be characterized by identifying the order of the corresponding growth function (e.g., linear, polynomial, exponential, factorial). In the context of data analysis, it is usually most important to assess how preprocessing and modeling procedures scale with respect to sample size and the number of variables. This relationship can be expressed using the "Big O notation," where O denotes the order of the function. Assuming that a data set is represented as an input matrix with the dimensions n (sample size) and p (number of variables), runtime can be expressed as a function of n and p. A linear relationship between sample size and runtime, for example, can be expressed as $O(n)$, whereas a polynomial relationship of degree 3 between the number of variables and runtime would be expressed as $O(p^3)$. The complexity of most procedures depends on both n and p, so most expressions contain both dimensions. The runtime for least squares fitting, for instance, would be $O(p^3 + np^2 / 2)$ or $O(np^2)$ depending on the choice of the decomposition algorithm (see Hastie et al., 2016). That means that computational costs for linear regression models are especially sensitive to the number of variables involved. Importantly, the same computational goal can often be achieved through different algorithms with different computational complexities. Knowing the runtime complexity of a given algorithm can give researchers an idea of how the algorithm scales and may therefore affect their decisions regarding the analytic strategy, the hardware configuration that is needed, and how much data should be used. The computational complexity of an algorithm is usually reported in the publication introducing the algorithm. For common algorithms, computational considerations are discussed in statistics and computer science textbooks such as *Elements of Statistical Learning* (Hastie et al., 2016). Additionally, because many techniques like hyperparameter tuning, cross-validation, or feature selection rely on fitting models repeatedly, it is advisable for researchers to first run a single model and then extrapolate execution time to the overall learning task.

Processing Power

Computational operations are usually performed by the CPU. Other things being equal, a faster CPU (or in the case of parallel processing a CPU with more cores) can perform more operations on data and will therefore improve the speed with which the data are processed. As the computational requirements for Big Data applications are generally very high, differences in processing power and parallelization can lead to appreciable differences in execution time. The term parallelization or parallel computing describes a paradigm where a computational task is divided into subtasks that can be run independently and simultaneously on different CPUs or CPU cores. The performance gain that can be achieved through parallelization depends not only on the hardware but also on properties of the computational task at hand. Highly parallelizable tasks can easily be split up into independent subtasks, whereas tasks that have a lot of interdependencies are harder to parallelize. Examples of easily parallelizable tasks in the context of machine learning are tasks that rely on repeated execution of subtasks, such as hyperparameter tuning, cross-validation, and variable selection, or ensemble models like random forests, where many base estimators are fitted independently. Tasks that are hard to parallelize include iterative processes where each computational subtask relies on the results of previous subtasks. This is the case for sequential methods, such as Markov Chain Monte Carlo algorithms, which play an important role in Bayesian machine learning. Both Python and R run on a single CPU core by default, which is often insufficient when large quantities of data need to be processed. However, standard machine learning software usually offers some multicore capabilities (e.g., *n_jobs* in Scikit-Learn). Additionally, both languages provide libraries that facilitate parallel computing (e.g., *doParallel* and *foreach* in R; *Dask* and *concurrent* in Python) and enable users to distribute tasks across multiple cores. Using multiple cores on a local machine is certainly preferable to running code on a single core while the other cores are idle, but truly large data sets may require more processing power than any laptop or desktop computer can provide. In these cases, it is advisable to use cloud computing and run code on a powerful server or on a high-performance computing cluster. A *cluster* is a set of servers (i.e., nodes) that works together to tackle a computational task. Researchers can flexibly determine the structure and the size of a cluster when using cloud services like Amazon Web Services, Google Cloud Platform, or Microsoft Azure. To make efficient use of a cluster, special software such as Apache Spark is required. Spark offers data manipulation and machine learning libraries optimized for parallelized distributed computing. Both R and Python can be used to interact with the Spark API. An additional advantage of using cloud services is that researchers can opt for specialized hardware such as graphics processing units or tensor processing units, which are optimized for machine learning tasks, especially in the context of deep learning.

Memory

Memory is used to temporarily store data for quick access so that the CPU can perform computations on them. In most standard applications the data that are being analyzed are fully loaded into memory. While the memory capacity of laptops and desktop machines has dramatically increased over the past decades, Big Data will easily push the memory of standalone machines to its limits. In such cases it is advisable to resort to more powerful remote machines or distributed computing. But even if memory can easily be expanded, as in cloud computing, researchers should be aware of how they use memory, because inefficient memory use can considerably slow down data processing. First, it is advisable to drop unnecessary data early on in the data analysis process. Second, researchers should avoid creating unnecessary copies of the data they are working with. A common mistake is the creation of copies when performing operations in a loop. Third, objects that are not needed anymore can be deleted from memory. Fourth, it is important to pay close attention to data structures and data types that are being used because different data structures and data types may have different memory requirements. For example, the use of the sparse matrix format can improve memory efficiency. Related to this point, researchers should be aware of how the transformations that they apply to the data interact with data structures and data types. Naively standardizing a sparse matrix, for instance, can dramatically increase memory requirements by replacing zeros with nonzero values.

If memory is still an issue after optimizing the code for memory efficiency and buying more memory or renting cloud memory is not an option, researchers can resort to several workarounds. First, it is possible to use secondary storage as "virtual memory." In practice, that means that disc space is treated like memory, which effectively expands memory capacity but will slow down operations that are performed on data that is stored on the disc. Second, it is often possible to process data in chunks, such that only a subset of data (a chunk) is loaded into memory. Once the data are processed, they are released from memory, so that the next chunk can be processed. The results of the operations performed on each of the chunks can then be assembled for further processing. This idea can be illustrated in the context of text analysis, where individual documents or groups of documents can be processed separately and derived metrics like word counts can be appended to a results file. Both R and Python offer tools that can help to overcome memory limitations. The R packages *ff*, *filehash* or *disc.frame* may be a reasonable starting point. In Python the popular *pandas* library enables researchers to read and process files in chunks. However, as mentioned previously, it is advisable to resort to cloud solutions and rent additional resources if memory capacity is a limiting factor.

CONCLUSION

Big data opens up a broad range of interesting new research opportunities for psychologists. Here we have discussed some of the opportunities and challenges associated with different kinds of Big Data (Big N, Big V, Big T) for psychological research. Furthermore, we have assembled a collection of current best practices for data collection (APIs, web-scraping, industry collaborations), data storage (RDBMSs, distributed file systems, cloud storage), preprocessing (filtering and merging, missing and sparse data, feature engineering) and modeling (explanatory and predictive) of large, complex data. While the current chapter is certainly too short to cover the intricacies of Big Data research, it was our goal to introduce the reader to the most important concepts and provide starting points for further reading. We hope that in doing so this chapter will facilitate innovative work using Big Data in the field of psychology.

REFERENCES

Adjerid, I., & Kelley, K. (2018). Big data in psychology: A framework for research advancement. *American Psychologist*, *73*(7), 899–917. https://doi.org/10.1037/amp0000190

Bergstra, J., & Bengio, Y. (2012). Random search for hyper-parameter optimization. *Journal of Machine Learning Research*, *13*, 281–305. https://www.jmlr.org/papers/volume13/bergstra12a/bergstra12a

Blei, D. M., Ng, A. Y., & Jordan, M. I. (2003). Latent Dirichlet allocation. *The Journal of Machine Learning Research*, *3*, 993–1022. https://doi.org/10.5555/944919.944937

Borgman, C. L. (2015). *Big data, little data, no data: Scholarship in the networked world*. The MIT Press. https://doi.org/10.7551/mitpress/9963.001.0001

Breiman, L. (2001). Random forests. *Machine Learning*, *45*(1), 5–32. https://doi.org/10.1023/A:1010933404324

Camerer, C. F., Dreber, A., Holzmeister, F., Ho, T.-H., Huber, J., Johannesson, M., Kirchler, M., Nave, G., Nosek, B. A., Pfeiffer, T., Altmejd, A., Buttrick, N., Chan, T., Chen, Y., Forsell, E., Gampa, A., Heikensten, E., Hummer, L., Imai, T., . . . Wu, H. (2018). Evaluating the replicability of social science experiments in Nature and Science between 2010 and 2015. *Nature Human Behaviour*, *2*(9), 637–644. https://doi.org/10.1038/s41562-018-0399-z

Cattell, R. B. (1953). The three basic factor-analytic research designs—Their interrelations and derivatives. *Psychological Bulletin*, *49*(5), 499–520. https://doi.org/10.1037/h0054245

Chollet, F. (2015). *Keras: The Python deep learning API*. https://keras.io/

Cortes, C., & Vapnik, V. (1995). Support-vector networks. *Machine Learning*, *20*(3), 273–297. https://doi.org/10.1007/BF00994018

Ebert, T., Götz, F. M., Obschonka, M., Zmigrod, L., & Rentfrow, P. J. (2019). Regional variation in courage and entrepreneurship: The contrasting role of courage for the emergence and survival of start-ups in the United States. *Journal of Personality*, *87*(5), 1039–1055. https://doi.org/10.1111/jopy.12454

Enamorado, T., Fifield, B., & Imai, K. (2019). Using a probabilistic model to assist merging of large-scale administrative records. *The American Political Science Review*, *113*(2), 353–371. https://doi.org/10.1017/S0003055418000783

Facebook Data for Good. (2020). *Our work on COVID-19*. https://dataforgood.fb.com/docs/covid19

Farrell, M. H., Liang, T., & Misra, S. (2021). Deep neural networks for estimation and inference. *Econometrica*, *89*(1), 181–213. https://doi.org/10.3982/ECTA16901

Friedman, J. H. (2000). Greedy function approximation: A gradient boosting machine. *Annals of Statistics*, *29*(5), 1189–1232. https://www.jstor.org/stable/2699986

Gelman, A. (2005). Two-stage regression and multilevel modeling: A commentary. *Political Analysis*, *13*(4), 459–461. https://doi.org/10.1093/pan/mpi032

Gosling, S. D., Vazire, S., Srivastava, S., & John, O. P. (2004). Should we trust web-based studies? A comparative analysis of six preconceptions about Internet questionnaires. *American Psychologist*, *59*(2), 93–104. https://doi.org/10.1037/0003-066X.59.2.93

Hastie, T., Tibshirani, R., & Friedman, J. (2016). *The elements of statistical learning: Data mining, inference, and prediction* (2nd ed.). Springer.

Hutter, F., Hoos, H. H., & Leyton-Brown, K. (2011). Sequential model-based optimization for general algorithm configuration. In C. A. C. Coello (Ed.), *Learning and intelligent optimization* (pp. 507–523). Springer. https://doi.org/10.1007/978-3-642-25566-3_40

Isaak, J., & Hanna, M. J. (2018). User data privacy: Facebook, Cambridge Analytica, and privacy protection. *Computer*, *51*(8), 56–59. https://doi.org/10.1109/MC.2018.3191268

Kosinski, M., Matz, S. C., Gosling, S. D., Popov, V., & Stillwell, D. (2015). Facebook as a research tool for the social sciences: Opportunities, challenges, ethical considerations, and practical guidelines. *American Psychologist*, *70*(6), 543–556. https://doi.org/10.1037/a0039210

Kosinski, M., Stillwell, D., & Graepel, T. (2013). Private traits and attributes are predictable from digital records of human behavior. *Proceedings of the National Academy of Sciences, 110*(15), 5802–5805. https://doi.org/10.1073/pnas.1218772110

Kulkarni, V., Kern, M. L., Stillwell, D., Kosinski, M., Matz, S., Ungar, L., Skiena, S., & Schwartz, H. A. (2018). Latent human traits in the language of social media: An open-vocabulary approach. *PLOS ONE, 13*(11), e0201703. https://doi.org/10.1371/journal.pone.0201703

Lachenbruch, P. A., & Mickey, M. R. (1968). Estimation of error rates in discriminant analysis. *Technometrics, 10*(1), 1–11. https://doi.org/10.1080/00401706.1968.10490530

Li, L., Jamieson, K., DeSalvo, G., Rostamizadeh, A., & Talwalkar, A. (2016). *Hyperband: A novel bandit-based approach to hyperparameter optimization.* arXiv. http://arxiv.org/abs/1603.06560

Maas, C. J. M., & Hox, J. J. (2004). The influence of violations of assumptions on multilevel parameter estimates and their standard errors. *Computational Statistics & Data Analysis, 46*(3), 427–440. https://doi.org/10.1016/j.csda.2003.08.006

Matz, S. C., Gladstone, J. J., & Stillwell, D. (2017). In a world of Big Data, small effects can still matter: A reply to Boyce, Daly, Hounkpatin, and Wood (2017). *Psychological Science, 28*(4), 547–550. https://doi.org/10.1177/0956797617697445

McInnes, L., Healy, J., & Melville, J. (2018). *UMAP: Uniform Manifold Approximation and Projection for dimension reduction.* arXiv. http://arxiv.org/abs/1802.03426

Metzler, K., Kim, D. A., Allum, N., & Denman, A. (2016). *Who is doing computational social science? Trends in Big Data research* [White paper]. SAGE Publishing. https://doi.org/10.4135/wp160926

Müller, S. R., Peters, H., Matz, S. C., Wang, W., & Harari, G. M. (2020). Investigating the relationships between mobility behaviours and indicators of subjective well-being using smartphone-based experience sampling and GPS tracking. *European Journal of Personality, 34*(5), 714–732. https://doi.org/10.1002/per.2262

Open Science Collaboration. (2015). Estimating the reproducibility of psychological science. *Science, 349*(6251), aac4716. https://doi.org/10.1126/science.aac4716

Pennington, J., Socher, R., & Manning, C. (2014). Glove: Global vectors for word representation. In A. Moschitti, B. Pang, & W. Daelemans (Eds.), *Proceedings of the 2014 Conference on Empirical Methods in Natural Language Processing (EMNLP)* (pp. 1532–1543). Association for Computational Linguistics. https://doi.org/10.3115/v1/D14-1162

Peters, H., Götz, F., Ebert, T., Müller, S., Rentfrow, J., Gosling, S., Obschonka, M., Ames, D., Potter, J., & Matz, S. (2020). *Regional personality differences predict variation in COVID-19 infections and social distancing behavior.* PsyArXiv. https://doi.org/10.31234/osf.io/sqh98

Redmon, J., Divvala, S., Girshick, R., & Farhadi, A. (2016). You only look once: Unified, real-time object detection. In *2016 IEEE Conference on Computer Vision and Pattern Recognition (CVPR)* (pp. 779–788). IEEE. https://doi.org/10.1109/CVPR.2016.91

Rentfrow, P. J., Jokela, M., & Lamb, M. E. (2015). Regional personality differences in Great Britain. *PLOS ONE, 10*(3), e0122245. https://doi.org/10.1371/journal.pone.0122245

Rumelhart, D. E., Hinton, G. E., & Williams, R. J. (1986). Learning representations by back-propagating errors. *Nature, 323*(6088), 533–536. https://doi.org/10.1038/323533a0

Schneble, C. O., Elger, B. S., & Shaw, D. (2018). The Cambridge Analytica affair and Internet-mediated research. *EMBO Reports, 19*(8), e46579. https://doi.org/10.15252/embr.201846579

Shi, C., Blei, D., & Veitch, V. (2019, December 8–14). *Adapting neural networks for the estimation of treatment effects* [Paper presentation]. 33rd Conference on Neural Information Processing Systems (NeurIPS 2019), Vancouver, BC, Canada.

Shmueli, G. (2010). To explain or to predict? *Statistical Science, 25*(3), 289–310. https://doi.org/10.1214/10-STS330

Stachl, C., Pargent, F., Hilbert, S., Harari, G. M., Schoedel, R., Vaid, S., Gosling, S. D., & Bühner, M. (2020). Personality research and assessment in the era of machine learning. *European Journal of Personality, 34*(5), 613–631. https://doi.org/10.1002/per.2257

Steinkraus, A. (2019). Estimating treatment effects with artificial neural nets: A comparison to synthetic control method. *Economic Bulletin, 39*(4), 2778–2791.

Stone, M. (1974). Cross-validatory choice and assessment of statistical predictions. *Journal of the Royal Statistical Society: Series B. Methodological, 36*(2), 111–133. https://doi.org/10.1111/j.2517-6161.1974.tb00994.x

Tibshirani, R. (1996). Regression shrinkage and selection via the Lasso. *Journal of the Royal Statistical Society: Series B. Methodological, 58*(1), 267–288. https://doi.org/10.1111/j.2517-6161.1996.tb02080.x

Varol, O., Ferrara, E., Davis, C. A., Menczer, F., & Flammini, A. (2017). *Online human-bot interactions: Detection, estimation, and characterization.* arXiv. http://arxiv.org/abs/1703.03107

Xu, Q.-S., & Liang, Y.-Z. (2001). Monte Carlo cross validation. *Chemometrics and Intelligent Laboratory Systems*, *56*(1), 1–11. https://doi.org/10.1016/S0169-7439(00)00122-2

Yarkoni, T., & Westfall, J. (2017). Choosing prediction over explanation in psychology: Lessons from machine learning. *Perspectives on Psychological Science*, *12*(6), 1100–1122. https://doi.org/10.1177/1745691617693393

Youyou, W., Kosinski, M., & Stillwell, D. (2015). Computer-based personality judgments are more accurate than those made by humans. In D. Funder (Ed.), *Proceedings of the National Academy of Sciences*, *112*(4), 1036–1040. https://doi.org/10.1073/pnas.1418680112

4

The Psychology of Mobile Technology and Daily Mobility

Morgan Quinn Ross, Sandrine R. Müller, and Joseph B. Bayer

The fact that mobile technologies can be used almost anywhere makes them seem almost placeless, as if they were independent from their surrounding environments. Theorists, designers, and researchers have long heralded the possibilities of mobile media to circumvent spatial constraints (see Campbell, 2019; McLuhan, 1964). Yet, over time it has become clear that this projection of mobile capabilities is overstated (Frith, 2015; Humphreys et al., 2018). In this chapter, we review how the use of mobile media is deeply embedded in and shaped by the spatial contexts that we inhabit in our daily lives. Consequently, our understanding of mobile technologies, including their psychological effects, must be grounded in an understanding of daily mobility: how people move through everyday spaces.

To illustrate, consider an individual, Jerry, who plays the popular (as of 2020) mobile game *Candy Crush*. Jerry's total usage of *Candy Crush* can be considered an aggregation of playing the game in a variety of spatial

---

https://doi.org/10.1037/0000290-005
The Psychology of Technology: Social Science Research in the Age of Big Data, S. C. Matz (Editor)
Copyright © 2022 by the American Psychological Association. All rights reserved.

contexts. Perhaps he especially likes to play the game at home before bed (*in-place*) and on the bus before work (*on-the-go*). The ways that Jerry experiences *Candy Crush* are thereby contingent on whether he is in-place or on-the-go (Engl & Nacke, 2013), as well as whether he is in a particular place or moving in a particular way (Mehrotra et al., 2017). For example, as Jerry swipes and swirls his thumbs across the screen, he might talk with his partner in bed or overhear some juicy gossip on the bus. He might find the game relaxing after work in the comfort of his own bed or find it challenging when the bus is crowded. In turn, how Jerry makes sense of his everyday surroundings may also depend on his *Candy Crush* habits (Figeac & Chaulet, 2018).

Here, we contend that the psychological effects of apps such as *Candy Crush*—and the overall psychological connection between user and device—are based in part on the spatial contexts and mobile patterns that comprise daily life. The above example foreshadows the structure of the current chapter by demonstrating how the psychology of mobile technology can be interwoven with (a) in-place contexts, (b) on-the-go contexts, and (c) daily mobility overall. Ultimately, we demonstrate how a contextual perspective can advance our understanding of the psychological implications of mobile devices.

Crucially, more contextual studies often rely on novel capabilities tied to everyday mobile technologies (e.g., smartphones, fitness trackers). Sensing tools provide opportunities to capture accurate log data of phone use (Stachl et al., 2020) and notifications (Mehrotra et al., 2017). In addition, smartphone sensing provides rich information (e.g., GPS, acceleration) to ascertain mobility features (e.g., places visited, transportation modes; Müller et al., 2020). In these ways, novel mobile methods can provide deeper insights into the psychological terrain of mobile technology and daily mobility, especially when used in tandem with one another and infused with theoretical rigor.

Given these advancements, we suggest that efforts are needed to lay an integrative agenda for the psychology of mobile technology and daily mobility. Recognizing that "technology" and "mobility" can refer to a wide range of phenomena (Fortunati & Taipale, 2017), we first clarify the scope of our review. Most notably, we treat the psychology of mobile technology as the mental connection between user and mobile device, along with its downstream implications (e.g., well-being). We place special emphasis on smartphones and mobile apps due to the deep literature on them. However, we anticipate other mobile technologies used between and beyond places of destination (e.g., wearables) to involve similar mobility-based psychological processes and receive more attention in the future (see Bayer & LaRose, 2018). Concurrently, we approach the psychology of mobility as the psychological processes underlying how we move and how different spaces (i.e., locations, places) impact our behavior. We thus focus on "daily spatial mobilities" (Kellerman, 2012), which involve "the corporeal travel of people for work, leisure, family life, [and] pleasure" (Büscher & Urry, 2009, p. 101).

In this chapter, we chart parallel lines of research on mobile technology and daily mobility. Specifically, we review how people engage with mobile technologies in-place and on-the-go, as well as their broader connection with their mobile devices. In each section, we link perspectives from psychology and mobility studies to an integrative understanding of mobile media use. We then survey emerging mobile methodologies with potential for interdisciplinary work. To conclude, we collapse these boundaries—between being in-place and on-the-go, and mobile technology and daily mobility—to consider the trajectory of future research on the psychology of mobile technology.

USING MOBILE TECHNOLOGY IN-PLACE

An avid gamer like Jerry will likely play *Candy Crush* in a wide variety of in-place spatial contexts. We use the broad term "spatial context" to refer to physical spaces (i.e., locations, places; e.g., in bed, on the bus), concentrating on the central spaces that make up daily mobility patterns given their attention in prior work. Importantly, such spaces often carry strong psychological associations in people's minds, in addition to providing a more regular influence on their technology behavior. As mobile devices afford the possibility of use in a variety of regular locations (e.g., at home, at work), usage is frequently tied to particular spaces (Humphreys et al., 2018). Jerry may play *Candy Crush* differently—and react differently—when alone at home versus with friends at a café. Spatial contexts can also be shaped by mobile technology; the surrounds of a café may be experienced differently while playing *Candy Crush* versus talking to friends. In this section, we first examine how people use their devices in varied ways across in-place contexts. We then review research from psychology and mobility studies on how spatial contexts may have distinct psychological effects, before integrating this work towards understanding mobile media use in-place.

A considerable body of work has demonstrated that people engage with their mobile devices differently across spatial contexts. This work has often distinguished between four major locations: home, work, "other meaningful places" (see also "third places"; Oldenburg, 1996), and "elsewhere," where the latter two locations are distinguished in terms of visit frequency. Usage sessions are consistently longer at home than at work, with mixed results in comparison to other meaningful places and elsewhere (Hintze et al., 2017; Soikkeli et al., 2011; Verkasalo, 2009). For example, a person might spend more time per phone check at home due to the (relative) lack of constraints on their time. In contrast, users display the highest usage frequency—the number of usage sessions per time spent in a certain place—in other meaningful places (Hintze et al., 2017); a person might check their phone frequently yet briefly while at their favorite watering hole. Although people may spend significant time on their phones at home, this usage is relatively less frequent due to the overall amount of time spent at home, even when accounting for sleep (Soikkeli et al., 2011). Due to this lower intensity, locked usage sessions (e.g., checking the notification dashboard) are longest at home though similarly frequent at other places, perhaps reflecting an accumulation of notifications when not in use (Hintze et al., 2017).

The content of these usage sessions also differs across spatial context. Verkasalo (2009) found that at home (vs. at work or elsewhere), people were more likely to make long voice calls, send text messages, and use alarms, but less likely to use multimedia and internet browsers. Other key spatial contexts have their own tendencies. Karnowski and Jandura (2014) found that people were more likely to use multimedia services (e.g., cameras) and less likely to communicate with others via calling or messaging apps when hanging out with peers than when at home or on-the-go. Mehrotra and colleagues (2017) demonstrated that music and reading apps were most frequently used at bus stops and train stations. In sum, mobile media use changes in combination with spatial context, including average session durations, session frequencies, and the types of activities engaged. These patterns of mobile technology usage suggest psychological implications as a function of the surrounding spatial context.

Perhaps the most fundamental in-place context is the home, where usage sessions are longer but occur less frequently. People have more freedom to do a variety of activities (e.g., play the boardgame *Candy Land*) and are less likely to need to be attentive or receptive to notifications (Mehrotra et al., 2017). As a result, phone use at home may represent a deviation from a laid-back environment; posting on Facebook at home—but not elsewhere—is associated with higher emotional arousal 30 minutes later (Bayer et al., 2018). Beyond particular places, Mehrotra et al. (2017) found connections between different qualities of places (e.g., relaxing vs. distressing places, as rated by independent coders) and notification receptivity, notification attentiveness, and phone use. Most notably, app use was lower in more distressing and more natural places, and attentiveness to notifications was higher in more productive places (Mehrotra et al., 2017).

The findings above can be understood through theoretical perspectives on the role of spatial context in media use (see Campbell, 2019). The constant accessibility of information and communication afforded by mobile technology spreads a digital layer over physical space, creating "hybrid spaces" (de Souza e Silva, 2006, p. 272). On the surface, this might suggest that mobile technology can be used independent of place, leading to an "annihilation of space and place" (Frith, 2015, p. 3). But conversely, it can be argued that mobile devices are actually more coupled with physical space than stationary media. Whereas the spatial context is constant for the television in the living room, the context constantly shifts for mobile media, implying that context may be more impactful for the latter. Furthermore, different spatial contexts may even enable certain uses (e.g., to catch a Magikarp in Pokémon Go) or trigger notifications (e.g., to give feedback on the crowdedness of a subway). Humphreys et al. (2018) mapped a theoretical framework to substantiate these notions, describing how context shapes our uses and gratifications of mobile technology, as well as our perception of the affordances of the device. This perspective echoes Farman's (2012) mobile interface theory, which characterizes mobile phone use as the negotiation of space. In these ways, the physical space surrounding our usage is intertwined with the digital spaces we explore through mobile interfaces.

Why Mobility Matters to In-Place Contexts

Prior work on the psychology of mobility is concerned with being in-place to the extent that (a) mobile people visit different places and (b) experiences with places shape subsequent movement (Di Masso et al., 2019). To the former, in a typical day, many people will spend a majority of their time in their homes or workplaces. However, people also choose to go to a variety of so-called "third places," such as cafés, shops, or parks (Oldenburg, 1996). Such places have been associated with relaxation and social activities (Mehta & Bosson, 2010), suggesting that the places we spend time in reflect our needs. People derive a sense of security (Scannell & Gifford, 2017), bolster their personal goals and identities (Ratcliffe & Korpela, 2016), and feel a sense of continuity in life (e.g., through repeated visits; Froehlich et al., 2006) through their attachment to particular places (Scannell & Gifford, 2010), and people select and manipulate their environments accordingly (Buss, 1987). These processes have been sometimes examined through the lens of place attachment, which broadly refers to the "emotional bonds which people develop with various places" (Lewicka, 2011, p. 219). Although most research on place attachment concerns itself with geographic regions (e.g., a country), some work has described the home in particular as a symbol of centrality, comfort, and security—and people often seek it out for these reasons (see Lewicka, 2011).

To the latter, places themselves have psychological implications that can impact downstream mobility decisions. Recent work has harnessed advancements in mobile capabilities to capture the affective and cognitive implications of particular places (Harari et al., 2017). On the affective front, Sandstrom et al. (2017) found that participants reported more positive moods in social places (compared with being at home) and at home (compared with being at work). From a cognitive standpoint, Slepian et al. (2015) found that experiences of high verticality (e.g., looking up) prime high construal level (i.e., more abstract, superordinate processing; see also Meyers-Levy & Zhu, 2007). Consequently, the affective and cognitive implications of particular places may be related to subsequent mobility choices, as people seek environments more conducive to their current mood or mental state (see Meagher & Marsh, 2017).

Integrating Research on Mobility and Mobile Media In-Place

The above work on daily mobility provides ample possibilities for explicating the psychology of mobile technology in-place. People move to spatial contexts in ways that reflect their identities and form attachments to particular places. These processes likely shape whether individuals view a certain location as a meaningful locale or a simple pitstop and then shape how people use mobile media. Furthermore, mobile technology can structure the importance of particular places (Schwartz, 2014): A previously overlooked street corner can become imbued with meaning to Pokémon Go users if it is the site of a virtual gym that offers in-game rewards. In these ways, hybrid spaces can underpin spatial contexts.

Moreover, the effects of mobile technology are likely contingent on the psychological contours of spatial contexts. By contextualizing the ways that our homes and other places structure our emotion and cognition, we will be better positioned to understand the unique role of mobile media use at home. Moreover, understanding how certain spaces and technology behaviors are interwoven may clarify when individuals choose to move. In the process, the perception of different types of spaces must be explored to reveal a deeper understanding of how place matters to media effects (and vice versa).

USING MOBILE TECHNOLOGY ON-THE-GO

Next, Jerry may play *Candy Crush* while going from place to place. These usage sessions can be seen as the "most mobile" of spatial contexts: when users are on-the-go. Being on-the-go is a unique state of behavior. People must navigate their physical surroundings, often with fast-changing scenes and dynamic decision-making. Thus, Jerry may play *Candy Crush* differently when in a Lyft versus at home. As above, spatial context can also be shaped by mobile technology; commuting is experienced differently when playing *Candy Crush* versus looking out the car window. In this section, we examine how people use their devices in varied ways while navigating on-the-go contexts, highlight work on how such actively mobile contexts may have distinct psychological ramifications, and conclude by integrating these perspectives.

Theoretical frameworks of mobile media are grounded in spatial mobility: the idea that users can use their devices to stay permanently online and permanently connected as they move (Vorderer et al., 2017). According to the theory of the niche (Dimmick et al., 2011), mobile media have a competitive advantage over other communication technologies in these moments. For example, on public transportation, playing *Candy Crush* is more easily accomplished on a phone rather than a computer or even a tablet; it is more socially acceptable, physically comfortable, and technologically possible. As such, mobile media are often used in on-the-go contexts (Karnowski & Jandura, 2014; Leung & Wei, 2000).

Several early studies established that people utilize different functions of mobile technology while on-the-go. Böhmer et al. (2011) found that users were more likely to use music apps and, perhaps counterintuitively, less likely to be using travel apps (e.g., Google Maps). Likewise, Verkasalo (2009) showed that on-the-go users were more likely to use productivity apps and Internet browsers, perhaps because the computer would have a competitive advantage for these functions in the home or at work. A brief caveat, though, lies in how "on-the-go" was operationalized in these studies:

moving faster than 25 kilometers per hour (Böhmer et al., 2011) and not being at home or at work (Verkasalo, 2009). More work is needed that adopts a broader measurement approach, in order to include slower modes of transportation (e.g., walking) and exclude being in other places (e.g., a café).

Perhaps due to these difficulties, most of the work regarding on-the-go mobile media use has been experimental or qualitative in nature. Numerous studies have demonstrated that people walk differently when using their phones (Argin et al., 2020). For example, Hyman et al. (2010) found that people on their phones were more likely to walk slowly, change directions frequently, and not acknowledge other people (compared with control conditions of walking with an MP3 player, with another person, or alone). This suggests that people split their attention between the physical and virtual domains. Figeac and Chaulet (2018) provided more nuance to this attention split via a video-ethnography. At the beginning of trips on public transportation, people are often visually engaged with their mobile devices to check for notifications and other updates. After this initial intensive period, users proceed to browse social media, play games, and do other activities that require less attention, with gaze switches increasing over the course of the trip (e.g., after a scroll, during a game transition). Hence, although the beginning of the trip may be more immersive due to the user's focus on checking notifications or completing a task, awareness of the physical environment increases over time. Schneier and Kudenov (2018) showed that the degree of this awareness is indeed based on the intensity of the task. When participants were texted more difficult questions, they displayed heightened keystroke activity (e.g., deletions, revisions) and had more difficulty navigating their physical environment. Notably, this study was grounded in mobile interface theory, affirming how an interplay exists between mobile technology and mobility because both our virtual interactions and physical engagements involve the senses (e.g., the "sensory-inscribed" body; Farman, 2012).

Such studies provide insights into our understanding of the psychological implications of using mobile technology while on-the-go. In terms of affect, usage of mobile technology is often driven by boredom, but perhaps more so while on-the-go due to the lack of available alternatives

(Figeac & Chaulet, 2018). Furthermore, Reichow and Friemel (2020) argued that perceived insecurity is more likely on public transportation, and users may utilize mobile technology to create a sense of social presence that dispels insecurity. The phone's status as a symbol of security, which was particularly relevant in the early days of mobile phone adoption (Aoki & Downes, 2003), may remain especially relevant in transit. In terms of cognition, being on-the-go presents a changing physical environment, with ramifications for attention. Some users may ignore external constraints and quickly immerse themselves in their mobile devices (Bayer et al., 2016); indeed, Engl and Nacke (2013) found that playing a mobile game on-the-go versus in-place was associated with more immersion. However, others seem to structure their device usage around their physical environment (Figeac & Chaulet, 2018). As such, on-the-go mobile media use may be particularly habitual, in that it is cued and contextualized by one's surroundings, as well as incorporated into the flow of everyday mobility routines (Schnauber-Stockmann & Naab, 2019).

Why Mobility Matters to On-The-Go Contexts

Research on mobility provides a fruitful avenue to explicate the psychological mechanisms that underlie on-the-go contexts. For example, a growing body of work investigates the implications of daily mobility experiences for well-being. Ettema et al. (2011) introduced travel satisfaction as a precursor to subjective well-being (Olsson et al., 2013). Satisfaction with travel is shaped by perceptions of travel and waiting time (Carrel et al., 2016), and also drives future transportation decisions (Le et al., 2020). Other recent work relates travel to a variety of well-being dimensions (De Vos et al., 2013; Mokhtarian, 2019). For instance, Singleton (2019) found that walking and cycling were both perceived as conducive to subjective well-being, but that cycling was also associated with distress, fear, and low security. Hence, daily mobility is inextricably linked to daily well-being.

Daily mobility patterns most often take the form of trips (i.e., going from A to B; Xu et al., 2016), and scholars have thus investigated different types of trips and their psychological consequences. For example, Zhang and Li (2017) distinguished between spontaneous and nonspontaneous trips, finding that spontaneous trips involve an activity range that is larger, more heterogenous, and less periodic. Glasgow et al. (2019) demonstrated that trips that were active (e.g., walking, biking), accompanied, or in natural environments were associated with higher moods, whereas neighborhoods with higher walk scores, an objective measure of walkability, negatively impacted mood (likely due to being in urban environments). The negative impact of walk scores suggests a contrast between subjective perceptions and objective aspects of one's environment while on-the-go. To that end, Chan et al. (2021) showed that positive perceptions of an environment can override low walkability. Thus, the implications of mobility depend on the type of trip as well as perceptions of the environment.

The importance of how one perceives the environment aligns with basic cognitive research on spatial perception. Schnall et al. (2008) demonstrated that social support can influence perceptions of geographical slants. Participants who walked up a hill with a friend or while thinking about a friend perceived a shallower slope, demonstrating that cognitive factors can impact spatial perception. Notably, Luo et al. (2021) replicated these findings on social support and spatial perception for mobile media, demonstrating that texting reduced perceptions of geographic slant. Similarly, Balcetis and Dunning (2007) found that hills were perceived as shallower when participants were given more choice about their environment. To reduce cognitive dissonance, people perceive their chosen environments as less aversive. In contrast, cross-country athletes and marine recruits likely view the hill that their coach forces them to climb as particularly steep. The previous findings have implications beyond hills for the psychological ramifications of mobility, providing initial evidence that psychological states (e.g., social support, perceived autonomy) can impact spatial perception and potentially broader perceptions of mobility (cf. Durgin et al., 2009; Schnall, 2017).

Integrating Research on Mobility and Mobile Media On-The-Go

The research threads covered thus far suggest that the psychology of mobile technology and that of mobility need to be considered in tandem. Indeed, a large proportion of mobile technology usage occurs on-the-go, and on-the-go contexts are increasingly marked by mobile technology. After long conceptualizing travel time to be a waste that was only worthwhile to reach a destination, scholarship in mobility studies over the last 2 decades now recognize that media use occupies a central part of contemporary mobile experiences (Lyons, 2019). For instance, research has demonstrated that perceptions of the value of travel time are pinned to technology use while traveling as well as downstream mobility choices (e.g., Ettema & Verschuren, 2007; Mokhtarian, 2019; Singleton, 2018). Our mobile devices even anchor how we move (e.g., to look up information about a destination or aid navigation), with psychological ramifications for how we experience mobile technology and mobility (Delbosc & Mokhtarian, 2018).

Unfortunately, these studies remain overlooked within the literature on mobile media, despite their clear relevance and psychological bent. Work on the psychological implications of mobile technology demonstrates that on-the-go usage is often driven by boredom and the need for security, which are likely based in part on characteristics of the mobility itself (e.g., travel time, co-presence of others). In turn, perceptions of mobility are likely shaped by the psychological effects of mobile media use. An example regarding perceived security illustrates this dynamic relationship. Reichow and Friemel (2020) found that traveling on an underground train was associated with perceived insecurity, hypothesizing that travelers would use their mobile devices to increase perceptions of social presence in order to reduce such feelings of insecurity (e.g., texting; Luo et al., 2021). But this behavior feeds back to influence perceptions of the environment itself. In other words, the environment could come to be seen as a lower threat of insecurity, whetting the initial reason to text in the first place. Over time, a malleable but stable balance between security and insecurity likely forms, perhaps codified in the gaze switches between the device and the environment (Figeac & Chaulet, 2018). This example displays the potential for mobile technology and mobility to become tightly interwoven in individuals' minds and depicts their relevance to psychological well-being in daily life.

USERS AND MOBILE DEVICES

In the previous sections, we explicated how the psychology of mobile technology is interwoven with the psychology of mobility across in-place and on-the-go contexts. We now compile these contexts into *daily mobility patterns*, seeking to untangle the overall psychological connection between users and their personal mobile technologies. To illustrate, we return once more to our obsessive *Candy Crush* user, Jerry. Because he plays *Candy Crush* in his favorite places as well as out-and-about, clarifying the effects of playing *Candy Crush*—or using smartphones in general—requires evaluating the interaction of media and space. Furthermore, daily mobility patterns are highly regular over time (González et al., 2008), suggesting that they serve an outsized role in shaping the mental connection between users and their mobile devices. Next, we review existing research on the user–device connection, overall daily mobility patterns, and their intersections.

People often come to feel psychologically close to their mobile devices, as they use these always-present tools for myriad purposes (Walsh & White, 2007). This phenomenon has been described by a multiplicity of terms, including digital companionship (Carolus et al., 2019), virtual friendship (Fullwood et al., 2017), extension of the self (Ross & Bayer, 2021), and attachment object (Trub & Barbot, 2016), among others. A large body of work considers this phenomenon from the lens of behavioral addiction, though this designation remains debated (cf. Bayer et al., 2016; Pivetta et al., 2019). Although the exact nature of the mental connection is complicated and contested, a range of perspectives have defined the strong psychological link that can exist between user and device.

A variety of processes drive this global psychological connection (Ross & Campbell, 2021). In terms of cognition, people can offload information and navigation to their devices, which can enhance access to knowledge (Clark, 2008; Ishikawa, 2019). Personal devices offer a wide array of cues, contexts, and outcomes that can be linked together to form complex mental habits (Bayer & LaRose, 2018). In terms of affect, feelings of pleasure, connection, control, and safety are always within arm's reach

(Cumiskey & Brewster, 2012; Fullwood et al., 2017). As a result, mobile technology is able to quell stress and anxiety at key times—and drive such feelings at other times (Carolus et al., 2019; Cheever et al., 2014; Melumad & Pham, 2020). In many cases, the resulting user–device connection is strong enough to disrupt or displace other cognitive and affective processes (e.g., memory, face-to-face interaction; Dwyer et al., 2018; Ward, 2013). This connection is also likely shaped by individual differences (e.g., Big Five; see Stachl et al., 2020).

The above research suggests that the psychological link between user and device—whether framed as "companionship" or "addiction"—is underpinned by daily cognitive and affective processes. Critically, these processes are often contextual. Jerry may develop the habit of playing *Candy Crush* to avoid boredom experienced while waiting in line at the bus stop. This contextual perspective, however, is often glossed over when considering the overall user–device connection. Consequently, we argue that the complex connection between user and device has the potential to be deciphered through a dual focus on mobile technology and daily mobility.

Why Mobility Matters to the User–Device Connection

The notion that the user–device connection represents a combination of contextual usage sessions is further supported by the fact that daily mobility patterns are highly regular. Although people move in a variety of ways (e.g., walking, scootering), mobility is marked by clear intraindividual regularity (González et al., 2008; Song et al., 2010). Most people walk or ride the same set of paths on daily loops, frequenting a few places (i.e., one's "activity space"; Smith et al., 2019) with clear temporal patterns. As a result of this stability, researchers are able to predict an individual's location at a given moment with a remarkably high degree of accuracy (de Montjoye et al., 2013). The regularity of mobility likely reflects consistent motivations for movement. Mokhtarian et al. (2015) argued that people move for broadly similar reasons, including intrinsic pushes (e.g., wanting to roam about) and extrinsic pulls (e.g., wanting to go to an event), and these motivations are largely stable within individuals.

In turn, researchers have explored the psychological implications of stable movement patterns by relating mobility to sociability, personality, and health. Alessandretti et al. (2018) found that people who visit more places tend to connect with more people, cementing an association between mobility and sociability. This lends evidence to the idea that people are driven to move because they are driven by a desire to interact (van der Waerden et al., 2019). Furthermore, extraverts tend to visit more places and travel farther than introverts (Alessandretti et al., 2018). Mobility patterns also correspond with mental health profiles (see Müller et al., 2020). For example, a wide variety of mobility features (e.g., lower regularity of movement) have been used to predict depression with relatively high levels of accuracy (Canzian & Musolesi, 2015; Saeb et al., 2016), at least in homogenous samples (Müller et al., 2021). Research has also linked movement patterns to other indicators of mental health, such as stress (Ben-Zeev et al., 2015). Overall, individuals move in regular patterns, and these patterns are related to sociability, personality, and health, illustrating how mobility is integral to mental and physical well-being.

Integrating Research on Mobile Media Users and Daily Mobility

The above precursors of mobility can dovetail or conflict with the precursors of mobile technology usage. Kellerman (2012), Taipale (2014), and others have argued that physical mobilities (e.g., spatial movement) and "virtual mobilities" (e.g., communication- or information-based mobile technology usage) are driven by the same core psychological motivations (e.g., desire for proximity to others; Mokhtarian et al., 2015). That is, the same motivation could shape both spatial and virtual behavior, but also potentially be satisfied by either one. Typically, the relationship between physical and virtual mobilities has been examined at the individual level (Dal Fiore et al., 2014; Mokhtarian, 2002; van der Waerden et al., 2019), such as investigating whether frequent mobile users spend more time at home (Thulin & Vilhelmson, 2019) or visit fewer locations (De Nadai et al., 2019). But this relationship could also manifest at the situational level (see Ling & Haddon, 2003). For example, extroverts may message their friends on the way to a party, whereas introverts may either message their friends *or* go to the party (and instead play *Candy Crush* on-the-way). Untangling the shared and distinct precursors of daily mobility and technology usage can advance the psychology of mobile technology.

Taken together, these perspectives provide an underutilized lens to understand Jerry's mobile device usage. Spatial contexts shape his experience of *Candy Crush*, whether he is in bed or on the bus, and these media-context pairings coalesce into an overall connection to his device. Given the regularity of daily mobility, Jerry likely spends the most time in these spaces and other key locations that form the foundation of his smartphone usage. Hence, the user–device connection cannot be understood without identifying the interactive effects of technology and mobility behaviors, including how the precursors—and effects—of these usage patterns and spaces operate in tandem. If mobile gaming relieves stress during his work commute, yet also amplifies stress during his sleep routine, then the underlying mechanisms of digital well-being may be overlooked (Vanden Abeele, 2020). For the above reasons, unpacking how users are psychologically synced with their technologies increasingly demands that researchers delineate the spatial contours of daily mobile behavior.

METHODOLOGICAL APPROACHES AND ADVANCES

Adopting a more integrative approach to studying mobile technology and mobility requires reflection on the methodological toolkit at hand. Much of the work mentioned in this chapter utilized cross-sectional self-reports as part of a survey or experimental approach. Of course, self-reports are the most appropriate way to measure many psychological constructs, as they can uniquely tap into user perceptions (e.g., attachment towards mobile devices, contours of social environments). However, self-reports of mobile media use have faced criticism for being prone to a number of biases. For example, self-reports of constructs that have verifiable answers (e.g., time spent on smartphones) have been found to be fairly inaccurate, if not systematically biased (e.g., confounded with well-being and mental health; Parry et al., 2021). For self-reports of other constructs (e.g., habitual phone use), it can still be challenging to determine validity, especially when measured in a cross-sectional survey (Ohme et al., 2016).

In addition to general self-report issues, a large share of the work on the psychology of mobile technology and mobility has relied on surveys that were not administered in naturalistic contexts. Hence, our current knowledgebase overlooks crucial aspects of in vivo spatial contexts. Going forward, more studies can be designed to make use of the family of approaches (and terms) falling under the umbrella of experiencing sampling method (ESM, also referred to as ecological momentary or ambulatory assessment). A variety of means have been adapted to achieve experience sampling over the years (e.g., pagers; Larson & Csikszentmihalyi, 1978), but the widespread diffusion of mobile media such as smartphones and wearables enables ESM studies to be conducted with relative ease (Schnauber Stockmann & Karnowski, 2020).

Contemporary media also provide new opportunities to obtain insights about the psychology of mobile technology and daily mobility. Researchers can be in direct contact with participants as they are actively mobile (Kaufmann & Peil, 2020). Alternatively, mobile technology can be used to record or log in vivo technology behavior (Deng et al., 2019). Some studies have used log data to infer psychological processes; for example, Oulasvirta et al. (2012) measured habitual smartphone use as the frequency of brief checking behaviors instead of using a more traditional self-report measure. Other approaches include video-ethnography, in which participants record their interactions with mobile technology while on-the-go (Figeac & Chaulet, 2018), and keystroke logging, in which researchers analyze how participants type in different spatial contexts (Schneier & Kudenov, 2018). Such data can serve as the basis for interviews and focus groups, as participants can retroactively discuss their phone use and mobility patterns (see Schoenebeck et al., 2016).

Beyond offering digital traces of emerging media use, mobile technologies represent a new frontier by enabling access to geospatial and transportation data. Scholars have utilized a variety of data sources to acquire spatial insights about mobile behavior, including call detail records (CDRs), cell tower identification, WiFi, smartphone GPS sensors (vis-à-vis native and downloaded apps), and geotagged social media (Harari et al., 2017). These methods can passively affix GPS coordinates to a variety of activities and allow researchers to link their datasets to other location-based data sources (e.g., maps, zip codes; Andris et al., 2019). Despite these advances, there can still be problems with data accuracy due to satellite geometry, receiver quality, and whether an obstacle (e.g., a building) blocks the direct line of sight between the receiver and the satellite. These problems may motivate researchers to use multiple data sources, if possible (Tu et al., 2017), or rely on broad typologies of places (e.g., home, work, "other meaningful," and "elsewhere") rather than more specific categorizations.

Last, an increasing number of studies use in vivo self-reports and track participants' spatial context in tandem (Harari et al., 2020; Rhee et al., 2020). Participant self-reports can be used to establish a ground truth, thus improving the accuracy of location sensing (Froehlich et al., 2006). Moreover, participants can self-report their psychological states (e.g., Sandstrom et al., 2017), aspects of their environment (e.g., Rauthmann et al., 2015), and their mobility patterns (e.g., Glasgow et al., 2019) while in-place or on-the-go. Looking forward, future work even has the potential to leverage mobile eye-tracking (Pérez-Edgar et al., 2020) and brain imaging (Huckins et al., 2019) to deepen our understanding of mobile processes.

FUTURE DIRECTIONS AND CONCLUSIONS

This chapter has advanced a contextual, spatial, and mobile perspective on the psychological implications of technology usage. We first affirmed the importance of central spatial contexts (e.g., home) by examining in-place use of mobile technologies. Next, we considered the contexts linking these focal places by reviewing on-the-go usage. These sections laid the groundwork for recentering how we study the fundamental psychological connection between user and device. We suggest that daily mobility patterns—which take the individual to and through these in-place and on-the-go contexts—play a pivotal role in shaping user–device connections, as well as the effects of technology usage. In other words, Jerry's attachment to his smartphone can also be seen as a mental connection between Jerry, the device, *and* the spaces where he uses apps such as *Candy Crush*.

Future research will be challenged to pinpoint the direct and interactive effects of spatial context on the psychology of technology usage. As noted above, in-place and on-the-go contexts are often studied separately or simply collapsed into "elsewhere," especially in studies on the psychology of mobile technology. However, this obfuscates the fact that being in-place and on-the-go, and using mobile media in these contexts, can have significantly different psychological implications for users. Emergent research should not only theoretically and methodologically tease apart these contexts, but also explore how place interacts with movement (Di Masso et al., 2019). Indeed, prior work illustrates how place attachment is related to mobility patterns (Gustafson, 2009) and how mobility can "spill over" to impact experience with places (Gärling, 2019). Following these ideas, technology researchers should acknowledge and assess the bidirectional, dynamic relationship between being in-place and on-the-go. For example, using mobile technologies to navigate to a location—or entertain oneself on the way—may shape the psychological effects of such locations. Alternatively, using mobile media in particular places may have downstream effects for subsequent mobility behavior. Notably, study designs to this end could draw on the state-of-the-art methods reviewed above, such as passively tracking individuals' mobility patterns and collecting subjective perceptions of in-place and on-the-go spatial contexts in parallel. Looking forward, the increasing number of psychological methods being translated for mobile data collection suggests that connections between psychology and mobility are just beginning.

There is also a need to bridge the disciplinary gaps between research on mobile technology and spatial mobility, which are seldom considered in tandem. In psychology, communication, human–computer interaction, and other research spaces that are focused on the psychological implications of mobile technologies; there has been scant recognition of the work in mobility studies. We argue that this tech-centric approach is myopic. Mobility researchers provide an essential viewpoint on the effects of mobile media by theoretically and empirically foregrounding the question of *why* people move. Moreover, mobility scholars have increasingly studied technology use in transit and incorporated psychological approaches (e.g., Le et al., 2020). Likewise, future research on daily mobility and human movement may draw on theoretical models of technology behavior, as contemporary mobility is often interwoven with mobile media. Ultimately, scholars in all domains should explicate the theoretical similarities between transportation and communication (Hildebrand, 2018; von Pape, 2020).

As everyday technologies become more and more mobile, their usage will interact with numerous spatial contexts in novel ways—creating complex combinations of potential psychological effects. Although this represents a formidable challenge to researchers of mobile technology, mobility, and psychology alike, it also represents an opportunity to merge energies and integrate agendas. The array of spaces in which mobile media are used today affirms the need to pull theoretical perspectives on movement and place into the central fold of technology research. In parallel, these same technologies surface new opportunities to assess the fundamental role of spatial context in human behavior and cognition. Though linking this constellation of research perspectives is certain to be complicated, it is likely necessary to fully decode our increasingly mobile world.

REFERENCES

Alessandretti, L., Lehmann, S., & Baronchelli, A. (2018). Understanding the interplay between social and spatial behaviour. *EPJ Data Science, 7*(1), 36. https://doi.org/10.1140/epjds/s13688-018-0164-6

Andris, C., Godfrey, B., Maitland, C., & McGee, M. (2019). The built environment and Syrian refugee integration in Turkey: An analysis of mobile phone data. In B. Martins, L. Moncia, & P. Murrieta-Flores (Eds.), *GeoHumanities '19: Proceedings of the 3rd ACM SIGSPATIAL International Workshop on Geospatial Humanities* (pp. 1–7). Association for Computing Machinery. https://doi.org/10.1145/3356991.3365472

Aoki, K., & Downes, E. J. (2003). An analysis of young people's use of and attitudes toward cell phones. *Telematics and Informatics, 20*(4), 349–364. https://doi.org/10.1016/S0736-5853(03)00018-2

Argin, G., Pak, B., & Turkoglu, H. (2020). Between post-flâneur and smartphone zombie: Smartphone users' altering visual attention and walking behavior in public space. *ISPRS International Journal of Geo-Information, 9*(12), 700. https://doi.org/10.3390/ijgi9120700

Balcetis, E., & Dunning, D. (2007). Cognitive dissonance and the perception of natural environments. *Psychological Science, 18*(10), 917–921. https://doi.org/10.1111/j.1467-9280.2007.02000.x

Bayer, J. B., Dal Cin, S., Campbell, S. W., & Panek, E. (2016). Consciousness and self-regulation in mobile communication. *Human Communication Research, 42*(1), 71–97. https://doi.org/10.1111/hcre.12067

Bayer, J. B., Ellison, N., Schoenebeck, S., Brady, E., & Falk, E. B. (2018). Facebook in context(s): Measuring emotional responses across time and space. *New Media & Society, 20*(3), 1047–1067. https://doi.org/10.1177/1461444816681522

Bayer, J. B., & LaRose, R. (2018). Technology habits: Progress, problems, and prospects. In B. Verplanken (Ed.), *The psychology of habit* (pp. 111–130). Springer. https://doi.org/10.1007/978-3-319-97529-0_7

Ben-Zeev, D., Scherer, E. A., Wang, R., Xie, H., & Campbell, A. T. (2015). Next-generation psychiatric assessment: Using smartphone sensors to monitor behavior and mental health. *Psychiatric Rehabilitation Journal, 38*(3), 218–226. https://doi.org/10.1037/prj0000130

Böhmer, M., Hecht, B., Schöning, J., Krüger, A., & Bauer, G. (2011). Falling asleep with Angry Birds, Facebook and Kindle: A large scale study on mobile application usage. In *MobileHCI '11: Proceedings of the 13th International Conference on Human–Computer Interaction With Mobile Devices and Services* (pp. 47–56). Association for Computing Machinery. https://doi.org/10.1145/2037373.2037383

Büscher, M., & Urry, J. (2009). Mobile methods and the empirical. *European Journal of Social Theory, 12*(1), 99–116. https://doi.org/10.1177/1368431008099642

Buss, D. M. (1987). Selection, evocation, and manipulation. *Journal of Personality and Social Psychology, 53*(6), 1214–1221. https://doi.org/10.1037/0022-3514.53.6.1214

Campbell, S. W. (2019). From frontier to field: Old and new theoretical directions in mobile communication studies. *Communication Theory, 29*(1), 46–65. https://doi.org/10.1093/ct/qty021

Canzian, L., & Musolesi, M. (2015). Trajectories of depression: Unobtrusive monitoring of depressive states by means of smartphone mobility traces analysis. In *UbiComp '15: Proceedings of the 2015 ACM International Joint Conference on Pervasive and Ubiquitous Computing* (pp. 1293–1304). Association for Computing Machinery. https://doi.org/10.1145/2750858.2805845

Carolus, A., Binder, J. F., Muench, R., Schmidt, C., Schneider, F., & Buglass, S. L. (2019). Smartphones as digital companions: Characterizing the relationship between users and their phones. *New Media & Society, 21*(4), 914–938. https://doi.org/10.1177/1461444818817074

Carrel, A., Mishalani, R. G., Sengupta, R., & Walker, J. L. (2016). In pursuit of the happy transit rider: Dissecting satisfaction using daily surveys and tracking data. *Journal of Intelligent Transportation Systems, 20*(4), 345–362. https://doi.org/10.1080/15472450.2016.1149699

Chan, E. T. H., Schwanen, T., & Banister, D. (2021). The role of perceived environment, neighbourhood characteristics, and attitudes in walking behaviour: Evidence from a rapidly developing city in China. *Transportation, 48*(1), 431–454. https://doi.org/10.1007/s11116-019-10062-2

Cheever, N. A., Rosen, L. D., Carrier, L. M., & Chavez, A. (2014). Out of sight is not out of mind: The impact of restricting wireless mobile device use on anxiety levels among low, moderate and high users. *Computers in Human Behavior, 37*, 290–297. https://doi.org/10.1016/j.chb.2014.05.002

Clark, A. (2008). *Supersizing the mind: Embodiment, action, and cognitive extension.* Oxford University Press. https://doi.org/10.1093/acprof:oso/9780195333213.001.0001

Cumiskey, K. M., & Brewster, K. (2012). Mobile phones or pepper spray? Imagined mobile intimacy as a weapon of self-defense for women. *Feminist Media Studies, 12*(4), 590–599. https://doi.org/10.1080/14680777.2012.741893

Dal Fiore, F., Mokhtarian, P. L., Salomon, I., & Singer, M. E. (2014). "Nomads at last"? A set of perspectives on how mobile technology may affect travel. *Journal of Transport Geography, 41*, 97–106. https://doi.org/10.1016/j.jtrangeo.2014.08.014

de Montjoye, Y. A., Hidalgo, C. A., Verleysen, M., & Blondel, V. D. (2013). Unique in the crowd: The privacy bounds of human mobility. *Scientific Reports, 3*(1), 1376. https://doi.org/10.1038/srep01376

De Nadai, M., Cardoso, A., Lima, A., Lepri, B., & Oliver, N. (2019). Strategies and limitations in app usage and human mobility. *Scientific Reports, 9*(1), 10935. https://doi.org/10.1038/s41598-019-47493-x

de Souza e Silva, A. (2006). From cyber to hybrid: Mobile technologies as interfaces of hybrid spaces. *Space and Culture, 9*(3), 261–278. https://doi.org/10.1177/1206331206289022

De Vos, J., Schwanen, T., van Acker, V., & Witlox, F. (2013). Travel and subjective well-being: A focus on findings, methods and future research needs. *Transport Reviews, 33*(4), 421–442. https://doi.org/10.1080/01441647.2013.815665

Delbosc, A., & Mokhtarian, P. (2018). Face to Facebook: The relationship between social media and social travel. *Transport Policy, 68*, 20–27. https://doi.org/10.1016/j.tranpol.2018.04.005

Deng, T., Kanthawala, S., Meng, J., Peng, W., Kononova, A., Hao, Q., Zhang, Q., & David, P. (2019). Measuring smartphone usage and task switching with log tracking and self-reports. *Mobile Media and Communication*, *7*(1), 3–23. https://doi.org/10.1177/2050157918761491

Di Masso, A., Williams, D. R., Raymond, C. M., Buchecker, M., Degenhardt, B., Devine-Wright, P., Hertzog, A., Lewicka, M., Manzo, L., Shahrad, A., Stedman, R., Verbrugge, L., & von Wirth, T. (2019). Between fixities and flows: Navigating place attachments in an increasingly mobile world. *Journal of Environmental Psychology*, *61*, 125–133. https://doi.org/10.1016/j.jenvp.2019.01.006

Dimmick, J., Feaster, J. C., & Hoplamazian, G. J. (2011). News in the interstices: The niches of mobile media in space and time. *New Media & Society*, *13*(1), 23–39. https://doi.org/10.1177/1461444810363452

Durgin, F. H., Baird, J. A., Greenburg, M., Russell, R., Shaughnessy, K., & Waymouth, S. (2009). Who is being deceived? The experimental demands of wearing a backpack. *Psychonomic Bulletin & Review*, *16*(5), 964–969. https://doi.org/10.3758/PBR.16.5.964

Dwyer, R. J., Kushlev, K., & Dunn, E. W. (2018). Smartphone use undermines enjoyment of face-to-face social interactions. *Journal of Experimental Social Psychology*, *78*, 233–239. https://doi.org/10.1016/j.jesp.2017.10.007

Engl, S., & Nacke, L. E. (2013). Contextual influences on mobile player experience—A game user experience model. *Entertainment Computing*, *4*(1), 83–91.

Ettema, D., Gärling, T., Eriksson, L., Friman, M., Olsson, L. E., & Fujii, S. (2011). Satisfaction with travel and subjective well-being: Development and test of a measurement tool. *Transportation Research Part F: Traffic Psychology and Behaviour*, *14*(3), 167–175. https://doi.org/10.1016/j.trf.2010.11.002

Ettema, D., & Verschuren, L. (2007). Multitasking and value of travel time savings. *Transportation Research Record*, (2010), 19–25. https://doi.org/10.3141/2010-03

Farman, J. (2012). *Mobile interface theory: Embodied space and locative media*. Routledge.

Figeac, J., & Chaulet, J. (2018). Video-ethnography of social media apps' connection cues in public settings. *Mobile Media and Communication*, *6*(3), 407–427. https://doi.org/10.1177/2050157917747642

Fortunati, L., & Taipale, S. (2017). Mobilities and the network of personal technologies: Refining the understanding of mobility structure. *Telematics and Informatics*, *34*(2), 560–568. https://doi.org/10.1016/j.tele.2016.09.011

Frith, J. (2015). *Smartphones as locative media*. Polity Press.

Froehlich, J., Chen, M. Y., Smith, I. E., & Potter, F. (2006). Voting with your feet: An investigative study of the relationship between place visit behavior and preference. In P. Dourish & A. Friday (Eds.), *UbiComp 2006: International*

Conference on Ubiquitous Computing (pp. 333–350). Springer. https://doi.org/10.1007/11853565_20

Fullwood, C., Quinn, S., Kaye, L. K., & Redding, C. (2017). My virtual friend: A qualitative analysis of the attitudes and experiences of smartphone users: Implications for smartphone attachment. *Computers in Human Behavior, 75*, 347–355. https://doi.org/10.1016/j.chb.2017.05.029

Gärling, T. (2019). Travel-related feelings: Review, theoretical framework, and numerical experiments. *Transportation Letters, 11*(1), 54–62. https://doi.org/10.1080/19427867.2017.1300399

Glasgow, T. E., Le, H. T. K., Geller, E. S., Fan, Y., & Hankey, S. (2019). How transport modes, the built and natural environments, and activities influence mood: A GPS smartphone app study. *Journal of Environmental Psychology, 66*, 101345. https://doi.org/10.1016/j.jenvp.2019.101345

González, M. C., Hidalgo, C. A., & Barabási, A. L. (2008). Understanding individual human mobility patterns. *Nature, 453*, 779–782. https://doi.org/10.1038/nature06958

Gustafson, P. (2009). Mobility and territorial belonging. *Environment and Behavior, 41*(4), 490–508. https://doi.org/10.1177/0013916508314478

Harari, G. M., Müller, S. R., Aung, M. S., & Rentfrow, P. J. (2017). Smartphone sensing methods for studying behavior in everyday life. *Current Opinion in Behavioral Sciences, 18*, 83–90. https://doi.org/10.1016/j.cobeha.2017.07.018

Harari, G. M., Müller, S. R., Stachl, C., Wang, R., Wang, W., Bühner, M., Rentfrow, P. J., Campbell, A. T., & Gosling, S. D. (2020). Sensing sociability: Individual differences in young adults' conversation, calling, texting, and app use behaviors in daily life. *Journal of Personality and Social Psychology, 119*(1), 204–228. https://doi.org/10.1037/pspp0000245

Hildebrand, J. M. (2018). Modal media: Connecting media ecology and mobilities research. *Media Culture & Society, 40*(3), 348–364. https://doi.org/10.1177/0163443717707343

Hintze, D., Hintze, P., Findling, R. D., & Mayrhofer, R. (2017). A large-scale, long-term analysis of mobile device usage characteristics. *Proceedings of the ACM on Interactive, Mobile, Wearable and Ubiquitous Technologies, 1*(2), 1–21. https://doi.org/10.1145/3090078

Huckins, J. F., daSilva, A. W., Wang, R., Wang, W., Hedlund, E. L., Murphy, E. I., Lopez, R. B., Rogers, C., Holtzheimer, P. E., Kelley, W. M., Heatherton, T. F., Wagner, D. D., Haxby, J. V., & Campbell, A. T. (2019). Fusing mobile phone sensing and brain imaging to assess depression in college students. *Frontiers in Neuroscience, 13*, 248. https://doi.org/10.3389/fnins.2019.00248

Humphreys, L., Karnowski, V., & von Pape, T. (2018). Smartphones as metamedia: A framework for identifying the niches structuring smartphone use.

International Journal of Communication, 12, 2793–2809. https://ijoc.org/index.php/ijoc/article/view/7922

Hyman, I. E., Jr., Boss, S. M., Wise, B. M., McKenzie, K. E., & Caggiano, J. M. (2010). Did you see the unicycling clown? Inattentional blindness while walking and talking on a cell phone. *Applied Cognitive Psychology, 24*(5), 597–607. https://doi.org/10.1002/acp.1638

Ishikawa, T. (2019). Satellite navigation and geospatial awareness: Long-term effects of using navigation tools on wayfinding and spatial orientation. *The Professional Geographer, 71*(2), 197–209. https://doi.org/10.1080/00330124.2018.1479970

Karnowski, V., & Jandura, O. (2014). When lifestyle becomes behavior: A closer look at the situational context of mobile communication. *Telematics and Informatics, 31*(2), 184–193. https://doi.org/10.1016/j.tele.2013.11.001

Kaufmann, K., & Peil, C. (2020). The mobile instant messaging interview (MIMI): Using WhatsApp to enhance self-reporting and explore media usage in situ. *Mobile Media and Communication, 8*(2), 229–246. https://doi.org/10.1177/2050157919852392

Kellerman, A. (2012). *Daily spatial mobilities: Physical and virtual*. Ashgate Publishing.

Larson, R., & Csikszentmihalyi, M. (1978). Experiential correlates of time alone in adolescence. *Journal of Personality, 46*(4), 677–693. https://doi.org/10.1111/j.1467-6494.1978.tb00191.x

Le, H. T. K., Carrel, A. L., & Li, M. (2020). How much dissatisfaction is too much for transit? Linking transit user satisfaction and loyalty using panel data. *Travel Behaviour & Society, 20*, 144–154. https://doi.org/10.1016/j.tbs.2020.03.007

Leung, L., & Wei, R. (2000). More than just talk on the move: Uses and gratifications of the cellular phone. *Journalism & Mass Communication Quarterly, 77*(2), 308–320. https://doi.org/10.1177/107769900007700206

Lewicka, M. (2011). Place attachment: How far have we come in the last 40 years? *Journal of Environmental Psychology, 31*(3), 207–230. https://doi.org/10.1016/j.jenvp.2010.10.001

Ling, R., & Haddon, L. (2003). Mobile telephony, mobility, and the coordination of everyday life. In J. E. Katz (Ed.), *Machines that become us: The social context of personal communication technology* (pp. 245–266). Routledge. https://doi.org/10.4324/9780203786826-18

Luo, M., Falisi, A., & Hancock, J. (2021). Can text messaging influence perceptions of geographical slant? A replication and extension of Schnall, Harber, Stefanucci, Proffitt (2008). *Technology, Mind, and Behavior, 2*(2). https://doi.org/10.1037/tmb0000031

Lyons, G. (2019). The wicked problem of travel time use and the effects of ICTs. In E. Ben-Elia (Ed.), *The evolving impacts of ICT on activities and travel behavior* (pp. 145–170). Elsevier. https://doi.org/10.1016/bs.atpp.2019.04.001

McLuhan, M. (1964). *Understanding media: The extensions of man.* McGraw-Hill.

Meagher, B. R., & Marsh, K. L. (2017). Seeking the safety of sociofugal space: Environmental design preferences following social ostracism. *Journal of Experimental Social Psychology, 68*, 192–199. https://doi.org/10.1016/j.jesp.2016.07.004

Mehrotra, A., Müller, S. R., Harari, G. M., Gosling, S. D., Mascolo, C., Musolesi, M., & Rentfrow, P. J. (2017). Understanding the role of places and activities on mobile phone interaction and usage patterns. In S. Santini (Ed.), *Proceedings of the ACM on Interactive, Mobile, Wearable and Ubiquitous Technologies, 1*(3), 1–22. https://doi.org/10.1145/3131901

Mehta, V., & Bosson, J. K. (2010). Third places and the social life of streets. *Environment and Behavior, 42*(6), 779–805. https://doi.org/10.1177/0013916509344677

Melumad, S., & Pham, M. T. (2020). The smartphone as a pacifying technology. *The Journal of Consumer Research, 47*(2), 237–255. https://doi.org/10.1093/jcr/ucaa005

Meyers-Levy, J., & Zhu, R. (2007). The influence of ceiling height: The effect of priming on the type of processing that people use. *The Journal of Consumer Research, 34*(2), 174–186. https://doi.org/10.1086/519146

Mokhtarian, P. L. (2002). Telecommunications and travel: The case for complementarity. *Journal of Industrial Ecology, 6*(2), 43–57. https://doi.org/10.1162/108819802763471771

Mokhtarian, P. L. (2019). Subjective well-being and travel: Retrospect and prospect. *Transportation, 46*(2), 493–513. https://doi.org/10.1007/s11116-018-9935-y

Mokhtarian, P. L., Salomon, I., & Singer, M. E. (2015). What moves us? An interdisciplinary exploration of reasons for traveling. *Transport Reviews, 35*(3), 250–274. https://doi.org/10.1080/01441647.2015.1013076

Müller, S. R., Peters, H., Matz, S. C., Wang, W., & Harari, G. M. (2020). Investigating the relationships between mobility behaviours and indicators of subjective well-being using smartphone-based experience sampling and GPS tracking. *European Journal of Personality, 34*(5), 714–732. https://doi.org/10.1002/per.2262

Müller, S. R., Xi, L. C., Peters, H., Chaintreau, A., & Matz, S. C. (2021). Depression predictions from GPS-based mobility do not generalize well to large demographically heterogeneous samples. *Scientific Reports, 11*, 14007. https://doi.org/10.1038/s41598-021-93087-x

Ohme, J., Albaek, E., & de Vreese, C. (2016). Exposure research going mobile: A smartphone-based measurement of media exposure to political information in a convergent media environment. *Communication Methods and Measures*, *10*(2–3), 135–148. https://doi.org/10.1080/19312458.2016.1150972

Oldenburg, R. (1996). Our vanishing "third places." *Planning Commissioners Journal*, *25*, 6–10. https://plannersweb.com/wp-content/uploads/1997/01/184.pdf

Olsson, L. E., Gärling, T., Ettema, D., Friman, M., & Fujii, S. (2013). Happiness and satisfaction with work commute. *Social Indicators Research*, *111*(1), 255–263. https://doi.org/10.1007/s11205-012-0003-2

Oulasvirta, A., Rattenbury, T., Ma, L., & Raita, E. (2012). Habits make smartphone use more pervasive. *Personal and Ubiquitous Computing*, *16*(1), 105–114. https://doi.org/10.1007/s00779-011-0412-2

Parry, D. A., Davidson, B. I., Sewall, C. J. R., Fisher, J. T., Mieczkowski, H., & Quintana, D. S. (2021). A systematic review and meta-analysis of discrepancies between logged and self-reported digital media use. *Nature Human Behaviour*. Advance online publication. https://doi.org/10.1038/s41562-021-01117-5

Pérez-Edgar, K., MacNeill, L. A., & Fu, X. (2020). Navigating through the experienced environment: Insights from mobile eye tracking. *Current Directions in Psychological Science*, *29*(3), 286–292. https://doi.org/10.1177/0963721420915880

Pivetta, E., Harkin, L., Billieux, J., Kanjo, E., & Kuss, D. J. (2019). Problematic smartphone use: An empirically validated model. *Computers in Human Behavior*, *100*, 105–117. https://doi.org/10.1016/j.chb.2019.06.013

Ratcliffe, E., & Korpela, K. M. (2016). Memory and place attachment as predictors of imagined restorative perceptions of favourite places. *Journal of Environmental Psychology*, *48*, 120–130. https://doi.org/10.1016/j.jenvp.2016.09.005

Rauthmann, J. F., Sherman, R. A., Nave, C. S., & Funder, D. C. (2015). Personality-driven situation experience, contact, and construal: How people's personality traits predict characteristics of their situations in daily life. *Journal of Research in Personality*, *55*, 98–111. https://doi.org/10.1016/j.jrp.2015.02.003

Reichow, D., & Friemel, T. N. (2020). Mobile communication, social presence, and perceived security on public transport. *Mobile Media and Communication*, *8*(2), 268–292. https://doi.org/10.1177/2050157919878759

Rhee, L., Bayer, J. B., & Hedstrom, A. (2020). Experience sampling method. In J. van den Bulck (Ed.), *The international encyclopedia of media psychology* (pp. 1–5). John Wiley & Sons, Inc. https://doi.org/10.1002/9781119011071.iemp0030

Ross, M. Q., & Bayer, J. B. (2021). Explicating self-phones: Dimensions and correlates of smartphone self-extension. *Mobile Media & Communication*, *9*(3), 488–512. Advance online publication. https://doi.org/10.1177/2050157920980508

Ross, M. Q., & Campbell, S. W. (2021). Thinking and feeling through mobile media and communication: A review of cognitive and affective implications. *Review of Communication Research*, *9*, 147–166. https://doi.org/10.12840/ISSN.2255-4165.031

Saeb, S., Lattie, E. G., Schueller, S. M., Kording, K. P., & Mohr, D. C. (2016). The relationship between mobile phone location sensor data and depressive symptom severity. *PeerJ*, *4*, e2537. https://doi.org/10.7717/peerj.2537

Sandstrom, G. M., Lathia, N., Mascolo, C., & Rentfrow, P. J. (2017). Putting mood in context: Using smartphones to examine how people feel in different locations. *Journal of Research in Personality*, *69*, 96–101. https://doi.org/10.1016/j.jrp.2016.06.004

Scannell, L., & Gifford, R. (2010). Defining place attachment: A tripartite organizing framework. *Journal of Environmental Psychology*, *30*(1), 1–10. https://doi.org/10.1016/j.jenvp.2009.09.006

Scannell, L., & Gifford, R. (2017). Place attachment enhances psychological need satisfaction. *Environment and Behavior*, *49*(4), 359–389. https://doi.org/10.1177/0013916516637648

Schnall, S. (2017). Social and contextual constraints on embodied perception. *Perspectives on Psychological Science*, *12*(2), 325–340. https://doi.org/10.1177/1745691616660199

Schnall, S., Harber, K. D., Stefanucci, J. K., & Proffitt, D. R. (2008). Social support and the perception of geographical slant. *Journal of Experimental Social Psychology*, *44*(5), 1246–1255. https://doi.org/10.1016/j.jesp.2008.04.011

Schnauber-Stockmann, A., & Karnowski, V. (2020). Mobile devices as tools for media and communication research: A scoping review on collecting self-report data in repeated measurement designs. *Communication Methods and Measures*, *14*(3), 145–164. https://doi.org/10.1080/19312458.2020.1784402

Schnauber-Stockmann, A., & Naab, T. K. (2019). The process of forming a mobile media habit: Results of a longitudinal study in a real-world setting. *Media Psychology*, *22*(5), 714–742. https://doi.org/10.1080/15213269.2018.1513850

Schneier, J., & Kudenov, P. (2018). Texting in motion: Keystroke logging and observing synchronous mobile discourse. *Mobile Media and Communication*, *6*(3), 309–330. https://doi.org/10.1177/2050157917738806

Schoenebeck, S., Ellison, N. B., Blackwell, L., Bayer, J. B., & Falk, E. B. (2016). Playful backstalking and serious impression management: How young adults reflect on their past identities on Facebook. In *CSCW '16: Proceedings of the ACM Conference on Computer Supported Cooperative Work* (pp. 1475–1487). Association for Computing Machinery. https://doi.org/10.1145/2818048.2819923

Schwartz, R. (2014). Online place attachment: Exploring technological ties to physical places. In A. de Souza e Silva & M. Sheller (Eds.), *Mobility and locative media: Mobile communication in hybrid spaces* (pp. 105–120). Routledge.

Singleton, P. A. (2018). How useful is travel-based multitasking? Evidence from commuters in Portland, Oregon. *Transportation Research Record*, *2672*(50), 11–22. https://doi.org/10.1177/0361198118776151

Singleton, P. A. (2019). Walking (and cycling) to well-being: Modal and other determinants of subjective well-being during the commute. *Travel Behaviour & Society, 16*, 249–261. https://doi.org/10.1016/j.tbs.2018.02.005

Slepian, M. L., Masicampo, E. J., & Ambady, N. (2015). Cognition from on high and down low: Verticality and construal level. *Journal of Personality and Social Psychology, 108*(1), 1–17. https://doi.org/10.1037/a0038265

Smith, L., Foley, L., & Panter, J. (2019). Activity spaces in studies of the environment and physical activity: A review and synthesis of implications for causality. *Health & Place, 58*, 102113. https://doi.org/10.1016/j.healthplace.2019.04.003

Soikkeli, T., Karikoski, J., & Hammainen, H. (2011). Diversity and end user context in smartphone usage sessions. In K. Al-Begain (Ed.), *2011 Fifth International Conference on Next Generation Mobile Applications, Services, and Technologies* (pp. 7–12). IEEE. https://doi.org/10.1109/NGMAST.2011.12

Song, C., Qu, Z., Blumm, N., & Barabási, A. L. (2010). Limits of predictability in human mobility. *Science, 327*(5968), 1018–1021. https://doi.org/10.1126/science.1177170

Stachl, C., Au, Q., Schoedel, R., Gosling, S. D., Harari, G. M., Buschek, D., Völkel, S. T., Schuwerk, T., Oldemeier, M., Ullmann, T., Hussmann, H., Bischl, B., & Bühner, M. (2020). Predicting personality from patterns of behavior collected with smartphones. *Proceedings of the National Academy of Sciences, 117*(30), 17680–17687. https://doi.org/10.1073/pnas.1920484117

Taipale, S. (2014). The dimensions of mobilities: The spatial relationships between corporeal and digital mobilities. *Social Science Research, 43*, 157–167. https://doi.org/10.1016/j.ssresearch.2013.10.003

Thulin, E., & Vilhelmson, B. (2019). More at home, more alone? Youth, digital media and the everyday use of time and space. *Geoforum, 100*, 41–50. https://doi.org/10.1016/j.geoforum.2019.02.010

Trub, L., & Barbot, B. (2016). The paradox of phone attachment: Development and validation of the Young Adult Attachment to Phone Scale (YAPS). *Computers in Human Behavior, 64*, 663–672. https://doi.org/10.1016/j.chb.2016.07.050

Tu, W., Cao, J., Yue, Y., Shaw, S. L., Zhou, M., Wang, Z., Chang, X., Xu, Y., & Li, Q. (2017). Coupling mobile phone and social media data: A new approach to understanding urban functions and diurnal patterns. *International Journal of Geographical Information Science, 31*(12), 2331–2358. https://doi.org/10.1080/13658816.2017.1356464

Vanden Abeele, M. M. P. (2020). Digital wellbeing as a dynamic construct. *Communication Theory*, qtaa024. https://doi.org/10.1093/ct/qtaa024

van der Waerden, P., Bérénos, M., & Wets, G. (2019). Communication and its relationship with digital and physical mobility patterns—A review. In E. Ben-Elia

(Ed.), *The evolving impacts of ICT on activities and travel behavior* (pp. 3–27). Elsevier. https://doi.org/10.1016/bs.atpp.2019.05.001

Verkasalo, H. (2009). Contextual patterns in mobile service usage. *Personal and Ubiquitous Computing, 13*(5), 331–342. https://doi.org/10.1007/s00779-008-0197-0

von Pape, T. (2020). Autonomous vehicles in the mobility system. In R. Ling, L. Fortunati, G. Goggin, S. S. Lim, & Y. Li (Eds.), *The Oxford handbook of mobile communication and society* (pp. 500–514). Oxford University Press. https://doi.org/10.1093/oxfordhb/9780190864385.013.33

Vorderer, P., Hefner, D., Reinecke, L., & Klimmt, C. (Eds.). (2017). *Permanently online, permanently connected: Living and communicating in a POPC world*. Routledge. https://doi.org/10.4324/9781315276472

Walsh, S. P., & White, K. M. (2007). Me, my mobile, and I: The role of self- and prototypical identity influences in the prediction of mobile phone behavior. *Journal of Applied Social Psychology, 37*(10), 2405–2434. https://doi.org/10.1111/j.1559-1816.2007.00264.x

Ward, A. F. (2013). Supernormal: How the internet is changing our memories and our minds. *Psychological Inquiry, 24*(4), 341–348. https://doi.org/10.1080/1047840X.2013.850148

Xu, Y., Shaw, S. L., Zhao, Z., Yin, L., Lu, F., Chen, J., Fang, Z., & Li, Q. (2016). Another tale of two cities: Understanding human activity space using actively tracked cellphone location data. *Annals of the Association of American Geographers, 106*(2), 489–502.

Zhang, S. M., & Li, X. (2017). Mobility patterns of human population among university campuses. In *2016 IEEE Asia Pacific Conference on Circuits and Systems, APCCAS 2016* (pp. 50–53). IEEE. https://doi.org/10.1109/APCCAS.2016.7803893

5

The Psychology of Virtual Reality

Marijn Mado and Jeremy Bailenson

Virtual reality (VR) technologies are immersive systems that simulate the actual world by tracking the movements of the user and responding with visual, audio and at times even haptic or olfactory feedback. This generates a strong sense of psychological *presence* for the user, who forgets about the physical world and feels to be "in" a virtual environment (Slater & Wilbur, 1997). As a result, VR experiences can have profound psychological effects on people, and scholars have for decades been investigating how this technology influences people's attitudes and behaviors. This research is ever more important as VR is no longer only available to research institutions and governments. Whereas the VR systems of the 20th century were heavy, bulky, difficult to operate, and wildly expensive,[1] the past decade has seen a surge in the acquisition and development of VR departments by mainstream technology corporations, who have created a

---

[1] For instance, the costs associated with the entire set of the Eyephone VR system that Jaron Lanier sold in the late 1980s, are more than $250,000 (Jacobs, 2020).

https://doi.org/10.1037/0000290-006
The Psychology of Technology: Social Science Research in the Age of Big Data, S. C. Matz (Editor)
Copyright © 2022 by the American Psychological Association. All rights reserved.

market of VR headsets that are accessible to the general public (Bailenson, 2018a). Currently, about 20% of U.S. households have VR hardware at home (Aubrey et al., 2018), and the application of VR in health care, education, and social networks is rapidly increasing.

This chapter recounts three types of scholarly work on VR. In line with Fox, Arena, and Bailenson (2009), we review research that considers VR as an *object*, as a *method*, and as an *application*. First, the psychological effects of VR technology as an object are discussed, especially the role of presence, immersion, and embodiment. This chapter also touches upon gender differences that emerge when using VR. Second, this chapter synthesizes the publications that address how VR can push social psychology forward. VR technologies allow scholars to set up and replicate experiments that are dangerous, expensive, or impossible in the real world. On top of this, VR allows for the precise capturing of behavioral data. Besides its research application, we will last discuss real-world applications of VR technology and particularly outline the use of VR for promoting empathy and facilitating education.

WHAT IS VIRTUAL REALITY?

Virtual reality (VR) is a blanket term that is used to connote a realm of different technologies that display digital environments. These technologies range from desktop computers and mobile phones over flight and driving simulators to fully immersive cave automatic virtual environment (CAVE) systems and head-mounted displays (Tarr & Warren, 2002). For the purpose of this review, we employ the term VR[2] to indicate the two most sophisticated immersive systems, namely, the head-mounted display (HMD) and the CAVE that employs multiple walls to project imagery (de la Peña et al., 2010; Kinateder et al., 2018; Loomis et al., 1999). These two systems create immersive virtual environments (IVEs) that perceptually surround the user and respond to his or her movements by constantly updating visual and audio feedback (Blascovich et al., 2002).

---

[2] The terms VR and immersive VR are used interchangeably in this chapter. Both indicate immersive VR systems, most notably the HMD and the CAVE.

The technical process of the HMD and the CAVE systems are similar in that they follow a cycle of tracking, rendering, and display. *Tracking* refers to the measuring of the body position and head movements of the user. For instance, when a user walks backwards or looks right, the displacement in body position and rotation in head movements are measured. In three degrees of freedom systems (3DoF) only the head rotation (pitch, yaw, and roll) is measured, while the six degrees of freedom systems (6DoF) also allow for moving the entire body in the space. *Rendering* implies the translation of the tracking information to the updating of the virtual scene with the appropriate sights, sounds, and at times touch or smell. When users walk towards an object in an IVE, their viewpoint changes, and the object should come closer and closer. *Display* refers to the way in which digital information in the form of rendered sights, sounds, haptics, and smells are delivered to the user. An example is the headset of the HMD system that blocks out sights from the physical world and replaces them with stereoscopic images (Bailenson, 2018a).

The HMD is most commonly used and typically includes a display device worn on the head that employs LCD screens to provide a stereoscopic view of a computer-generated environment. The movements of the user are tracked by sensors such as cameras or magnets that coordinate with the headset. Many HMD systems also include hand controllers, which enable additional tracking of the users' movements and enhanced interactivity between the user and the rendering computer. Similar to the HMD, a CAVE system also generates an IVE that responds to the movements of the user by means of tracking, rendering, and display. The largest difference between HMD and CAVEs is the display aspect, since in the CAVE system, the user does not wear a headset but rather wears stereoscopic glasses and enters an enclosed space in which imagery is projected on multiple walls (Fox et al., 2009; Kinateder et al., 2018). Both systems are able to create IVEs because they keep stimuli from the physical world to a minimum and unobtrusively track the movements of the user (Bailenson et al., 2008).

IVEs can either be 360-degree videos or computer-generated (CG) experiences. In 360-degree videos, the user is surrounded by a panoramic video that shows recorded footage. Users can move their heads around to change their viewing angle, but they cannot walk around in the virtual environment as the vision is limited to the perspective of the omnidirectional camera. Thus, 360-degree videos only allow for tracking the head rotations with 3 degrees of freedom (3DoF) and are therefore less interactive than 6 degrees of freedom (6DoF) CG experiences, in which users can not only move their heads but also walk around in an artificial, three-dimensional world. Rather than relying on real-world footage, CG experiences are generated by computer graphics and are more comparable to video games. In CG experiences, the user can be represented by different kinds of avatars and interact with objects and virtual humans (Bailenson, 2018a). These virtual humans can either be agents, which means they are controlled by computer algorithms, or avatars, which implies that their behavior is controlled by real human users (Kinateder et al., 2018). There is also a type of IVE that blurs the line between 360-degree video and CG experiences, namely, volumetric video (VV) that employs multiple cameras to generate 3D models from recorded footage, providing the audience with interactive free navigation and the choice to select their own viewpoint within a scene (O'Dwyer et al., 2020).

VR AS A PSYCHOLOGICAL PHENOMENON

Media psychologists have been particularly interested in the potential of VR technologies to provide the user with an immersive environment in which they can embody a virtual self. These technological affordances are often studied in relation to users experiencing the sense of "being there" in the virtual environment, otherwise known as "presence." The psychological effects of VR are not similar for all users, as individual differences, most notably gender, may impact the processing of virtual environments. This section discusses these differences as well as explains the terms immersion, presence and embodiment.

Immersion and Presence

VR technologies allow users to explore IVEs. The term "immersion" describes the extent to which the technology provides users with a convincing illusion of reality (Slater & Wilbur, 1997). According to Slater and Wilbur (1997), in order for a virtual environment to be immersive, the technology should (a) shut physical reality out from the senses; (b) activate an extensive range of sensory modalities; (c) surround the user in a panoramic rather than narrow field of view; (d) provide high fidelity, resolution, and richness of the display; and (e) allow the user to represent themselves with a virtual body, and match their movements with the visual and auditory display of information.

The technological quality of immersion is intricately connected to the psychological effect of presence. Whereas the immersion that a technology provides can be more or less objectively described in terms of its inclusive, extensive, surrounding and vivid affordances, presence is the individual experience of being conscious in a certain environment (Cummings & Bailenson, 2016; Lombard & Ditton, 1997; Slater & Wilbur, 1997). While the terms "presence" and "immersion" have often been confounded, Slater and Wilbur (1997) drew a stark distinction between the quality of a system's technology, which provides immersion, and the user's psychological response to the technology, which can be described with the term presence. The immersive qualities of the medium provide a plausible space in which users have the possibility to act, which may facilitate their sense of being present in the environment (Cummings & Bailenson, 2016; Wirth et al., 2007).

The relationship between immersive technologies and psychological presence was reviewed in a meta-analysis by Cummings and Bailenson (2016). They asked, "How immersive is enough?" as they investigated the extent to which technologies need to be immersive in order to generate a sense of presence. The immersive features they include in their meta-analysis are (a) the tracking level, which implies the degrees of freedom with which a user is tracked; (b) monoscopic versus stereoscopic visuals; (c) image quality; (d) field of view; (e) sound quality; (f) update rate; and (g) first-person versus third-person perspective. They found that these features overall have a significant effect on presence, especially tracking level, stereoscopy, and field of view. Their findings echo Reeves and Nass's (1996) conclusion that the fidelity of visuals, or image quality, has less effect on user experience than other features. In order to increase presence, it appears more important to invest in adequate tracking, which allows the user to move around realistically in a virtual space, than to increase the quality of visuals and sound.

The concern over increasing presence through immersive technologies begs the question, what exactly is presence? And why is it such a sought-after quality in the design of virtual environments, as well as elaborately studied and discussed in the literature on virtual environments?[3] Presence was originally conceptualized as the subjective sense of "being" in a certain environment while one is physically located in a different environment (Steuer, 1992; Witmer & Singer, 1998). For Biocca (1997), this sense of "being there" can occur in real environments, virtual or mediated environments, and imaginal environments, such as daydreaming. Lee (2004) critiqued this position and expanded on previous definitions to argue that the work on presence falls squarely in the domain of virtual experience. For Lee, presence could be defined as "a psychological state in which virtual (para-authentic or artificial) objects are experienced as actual objects in either sensory or nonsensory ways" (p. 37). Presence thus applies to the experience of mediated objects and plays no role in actual or imaginary experience. As such, presence indicates an attentional shift from the physical environment to the virtual environment based on the suspension of disbelief (Slater et al., 1994).

Presence can be further divided into three subtypes. Even though scholars have proposed many different labels and definitions for these three types, they essentially agree upon a similar distinction. First, there is spatial presence (Wirth et al., 2007), which is also called environmental presence (Heeter, 1992), telepresence (Steuer, 1992), or physical presence (Biocca, 1997). Spatial presence implies the extent to which one believes oneself to be experiencing physical entities and environments when engaging with the virtual environment. When spatial presence is strong, users are no longer aware of the immediate physical environment and temporarily forget that the virtual environment is mediated (Lee, 2004;

---

[3] Even though presence is crucial for all mediated experiences, it has become tightly connected to immersive VR technologies that promise to optimize the users' sense of presence in comparison to less immersive media (IJsselsteijn & Riva, 2003).

Oh et al., 2018). Second, personal presence or self-presence refers to the experience of one's own self as existing in a virtual environment (Heeter, 1992; Lee, 2004). When users have a strong sense of personal presence, they experience the virtual self as the actual self (Aymerich-Franch et al., 2012). A third type of presence is social presence, the extent to which one perceives oneself to be with others and has access to the sensory experience and mental considerations of others (Biocca, 1997). Social presence is crucial for the mediation of people as social beings, and a vast amount of literature is interested in the technological, psychological, and social requirements for generating social presence (e.g., Cummings & Bailenson, 2016; Lee et al., 2006; Oh et al., 2018; Walther, 1992).

The subcategory of social presence in particular helps to explain why academia and industry alike have been interested in maximizing the experience of presence in their virtual environments. Social presence allows users to perceive others as "being there" with them, which boosts the potential for yielding social influence by increasing attraction and providing opportunities for persuasion (Oh et al., 2018). This explains why companies such as online shopping websites (Hassanein & Head, 2007), as well as news outlets that aim to promote understanding for the people they document (de la Peña et al., 2010), are interested in fostering a sense of social presence. In addition to social presence, spatial presence can also be a desired media effect. For instance, in the case of virtual reality exposure therapy, the more spatial presence the patient experiences, the more realistic their emotional reactions to the virtual stimulus will be, which allows for the treatment to be successful (Felnhofer et al., 2012).

Due to the multifaceted nature of the concept of presence, it has been challenging to develop an all-encompassing measure of presence, and researchers have designed a number of different standardized measures of presence. Many of these relate to presence as a subjective experience and measure whether people experience a virtual environment as real and suspend their disbelief. Subjective measures include qualitative approaches (e.g., ethnographic observation, focus groups, interviews), but the measures most often used are quantitative questionnaires (e.g., presence questionnaire, MEC Spatial Presence Questionnaire: Vorderer et al., 2004; IPO social presence questionnaire: De Greef & Ijsselsteijn, 2000).

In addition to subjective measures, presence can also be seen as objective when people behave in a virtual environment as if they were in the real world. Objective measures include testing for physiological indicators such as skin conductance or behavioral measures that test whether subjects respond adequately to virtual cues. These objective measures are perceived as less granular than subjective measures, since it remains unclear what aspect of the experience prompts certain behavior or physiological responses (Lombard et al., 2009; Slater & Wilbur, 1997).

Embodiment

The psychological effects of IVEs are not only associated with a sense of presence but are also fostered by the potential to embody a virtual self that can be radically different from one's own body. While many media allow for some form of self-presentation, often in the form of an avatar, immersive VR is able to encourage the user to strongly identify with the virtual body, thereby creating an embodiment illusion. Even when the virtual bodies people encounter in VR are different from their own—for instance, a human avatar with a different skin color (Hasler et al., 2017), a human avatar with a tail (Steptoe et al., 2013), or an avatar in the form of a cow (Ahn et al., 2016)—it is possible to feel a sense of ownership over these virtual self-representations.

How do people come to identify with a virtual body that is significantly different from one's own? There are two main strategies for inducing body transfer, which implies that the specific accordances and qualities of the virtual body shape people's mental models of their physical body (Bailey et al., 2016; Biocca, 1997; Slater et al., 2010). The first technique is signal correspondence between the virtual and actual body. The most common example of this technique is the rubber hand illusion, which involves the synchronous touching of a fake hand that the participant can see and the participant's real hand that is hidden from view. Research consistently confirms that people develop a sense of being touched and subsequently feel ownership over the fake hand (Blanke, 2012; Lenggenhager et al., 2007; Petkova & Ehrsson, 2008; Slater et al., 2009). By synchronizing signals such as touch with the actual and virtual body, people start to process virtual bodies as if they were their own (Bailey et al., 2016; Slater et al., 2010).

A second strategy to induce body transfer is sensorimotor correspondence between the actual and virtual body. This technique relies on synchrony between a person's physical movements and the movements of their virtual representation. People experience this synchrony, for instance, when seeing their avatar's arm move at the same time that they move their own physical arm. For this sensorimotor strategy, mirrors are often employed, which allows users to see their whole body respond to their movements in real time (Ahn et al., 2016; Fox, Bailenson, & Tricase, 2013; Guegan et al., 2016; Yee & Bailenson, 2007). This strategy may be even more psychologically impactful than the signal correspondence technique because rather than just ownership over the virtual body, people also experience a sense of agency over their avatar due to the correspondence of movements (Bailey et al., 2016; Kalckert & Ehrsson, 2012).

Since the body ownership illusion can be induced even when the virtual body is radically different from the actual body (Slater et al., 2010), virtual embodiment allows people to temporarily alter their personal identity. As personal identity is the source of personal perceptions about the world, virtual embodiment can be used to encourage "walking in the shoes of another" and allows people to perceive the world from a different point of view. This is why virtual embodiment is studied in relation to raising awareness of racial stereotyping (e.g., Banakou et al., 2016; Groom et al., 2009; Hasler et al., 2017; Peck et al., 2013), people with disabilities (Ahn et al., 2013), victims of domestic violence (Seinfeld et al., 2018), and even the natural environment (Ahn et al., 2016). These studies assess to what extent body transfer can aid the promotion of empathy and understanding towards their social targets. In contrast to communication media that do not support body transfer, a number of studies suggest that embodied experiences in VR are more likely to increase involvement or empathy (Ahn et al., 2013, 2016; Herrera et al., 2018). At the same time, embodied experiences could also backfire into strengthening resentment towards the social target (Groom et al., 2009), which we further explicate in the section on empathy applications.

Besides a change in attitudes, the temporary transformation of one's personal identity via virtual embodiment can also lead to behavioral change (e.g., Ahn et al., 2013; Herrera et al., 2018; Kilteni et al., 2013). One of the mechanisms that underlies behavioral change as a result of transformed self-representation is what Yee and Bailenson (2007) termed the *Proteus effect*. This effect prescribes that people will conform to the behavior they think is in line with their digital self-representation. When people believe that others would expect certain behavior as a consequence of the identity cues of their virtual self, they start to conform to that expected behavior, regardless of how others treat them. Yee and Bailenson showed that participants with more attractive avatars were more intimate in social interactions, while participants with taller avatars were more confident, regardless of their interaction partner. This Proteus effect has been replicated in a variety of contexts, and a meta-analysis of 46 studies by Ratan et al. (2019) showed it to be a reliable phenomenon.

The Proteus effect is not the only way in which the temporary transformation of one's personal identity via virtual embodiment generates behavioral change. Kilteni et al. (2013) found that the illusion of body ownership can change a participant's behavior even when the participant does not engage in substantial social interaction with others. The results of their experiment reveal that Caucasian participants who were asked to play West-African Djembe in an IVE showed more frequency and variation in their drumming movements when they embodied a casually dressed dark-skinned virtual body instead of a formally suited light-skinned body. This effect was stronger the greater the participants experienced an illusion of body ownership. The researchers argued that this enhancement in musical performativity can be explained by a temporary self-identification with the perceived social group to which the virtual body belongs, which in the case of the dark-skinned body is perceived to be more capable for the drumming task. This finding has implications for both empathy work and education, as embodying an "appropriate" body for the task at hand may enhance training outcomes (Kilteni et al., 2013).

VR can thus be a powerful psychological tool that allows people to temporarily experiment with different self-identities. While extreme self-transformations in physical reality are usually expensive and difficult to perform, VR offers this opportunity by simply customizing one's avatars. This could aid researchers, activists, and educators, who may aspire to increase empathy for social targets or train their students for difficult tasks.

Gender

Since the utility of VR relies on its strong psychological effects, such as presence and body ownership illusion, it is worthwhile to consider whether individual differences exist in the psychological response to using VR. One of these individual differences that may be most pronounced in the VR field is gender, considering that historically, women and men have had differential attitudes toward technology use more generally. Women have previously indicated feeling less confident and more anxious using technologies such as desktop computers than have men, which appears largely due to differences in prior experience that can be explained by sociocultural gender norms (Ausburn et al., 2009; He & Freeman, 2019; Huffman et al., 2013). As a result of differences in experience, female technology self-efficacy can be lower, and women experience less comfort when exploring virtual environments.

The gender gap in self-efficacy and prior experience potentially explains Felnhofer et al.'s (2012) findings that men experienced a stronger sense of presence in VR. In their experiment, Felnhofer et al. asked participants to either give a presentation in front of a virtual audience or merely imagine the audience. Women not only felt a weaker sense of being in the virtual environment as opposed to men but even felt a stronger sense of being present in the imaginative condition than in the virtual condition. As men reported significantly higher levels of spatial presence and perceived realism, as well as a more strongly felt sense of being in the environment, Felnhofer et al.'s study reinforced the tentative findings of previous studies that showed that men generally experience higher levels of presence in virtual environments (Nicovich et al., 2005; Slater & Wilbur, 1997).

Apart from gender differences in prior computer use and technology self-efficacy, Felnhofer et al. (2012) suggested that the results may be explained by differences in spatial abilities. This would imply biological differences between the sexes, which is disputed among scholars.

While some scholars propose that gender differences can be accounted for biologically, specifically in relation to men's superior spatial abilities (e.g., Hunt & Waller, 1999), Hoffman et al. (2011) showed that the gender gap in cognitive abilities is explained by nurture, rather than nature. However, biological differences that are unrelated to cognitive abilities do appear relevant when assessing another aspect of VR where a gender gap is detected—the degree of cybersickness.

Cybersickness is a visually induced motion sickness that can be felt by users of VR. The symptoms of discomfort include dizziness, eyestrain, headache, sweating, and vertigo (Weech et al., 2019). With regard to non-VR virtual environments, scholars have not come to an agreement on whether biological sex differences in the experience of cybersickness exist. Whereas some scholars failed to find any significant gender differences in cybersickness (Gamito et al., 2008; Knight & Arns, 2006; Ling et al., 2013), others found that women experienced significantly more cybersickness symptoms than did men (Jaeger & Mourant, 2001; Park et al., 2006). However, studies that assess cybersickness associated with VR exposure generally find that females appear more likely than males to suffer from cybersickness (De Leo et al., 2014; Häkkinen et al., 2002; Jun et al., 2020; Munafo et al., 2017; Stanney et al., 2003).

A study by Stanney et al. (2020) provided a physiological reason that explains these HMD-induced gender differences. The results from two experiments exposing men and women to a virtual rollercoaster reveal that interpupillary distance (IPD) nonfit is the main driver of gender differences in relation to cybersickness. Women whose distance between the center of their pupils was too small to fit properly to the VR headset took the longest to recover from VR exposure, while women who could fit their IPD to the headset experienced similar rates of cybersickness as men. The study also provides a table of the adjustable range of IPD per headsets, showing how 35% to 45% of women would not fit to the smallest IPD fit of some of the mainstream HMDs (HTC Vive and Oculus Rift S, respectively; Stanney et al., 2020). These findings imply that the future redesigning of VR displays can reduce gender differences in cybersickness.

VR AS A RESEARCH TOOL

In addition to studying the effects of VR technology on people's psychological states, researchers may also be drawn to VR as a research tool to study phenomena that are difficult to study in the physical world. A number of scholars have considered how VR can solve methodological problems that have traditionally troubled the field of social psychology (Blascovich et al., 2002; Fox et al., 2009; Kinateder et al., 2018; Loomis et al., 1999; Tarr & Warren, 2002; van Veen et al., 1998). These scholars' results highlight the advantages of employing VR for both the creation and control of empirical stimuli, as well as the collection of data.

Control Over Empirical Stimuli

One of the main issues in the field of social psychology is the development of and control over empirical stimuli that are used in experiments to measure the psychological reactions of participants. These stimuli can range widely in terms of complexity and elaboration, as they may be as simple as a written vignette or as complex as a scenario involving multiple confederates. Whereas more straightforward and simple empirical stimuli allow for precise manipulation and thus facilitate experimental control, more complex empirical stimuli are better at resembling everyday life and therefore score higher on mundane realism. In traditional experiments, a large amount of experimental control appears to be incompatible with high mundane realism. In order to ensure mundane realism, complex scenarios with confederates are warranted, which directly inhibits perfect experimental control, due to the difficulty of, for instance, keeping the performance constant for each confederate. As a result, many social psychologists face a trade-off between control and realism in designing their experiments (Blascovich et al., 2002; Kinateder et al., 2018; Loomis et al., 1999).

Technological advancements have played a role in diminishing this trade-off, and Blascovich et al. (2002) argued that IVEs provide the next progression in stimuli creation and control. IVEs generate mundane realism, as they provide immersive experiences that engender high levels of presence for the participant, as explained earlier. At the same time,

IVEs also allow the investigator almost perfect control over the experimental condition, as the empirical stimuli are pre-coded (Blascovich et al., 2002). An example of the decline of the classical trade-off is a study on stereotypes in social interaction. In IVEs, it is possible to manipulate the race and sex of the virtual confederates while making sure all their other characteristics, as well as their presentation, are the same across conditions. This supports researchers who intend to isolate the effects of demographic qualities on social interaction (Hasler et al., 2017; Loomis et al., 1999).

In addition to decreasing the biases of confounding variables by precisely manipulating the empirical stimuli, VR technology also enables researchers to develop scenarios that are costly, dangerous, or just plainly impossible in the real world. For instance, Kinateder et al. (2018) studied complex crowd dynamics by making use of virtual simulations of fire emergency evacuations. Without putting the participant at risk, the researchers can assess the particularities of crowd behavior and probe at questions such as how the probability that an individual will fall is related to crowd density and walking speed. An example of using VR to create "impossible" stimuli is a study by Tarr and Warren (2002), who break the laws of optics and physics in their virtual environments. To investigate how metric distances and angles contribute to cognitive navigation processes, their experiment employed a virtual hedge-maze environment that was distorted during the experiment by altering distances or angles. As participants walked through the maze, the world was stretched or sheared to assess the extent to which participants used metric distances and angles to navigate (Tarr & Warren, 2002).

The advantages of using VR for the development of and the control over experimental stimuli begs the question whether the results of virtual simulations can be extended to the real world. Van Veen et al. (1998) tackled this issue directly by recreating the real city of Tübingen in VR. They assessed whether navigation and orientation behavior are similar in the real and virtual city. One of the differences they found is that people perceive themselves to be moving slower in a virtual world than they actually are. This underestimation of one's velocity potentially points to differences in psychological processing in virtual and actual environments, which could be a limitation when using VR as a research tool (van Veen et al., 1998).

Kinateder et al. (2018) also investigated whether measured behavior in VR experiments applies to behavior in the real world. They found that there are biases in the perception of the size and distance of objects in VR, which appear reduced after spending more time in VR. They also found that wearing an HMD slows down a person's walking pace and alters their body posture in comparison to walking in a real environment. Another limitation of using virtual worlds is that participants are often aware that the environment is controlled by the experimenter (Kinateder et al., 2018). These limitations curtail the capacity of VR to create a perfect alternative to social psychological research in the real world. Nonetheless, the employment of VR for research retains enormous appeal due to the technology's potential to create cost effective, safe, extraordinary, and relatively realistic scenarios that can be tightly controlled.

Data Collection

The tightly controlled experimental manipulations in the shape of malleable executable files also provide advantages for data collection. In the social psychological sciences, some of the main challenges of data collection are the difficulty with replicating experimental conditions, the lack of access to a representative participant pool, and the meticulous tracking and coding of behavioral data. VR technology offers a potential solution to these issues.

Perfect replication of prior studies has been nearly impossible in traditional social psychology, since the exact experimental conditions are difficult to transport from lab to lab. This is because of the lack of details regarding the methods and procedures of experiments in journals, as well as the physically different laboratories that researchers work in (Blascovich et al., 2002). However, when using IVEs for the experimental setting, researchers can share software with other labs, who are then enabled to meticulously replicate the experimental setting (Blascovich et al., 2002; Kinateder et al., 2018). Instead of uncertainty about the confederates' dress or manner, or environmental details such as the lighting, the IVE software allows scholars to directly copy each other's experimental condition.

The other advantage of the easy sharing of IVEs across geographical locations is a potential increase in representative sampling. Whereas most studies in social psychology have used convenience samples in the form of undergraduate students obtaining course credit, which threatens the studies' external validity and generalizability, studies employing VR may access a much wider range of participants (Blascovich et al., 2002). Especially since VR is becoming more commonplace in people's living rooms (Bailenson, 2018a), the representative selection of participants can be facilitated by experimental IVE setups. Nonetheless, the advantages of easy replication and representative sampling depend on the ease with which scholars can share their software and their willingness to do so. To realize the vision of open access to IVEs that can be employed for psychological experiments, Li et al. (2017) made an attempt to develop a library of immersive VR clips. Another attempt to reach a larger and more representative participant pool was Ma et al.'s (2018) effort to recruit participants for VR studies via Amazon Mechanical Turk. They conceived of a validation method to check whether users indeed own a VR device and found 242 eligible participants who vary more in terms of age, education level, occupation, and household income than the population most VR studies draw from.

An additional advantage of using VR technology for the collection of data is that the technology allows one to capture and precisely record all the actions and movements of the participants. Not only does this save the time and money it would cost to employ human coders, but the technology is also more precise than humans and is able to measure additional variables that could not be measured otherwise. For example, the VR hardware could measure physiological responses such as the heart rate, skin conductance, or brain activity of the participants. This can be valuable when studying the neural responses to social interactions or other events (Kinateder et al., 2018; Loomis et al., 1999). At the same time, the meticulous capturing of behavioral information—up to 2 million unique instances of body language can be captured during a 20-minute VR simulation—could impede the privacy of participants. Since this nonverbal data can be used not only to identify individuals but also to predict future health, consuming, or romantic behavior, it is crucial to find ways to protect individuals from the easy sharing of their tracked data with companies and governments (Bailenson, 2018b).

VR APPLICATIONS

VR technology is not only a promising tool for scholars but is also used in a wide range of other fields, including health care (e.g., distracting patients from pain; Gold et al., 2007) or physical rehabilitation therapy (Schultheis & Rizzo, 2001), manufacturing industries (Mujber et al., 2004), well-being (Gutiérrez-Maldonado et al., 2006; Li & Bailenson, 2018), social networks (e.g., applications such as AltSpace, ENGAGE, and VRChat), and entertainment (e.g., applications such as Beat Saber, Half-Life: Alyx, and Superhot VR). This chapter discusses how VR is employed for education and for the promotion of empathy, which offers opportunities for scholars, educators, and activists alike.

Empathy and Perspective Taking

Similar to less immersive media, content developers design IVEs to share the experience of marginalized people with the aim of invoking empathy. Because VR allows users to viscerally experience a narrative from the first-person perspective, the technology has regularly been described as the "ultimate empathy machine" (Milk, 2015). Whereas for VR producers this empathetic function is a way to "hack" the human body into curing callousness, cultural critics have signaled the dangers with exploiting the pain of marginalized bodies for producing morally right feelings that may distract people from implementing institutional and structural change (Nakamura, 2020). The social-psychological response to this commentary is to study what kind of empathy VR is prone to evoke, whether this is coupled with behavioral change, and the duration of this increased sensitivity to the experience of others and to test whether VR is more effective in doing so than other media (e.g., Banakou et al., 2016; Herrera et al., 2018; Sri Kalyanaraman et al., 2010). We will discuss the research on VR as an empathy tool that can simulate impairments, encourage perspective taking, or provide direct empathy training.

Physical or Mental Impairment Simulation

Considering the affordance of VR to virtually embody someone else, the medium may be uniquely equipped to simulate uncomfortable and stressful physical or mental experiences. For instance, Sri Kalyanaraman et al. (2010) and Formosa et al. (2018) employed a VR experience that simulates schizophrenia. The VR simulator used by Sri Kalyanaraman et al. engages the user in a pharmacy visit and then exposes them to multiple visual and auditory hallucinations, such as anxious disembodied voices and a label reading "POISON" on the medicine offered by the pharmacist. The investigators found that the virtual simulation is only more effective in invoking empathy and generating positive perceptions towards people suffering from schizophrenia when the simulation is combined with a traditional perspective-taking task. In this traditional task, participants were asked to imagine what it is like to suffer from schizophrenia and write down their thoughts. The participants who engaged with both the writing task and the virtual simulation scored significantly higher on empathy and attitudes toward people with schizophrenia than did those who did only one of the tasks or a control. Interestingly, when participants would only engage with the virtual simulation, they would feel significantly more social distance to people with schizophrenia. The authors thus cautioned that virtual simulations should be accompanied with additional resources in other formats to prevent counterproductive learning experiences (Sri Kalyanaraman et al., 2010). This study points both to the potential of virtual simulations when integrated in a broader learning curriculum and the danger of exposing people to stand-alone immersive experiences.

As an extension to Sri Kalyanaraman et al.'s (2010) study, Formosa et al. (2018) tested the effects of exposure to a VR experience that simulates symptoms associated with both schizophrenia and other psychotic disorders. They found that the participants, who all read a background vignette and engaged with the interactive VR experience, scored higher on knowledge, attitudes, and empathic understanding of psychotic disorders after the experience than before. Another interesting finding is that the knowledge about the diagnosis increased, even though the intervention only simulated an experience rather than included any overt teaching (Formosa et al., 2018). However, since Formosa et al. did not include additional conditions that could compare the effects of the VR intervention with those of less immersive media or only reading the background vignette, it remains unclear what aspects of the intervention were most effective in bringing about the improvement in attitudes and knowledge.

Besides mental disorders, virtual simulations that imitate physical impairments have been studied by VR scholars. For instance, Ahn et al. (2013) assessed whether an embodied experience of colorblindness would elicit greater self-other merging with people with disabilities and promote more helping behavior than imagining being colorblind. Their results indicated that participants who embodied colorblindness in the virtual simulator spent twice as much time to help people with colorblindness in comparison with participants who only imagined what it was like to be colorblind (Ahn et al., 2013). Another example of a physical impairment simulation is the stroke simulator built and tested by Maxhall et al. (2004), which simulates the anomalies stroke patients experience, such as the inability to use one side of the body, rotating surroundings, and motion blur. Participants navigate the virtual apartment in a wheelchair and can engage in basic interactions like filling a glass with water, which become difficult due to the stroke anomalies. The qualitative observations suggest that the simulator influences social workers' empathy towards stroke patients (Maxhall et al., 2004). These simulations of physical or mental disabilities could function to relieve some of the burden of patients to explain their condition and daily life to others.

Taking the Perspective of Marginalized Groups

In addition to simulations, virtual reality perspective-taking (VRPT) tasks are often used to invoke empathy for specific social targets by nudging users to take on the first-person perspective of someone else. In 360-degree videos, this first-person perspective is generated by the immersion in a panoramic video and experiencing people looking and talking to you as if you were really there. An example is the Guardian VR film "The Party," which shows the thought process and sensory overload of a 16-year-old autistic girl at a surprise birthday party from a first-person perspective. This 360-degree film is both a VRPT experience and an impairment simulation of symptoms seen in autistic individuals, as faces become blurred, loud noises turn into overwhelming ringing, and bright lights distort overall vision (Bregman et al., 2017).

In computer-generated VR content, the VRPT task is regularly coupled with embodiment, which allows users to virtually embody an avatar that looks different from their own body—for instance by having a different gender, race, or age. The process of body transfer is explained above and is either induced through signal or sensorimotor correspondence. A number of empathy-related VRPT tasks that work with the latter strategy prompt first-person embodiment by means of virtual mirrors that allow the users to see their avatar moving synchronously with their own movements. The mirror strategy was first applied to empathy research by Yee and Bailenson (2006), who asked participants to embody an older avatar. They found that participants showed significantly less stereotyping of elderly persons in a word association task when they were placed in avatars of old people rather than in avatars of young people.

In addition to age discrimination, a body of scholarly work conducted by Mel Slater's research group focused on reducing racial prejudice by means of the virtual embodiment of black avatars (Banakou et al., 2016; Hasler et al., 2017; Peck et al., 2013). These studies confirm that when light-skinned participants embody a dark-skinned virtual body, their implicit racial bias against black people significantly decreases (Peck et al., 2013), an effect that is sustained over the course of 1 week (Banakou et al., 2016). The psychological effects of embodying an avatar of a different race not only include reductions in racial bias but can even involve a temporary reversal of ingroup and outgroup perceptions. In a clever experiment, Hasler et al. (2017) assessed whether a participant's Black and White virtual partners are treated as an ingroup or outgroup member by measuring the amount of mimicry in movements and gestures. In real-world interactions, people mimic people from their ingroup (often the same race) more than people from their outgroup. The study found that when White participants embodied a black virtual body, they would mimic Black virtual partners more than White virtual partners, suggesting the malleability of racial self-categorizations (Hasler et al., 2017).

However, the study by Hasler et al. (2017) showed no reduction in racial bias as a result of embodiment, in contrast to the previous studies (Banakou et al., 2016; Peck et al., 2013). There is only one study that could have predicted this otherwise surprising result, namely, a study by Groom et al. (2009), who found that participants who embodied a Black avatar in a mock job interview scored higher on implicit racial bias than those who embodied a White avatar. The explanation they offered for this counterproductive finding is that the context of the job interview may have induced performance anxiety accompanied by feelings of self-threat (Groom et al., 2009). This suggests the importance of context and activity in VR experiences that aim to reduce bias and promote empathy. Especially when empathizing with the social target becomes inconvenient in the social setting, virtual perspective-taking exercises may no longer be effective or may even be counterproductive.

The failure in improving attitudes when engaging participants with threatening situations is also indicated by Oh et al. (2016). The scholars built on Yee and Bailenson's (2006) work as they asked participants to take on the experience of an elderly person either through a VRPT task supported with a virtual mirror or by means of a traditional mental simulation task. In one of the conditions, participants were ostracized by two elderly partners in a game of Cyberball, which led to negative intergenerational attitudes that could not be mitigated by the VRPT task. Meanwhile, intergenerational attitudes did improve after the VRPT task when the threat by the elderly was indirectly presented in a news article, without any direct intergroup contact (Oh et al., 2016). These findings thus suggest the limits of virtual simulations for improving attitudes for threatening social targets. When the threat is indirect and impersonal, VRPT tasks still prove successful, while a concrete and experiential threat results in a failure to improve attitudes for the social target. The positive effect of VRPT tasks seems to be disrupted when the social target is perceived as directly and concretely threatening the participant.

For the VRPT tasks that do appear to increase understanding for marginalized others, the question remains how long such effects will last after the VR experience. Most VRPT studies measure for psychological effects immediately after the intervention, with the exception of a handful of studies that bring in the participants a week later for additional measurements (Ahn et al., 2013, 2016; Banakou et al., 2016). The most promising exception to the short-term VRPT effects studies is the work by Herrera et al. (2018), who included repeated measurements over time, up to 2 months after the intervention. Their study exposed participants to a first-person perspective of becoming homeless. They found that positive attitudes toward the homeless decrease more slowly over time for participants in the VRPT condition than for those engaging with narrative-based perspective-taking tasks (Herrera et al., 2018). Because the ultimate aim of most VRPT tasks is to create a sustainable change in attitudes, the field is in need of additional longitudinal studies that probe at the condition for durable effects to emerge.

Empathy Training

The perspective-taking potential of immersive VR makes it a promising medium for empathy training through role reversal, as offenders can take the role of victim (Seinfeld et al., 2018), parents can be put in the position of a child (Hamilton-Giachritsis et al., 2018), and health care students can explore the perspective of a patient (Dyer, Swartzlander, & Gugliucci, 2018). These pioneering studies suggest a range of positive learning outcomes. Male domestic violence offenders who were embodied in a virtual female body and experienced abuse in a first-person perspective improved their capacity to detect fearful female faces and were less likely to perceive fearful faces as happy (Seinfeld et al., 2018). Mothers who virtually embodied a 4-year-old child and experienced negative maternal behavior improved their levels of empathy for children (Hamilton-Giachritsis et al., 2018). Health care students who took the virtual perspective of a 74-year-old African American man suffering from hearing loss and macular degeneration show an increase in empathy and understanding for older adults suffering from age-related conditions (Dyer et al., 2018). All these effects were only measured in the short term, directly after the intervention, and were not paired with behavioral measures to investigate whether the VR intervention would have led to durable behavior change. The use of VR to train empathic skills is wildly promising to VR developers (e.g., Froyd, 2018; Milk, 2015) but can benefit from a rigorous research project into the type and durability of empathy that VR may enable.

Education

The studies on empathy training indicate how the application of VR to fostering empathy strongly resonates with another application of VR, namely, education. One of the earliest applications of the predecessors of present-day VR has been for training, as flight simulators were used to provide pilots with a cheaper and safer way to learn flying skills (Pausch et al., 1992). During the 21st century, the educational applications for VR have developed extensively, and VR has even emerged as a promising tool in the classroom. This has implications for children's use of VR, who may show different psychological responses than adults. This chapter addresses the use of VR with children, the potentials of classroom use, and the myriad of ways VR is used for training. In addition, this chapter briefly touches upon the role of augmented reality (AR) in education settings.

Training

In addition to empathy training, VR has been used as a training tool in a wide variety of contexts. Similar to its use as a research tool, VR's allure lies in its capacity to simulate situations that are expensive, dangerous, or plainly impossible in physical reality. Examples include the simulation of burning houses for firemen and space for astronauts (Tarr & Warren, 2002). Another large use case for VR training is found in health care. In addition to teaching students' empathetic communication skills (Dyer et al., 2018), a number of nonimmersive VR systems exist that allow students to practice talking with and diagnosing patients (Kenny et al., 2007; Lok et al., 2006; Stevens et al., 2006). Furthermore, immersive VR is well equipped to provide students with virtual models of the human body. This enables students to train for complex surgeries in a way that avoids risk to patients (Spitzer & Ackerman, 2008). In addition to anatomy, VR can also help prepare medical volunteers by training them to make correct decisions in health emergency situations (De Leo et al., 2003).

VR also forms a helpful training tool in less dramatic circumstances. For instance, when it comes to practicing athletic movements, the sheer hours of practice are what make the movements perfect. In order to reach those hours beyond the physical field, VR training simulations are used for football, amongst others (Bailenson, 2018a; Belch et al., 2020). The efficacy of training movements in VR was shown by a study in which participants learnt Tai Chi movements with greater accuracy when being taught by a 3D virtual instructor projected stereoscopically onto a screen than when watching a 2D video instructor (Patel et al., 2006). The VR for training field reaches a wide range of applications but is in need of more solid research to assess the conditions on which positive learning outcomes are reached.

Classrooms

A substantial body of work has been focusing on incorporating non-immersive virtual environments in the STEM (science, technology, engineering, and mathematics) classroom that are accessed via media such as computer desktops. These virtual environments are comparable to using immersive VR for education in that they focus on the "doing" part of science, increase student motivation and attendance, and allow students novel ways to experiment with their identity (Clarke & Dede, 2009; Dede, 2009; Dede et al., 2004; Ketelhut et al., 2010). These studies suggest that incorporating virtual environments in the classroom can increase student motivation and boost their self-efficacy, especially for students that generally perform less well.

The finding that virtual environments increase engagement and motivation resonate with the recent work on learning in immersive VR. For instance, Parong and Mayer (2018) found that students who watched a biology lesson in immersive VR reported higher engagement, motivation, and interest but learned less well than students who watched a PowerPoint slideshow on a desktop computer. Similarly, Moreno and Mayer (2002) discovered that students were more engaged in immersive VR rather than desktop computer but did not learn more. This suggests that VR can play a large role in improving student motivation but that student learning is still largely dependent on the curriculum that contextualizes the immersive experience.

Nonetheless, there is another way in which VR can be employed to directly enhance student learning in the classroom. Bailenson et al. (2008) experimented with different forms of transformed social interaction in an immersive virtual classroom by altering participants' social-sensory abilities, their self-representation, and the social environment. First, they found that participants who take on the role of a teacher and see their students fade out when they neglect to look at them are able to spread their attention more equally. This augmented perception helps teachers to fairly distribute their eye gaze, which in turn is shown to support students' learning. Second, immersive virtual classrooms can break the rule of spatial proximity by allowing every student to sit close by the teacher and within the teacher's field of view. They show that sitting at the center of the classroom and proximate to the teacher enhances students' learning. Third, inserting virtual colearners that are attentive to the teacher, rather than distracting, improved the learning abilities of participants. In this way, the transformed social interaction that is enabled by the use of IVEs can support student learning (Bailenson et al., 2008).

Another potential VR use case that supports students are virtual field trips. As mentioned before, the benefit of VR is its potential to simulate dangerous, expensive, or impossible field trips in the safe environment of the physical classroom (Bailenson, 2018a). In IVEs, students can explore the bottom of the ocean (Markowitz et al., 2018), take on the role of a gorilla (Allison & Hodges, 2000), or travel back in time to explore medieval Bruges (https://www.historium.be). The efficacy of these virtual field trips was studied by Klippel et al. (2019), who compared a group of students who experience a traditional field trip versus those experienced a virtual field trip of the same site. The virtual field trip provided advantages as students reported higher enjoyment as well as showed better learning outcomes than the students who engaged with the physical field trip. In addition, a qualitative follow-up study by Klippel et al. (2020) also showed that immersive virtual field trips provide an adequate preparation for the actual field trip. Similarly, Markowitz et al.'s (2018) findings show that an immersive VR experience that teaches about the causes and effects of ocean acidification led to a positive gain in knowledge and interest in learning more. In addition to the students, teachers also appear interested in VR, as Fauville et al. (2020) found that marine educators perceive VR as a valuable tool for teaching ocean acidification. The educators stress VR's capacity to support their teaching by visualizing the effect of CO_2 in the ocean, showing the impact of daily choices on the environment, and allowing students to take the perspective of someone who is directly affected by ocean acidification (Fauville et al., 2020).

Children

Since in these educational classroom settings VR is used not for adults but for children of a variety of ages, it is important to consider the differential psychological effects VR may have on children. Research shows that children under the age of 18 report higher levels of presence and consider the virtual environment to be more real than do adults above the age of 18 (Sharar et al., 2007). When it comes to young children, Segovia and Bailenson (2009) showed in an initial study that 6- to 7-year-old children may create "false memories" when interacting with immersive VR. They reported that young children were likely to confuse an IVE with real life when they watched a representation of themselves swim with orcas in immersive VR. An explanation for the intensity of effects of VR on children is that children are still developing the ability to differentiate symbols they encounter in media with reality (Bailey & Bailenson, 2017; Flavell et al., 1990). The blurring of real and virtual that is more prevalent in young children may be crucial to keep in mind when designing educational VR applications.

Augmented Reality

A promising future direction for education is the use of AR, a technology that is somewhat similar to VR in its use of tracking sensors and displays. However, whereas VR technology completely replaces the sensory information of the physical world, AR technology supplements physical reality. AR allows users to perceive the real world and superimposes or composes virtual objects in physical reality. Examples of AR technology range from the camera view of the mobile phone to the Microsoft Hololens (Scavarelli et al., 2021). The cheaper mobile AR options are suitable for education applications, as they allow students to, for instance, overlay a physical tree with its botanical characteristics or compare a city scene with a historical photograph (Dunleavy et al., 2009). In recent years, AR technology has been employed to enhance student learning in all sorts of school subjects, including geography (Carbonell Carrera & Bermejo Asensio, 2017), geology (Bursztyn et al., 2017), chemistry (Chen, 2006), biology (Kamarainen et al., 2013), art history (McNamara, 2011), and mathematics (Kaufmann & Schmalstieg, 2002). These studies, amongst others, show that students treat AR objects as real objects (Chen, 2006), AR-enhanced maps are more effective than 2D maps in developing students' spatial orientation skills (Carbonell Carrera & Bermejo Asensio, 2017), and AR generally increases student interest and engagement, especially for those students who generally perform less well (Bursztyn et al., 2017; Dunleavy et al., 2009; Kamarainen et al., 2013).

According to Dunleavy and Dede's (2014) review of the research assessing AR teaching tools, the main learning opportunities of using AR with students are enhanced collaboration, the leveraging of physical environments, and the access to additional resources. Because each student holds their own device, they can be presented with different pieces of information and have to collaborate in order to solve the problems at hand. Students can leverage the AR systems to layer physical space with rich layers of content, as well as use the mobile devices for access to the internet or for video recording. The main limitation of the use of AR in schools is that students can become cognitively overwhelmed with the complexity of the multifaceted information. In addition, curricula centered on standardized tests are not well-suited to the time-consuming collaborative activities provided by AR learning tools, and not all teachers are sufficiently technologically savvy to implement AR in their teaching. The technological issues associated with using AR, most prominently GPS error (Dunleavy & Dede, 2014; Klopfer & Squire, 2008; Perry et al., 2008), provide an additional challenge.

CONCLUSION

This chapter outlined the main psychological effects that are associated with VR use, as well as the applications of VR for conducting social psychological research, promoting empathy, and facilitating education.

We discussed how the technological quality of immersion is related to psychological presence, a sense of "being there," and how virtual embodiment is induced through signal or sensorimotor correspondence. There appear to be individual and demographic differences to the psychological effects of interacting with VR, as women, for instance, experience less presence than men in IVEs. The finding that VR does not have the same immersive effect on everyone has implications for its utility to be employed in social psychological research and other settings, which rely on VR's capacity to simulate reality with high fidelity and thereby suspend the disbelief of the user. In fact, since scholars have found some differences in how people perceive their environment and behave between virtual simulations and physical reality, caution is warranted when using VR as an alternative to the real world. Nonetheless, the appeal to apply VR to research, health care, perspective taking, and education remains because of the technology's potential to provide cheap, secure, exceptional, and relatively realistic scenarios that can be tightly controlled.

The VR empathy research points out that IVEs are not always effective in increasing empathy and understanding for marginalized social targets. When participants are exposed to a virtual situation that can be perceived as threatening to their self-esteem, they are less likely to empathize with the social target. VRPT exercises can even be counterproductive when they are presented as stand-alone experiences and increase the distance participants feel between themselves and the social target. A solution is to integrate VRPT tasks into a broader learning curriculum in which people get the chance to prepare for and reflect on the task using less immersive media such as text. The contextualization of VR experiences may also support student learning in educational applications of VR. Thus far, scholars have found that VR plays a large role in improving student engagement and motivation but does not necessarily lead to higher learning outcomes. Educators are recommended to adequately embed VR experiences in a larger curriculum and keep the intense psychological effects of VR on young children in mind. In this way, the field can work toward keeping the nefarious effects and manipulative use cases of VR to a minimum while stretching the technology to optimally benefit education, activism, and research.

REFERENCES

Ahn, S. J., Bostick, J., Ogle, E., Nowak, K., McGillicuddy, K., & Bailenson, J. N. (2016). Experiencing nature: Embodying animals in immersive virtual environments increases inclusion of nature in self and involvement with nature. *Journal of Computer-Mediated Communication, 21*(6), 399–419. https://doi.org/10.1111/jcc4.12173

Ahn, S. J., Le, A. M. T., & Bailenson, J. (2013). The effect of embodied experiences on self-other merging, attitude, and helping behavior. *Media Psychology, 16*(1), 7–38. https://doi.org/10.1080/15213269.2012.755877

Allison, D., & Hodges, L. F. (2000, October). Virtual reality for education? In *Proceedings of the ACM symposium on Virtual reality software and technology* (pp. 160–165). ACM Digital Library. https://dl.acm.org/doi/10.1145/502390.502420

Aubrey, J. S., Robb, M. B., Bailey, J., & Bailenson, J. (2018, April 4). *Virtual Reality 101: What you need to know about kids and VR*. Common Sense. https://www.commonsensemedia.org/research/virtual-reality-101

Ausburn, L. J., Martens, J., Washington, A., Steele, D., & Washburn, E. (2009). A cross-case analysis of gender issues in desktop virtual reality learning environments. *Journal of STEM Teacher Education, 46*(3), 6.

Aymerich-Franch, L., Karutz, C., & Bailenson, J. N. (2012). Effects of facial and voice similarity on presence in a public speaking virtual environment. In *Proceedings of the International Society for Presence Research Annual Conference* (pp. 24–26). Philadelphia, PA, United States.

Bailenson, J. (2018a). *Experience on demand: What virtual reality is, how it works, and what it can do*. W. W. Norton & Company.

Bailenson, J. (2018b). Protecting nonverbal data tracked in virtual reality. *JAMA Pediatrics, 172*(10), 905–906. https://doi.org/10.1001/jamapediatrics.2018.1909

Bailenson, J. N., Yee, N., Blascovich, J., Beall, A. C., Lundblad, N., & Jin, M. (2008). The use of immersive virtual reality in the learning sciences: Digital transformations of teachers, students, and social context. *Journal of the Learning Sciences, 17*(1), 102–141. https://doi.org/10.1080/10508400701793141

Bailey, J. O., & Bailenson, J. N. (2017). Immersive virtual reality and the developing child. In F. C. Blumberg & P. J. Brooks (Eds.), *Cognitive development in digital contexts* (pp. 181–200). Academic Press. https://doi.org/10.1016/B978-0-12-809481-5.00009-2

Bailey, J. O., Bailenson, J. N., & Casasanto, D. (2016). When does virtual embodiment change our minds? *Presence, 25*(3), 222–233. https://doi.org/10.1162/PRES_a_00263

Banakou, D., Hanumanthu, P. D., & Slater, M. (2016). Virtual embodiment of white people in a black virtual body leads to a sustained reduction in their implicit racial bias. *Frontiers in Human Neuroscience, 10*, 601. https://doi.org/10.3389/fnhum.2016.00601

Belch, D., Bailenson, J., Casale, M., & Manuccia, M. (2020). *U.S. Patent No. 10,586,469.* U.S. Patent and Trademark Office.

Biocca, F. (1997). The cyborg's dilemma: Progressive embodiment in virtual environments. *Journal of Computer-Mediated Communication, 3*(2), JCMC324.

Blanke, O. (2012). Multisensory brain mechanisms of bodily self-consciousness. *Nature Reviews Neuroscience, 13*(8), 556–571. https://doi.org/10.1038/nrn3292

Blascovich, J., Loomis, J., Beall, A. C., Swinth, K. R., Hoyt, C. L., & Bailenson, J. N. (2002). Immersive virtual environment technology as a methodological tool for social psychology. *Psychological Inquiry, 13*(2), 103–124. https://doi.org/10.1207/S15327965PLI1302_01

Bregman, A., Fernando, S., & Hawking, L. (2017, October 7). *The party: A virtual experience of autism—360 video* [Video]. The Guardian. https://www.theguardian.com/technology/2017/oct/07/the-party-a-virtual-experience-of-autism-360-video

Bursztyn, N., Walker, A., Shelton, B., & Pederson, J. (2017). Increasing undergraduate interest to learn geoscience with GPS-based augmented reality field trips on students' own smartphones. *GSA Today, 27*(5), 4–10. https://doi.org/10.1130/GSATG304A.1

Carbonell Carrera, C., & Bermejo Asensio, L. A. (2017). Augmented reality as a digital teaching environment to develop spatial thinking. *Cartography and Geographic Information Science, 44*(3), 259–270. https://doi.org/10.1080/15230406.2016.1145556

Chen, Y. C. (2006, June). A study of comparing the use of augmented reality and physical models in chemistry education. In *Proceedings of the 2006 ACM International Conference on Virtual Reality Continuum and Its Applications* (pp. 369–372). https://dl.acm.org/doi/proceedings/10.1145/1128923

Clarke, J., & Dede, C. (2009). Design for scalability: A case study of the River City curriculum. *Journal of Science Education and Technology, 18*(4), 353–365. https://doi.org/10.1007/s10956-009-9156-4

Cummings, J. J., & Bailenson, J. N. (2016). How immersive is enough? A meta-analysis of the effect of immersive technology on user presence. *Media Psychology, 19*(2), 272–309. https://doi.org/10.1080/15213269.2015.1015740

de la Peña, N., Weil, P., Llobera, J., Giannopoulos, E., Pomés, A., Spanlang, B., Friedman, D., Sanchez-Vives, M. V., & Slater, M. (2010). Immersive journalism: Immersive virtual reality for the first-person experience of news. *Presence, 19*(4), 291–301. https://doi.org/10.1162/PRES_a_00005

De Leo, G., Diggs, L. A., Radici, E., & Mastaglio, T. W. (2014). Measuring sense of presence and user characteristics to predict effective training in an online simulated virtual environment. *Simulation in Healthcare*, *9*(1), 1–6. https://doi.org/10.1097/SIH.0b013e3182a99dd9

De Leo, G., Ponder, M., Molet, T., Fato, M., Thalmann, D., Magnenat-Thalmann, N., Bermano, F., & Beltrame, F. (2003). A virtual reality system for the training of volunteers involved in health emergency situations. *Cyberpsychology & Behavior*, *6*(3), 267–274. https://doi.org/10.1089/109493103322011551

Dede, C. (2009). Immersive interfaces for engagement and learning. *Science*, *323*(5910), 66–69. https://doi.org/10.1126/science.1167311

Dede, C., Nelson, B., Ketelhut, D. J., Clarke, J., & Bowman, C. (2004). Design-based research strategies for studying situated learning in a multi-user virtual environment. In *Proceedings of the Sixth International Conference on the Learning Sciences* (pp. 158–165). Santa Monica, CA, United States.

De Greef, P., & IJsselsteijn, W. (2000). Social presence in the PhotoShare tele-application. *Proceedings of PRESENCE*, 27–28.

Dunleavy, M., & Dede, C. (2014). Augmented reality teaching and learning. In J. M. Spector, M. D. Merrill, J. Elen, & M. J. Bishop (Eds.), *Handbook of research on educational communications and technology* (pp. 735–745). Springer. https://doi.org/10.1007/978-1-4614-3185-5_59

Dunleavy, M., Dede, C., & Mitchell, R. (2009). Affordances and limitations of immersive participatory augmented reality simulations for teaching and learning. *Journal of Science Education and Technology*, *18*(1), 7–22. https://doi.org/10.1007/s10956-008-9119-1

Dyer, E., Swartzlander, B. J., & Gugliucci, M. R. (2018). Using virtual reality in medical education to teach empathy. *Journal of the Medical Library Association*, *106*(4), 498–500. https://doi.org/10.5195/JMLA.2018.518

Fauville, G., Bailenson, J. N., & Queiroz, A. C. M. (2020). Virtual reality as a promising tool to promote climate change awareness. *Technology and Health*, 91–108. https://doi.org/10.1016/B978-0-12-816958-2.00005-8

Felnhofer, A., Kothgassner, O. D., Beutl, L., Hlavacs, H., & Kryspin-Exner, I. (2012). Is virtual reality made for men only? Exploring gender differences in the sense of presence. *Proceedings of the International Society on Presence Research*, 103–112.

Flavell, J. H., Flavell, E. R., Green, F. L., & Moses, L. J. (1990). Young children's understanding of fact beliefs versus value beliefs. *Child Development*, *61*(4), 915–928. https://doi.org/10.2307/1130865

Formosa, N. J., Morrison, B. W., Hill, G., & Stone, D. (2018). Testing the efficacy of a virtual reality-based simulation in enhancing users' knowledge, attitudes, and empathy relating to psychosis. *Australian Journal of Psychology*, *70*(1), 57–65. https://doi.org/10.1111/ajpy.12167

Fox, J., Arena, D., & Bailenson, J. N. (2009). Virtual reality: A survival guide for the social scientist. *Journal of Media Psychology, 21*(3), 95–113. https://doi.org/10.1027/1864-1105.21.3.95

Fox, J., Bailenson, J. N., & Tricase, L. (2013). The embodiment of sexualized virtual selves: The Proteus effect and experiences of self-objectification via avatars. *Computers in Human Behavior, 29*(3), 930–938. https://doi.org/10.1016/j.chb.2012.12.027

Froyd, S. (2018). 100 Colorado Creatives 4.0: Romain Vak. *Westword.* https://www.westword.com/arts/romain-vak-is-erasing-hate-with-virtual-reality-10015134

Gamito, P., Oliveira, J., Santos, P., Morais, D., Saraiva, T., Pombal, M., & Mota, B. (2008). Presence, immersion and cybersickness assessment through a test anxiety virtual environment. *Annual Review of Cybertherapy and Telemedicine, 6*, 83–90.

Gold, J. I., Belmont, K. A., & Thomas, D. A. (2007). The neurobiology of virtual reality pain attenuation. *Cyberpsychology & Behavior, 10*(4), 536–544. https://doi.org/10.1089/cpb.2007.9993

Groom, V., Bailenson, J. N., & Nass, C. (2009). The influence of racial embodiment on racial bias in immersive virtual environments. *Social Influence, 4*(3), 231–248. https://doi.org/10.1080/15534510802643750

Guegan, J., Buisine, S., Mantelet, F., Maranzana, N., & Segonds, F. (2016). Avatar-mediated creativity: When embodying inventors makes engineers more creative. *Computers in Human Behavior, 61*, 165–175. https://doi.org/10.1016/j.chb.2016.03.024

Gutiérrez-Maldonado, J., Ferrer-García, M., Caqueo-Urízar, A., & Letosa-Porta, A. (2006). Assessment of emotional reactivity produced by exposure to virtual environments in patients with eating disorders. *Cyberpsychology & Behavior, 9*(5), 507–513. https://doi.org/10.1089/cpb.2006.9.507

Häkkinen, J., Vuori, T., & Puhakka, M. (2002). Postural stability and sickness symptoms after HMD use. In *IEEE International Conference on Systems, Man and Cybernetics*, 147–152. https://doi.org/10.1109/ICSMC.2002.1167964

Hamilton-Giachritsis, C., Banakou, D., Garcia Quiroga, M., Giachritsis, C., & Slater, M. (2018). Reducing risk and improving maternal perspective-taking and empathy using virtual embodiment. *Scientific Reports, 8*(1), 2975. https://doi.org/10.1038/s41598-018-21036-2

Hasler, B. S., Spanlang, B., & Slater, M. (2017). Virtual race transformation reverses racial in-group bias. *PLOS ONE, 12*(4), e0174965. https://doi.org/10.1371/journal.pone.0174965

Hassanein, K., & Head, M. (2007). Manipulating perceived social presence through the web interface and its impact on attitude towards online shopping.

International Journal of Human-Computer Studies, 65(8), 689–708. https://doi.org/10.1016/j.ijhcs.2006.11.018

He, J., & Freeman, L. A. (2019). Are men more technology-oriented than women? The role of gender on the development of general computer self-efficacy of college students. *Journal of Information Systems Education, 21*(2), 7.

Heeter, C. (1992). Being there: The subjective experience of presence. *Presence, 1*(2), 262–271. https://doi.org/10.1162/pres.1992.1.2.262

Herrera, F., Bailenson, J., Weisz, E., Ogle, E., & Zaki, J. (2018). Building long-term empathy: A large-scale comparison of traditional and virtual reality perspective-taking. *PLOS ONE, 13*(10), e0204494. https://doi.org/10.1371/journal.pone.0204494

Hoffman, M., Gneezy, U., & List, J. A. (2011). Nurture affects gender differences in spatial abilities. *Proceedings of the National Academy of Sciences, 108*(36), 14786–14788. https://doi.org/10.1073/pnas.1015182108

Huffman, A. H., Whetten, J., & Huffman, W. H. (2013). Using technology in higher education: The influence of gender roles on technology self-efficacy. *Computers in Human Behavior, 29*(4), 1779–1786. https://doi.org/10.1016/j.chb.2013.02.012

Hunt, E., & Waller, D. (1999). *Orientation and wayfinding: A review.* http://citeseerx.ist.psu.edu/viewdoc/summary?doi=10.1.1.46.5608

IJsselsteijn, W., & Riva, G. (2003). Being there: The experience of presence in mediated environments. In G. Riva, F. Davide, & W. A. IJsselsteijn (Eds.), *Being there: Concepts, effects and measurements of user presence in synthetic environments* (pp. 3–16). IOS Press.

Jacobs, M. (2020). *Here's what you didn't know about the history of virtual reality.* Delta2020. https://delta2020.com/blog/221-here-s-what-you-didn-t-know-about-the-history-of-virtual-reality

Jaeger, B. K., & Mourant, R. R. (2001, October). Comparison of simulator sickness using static and dynamic walking simulators. *Proceedings of the Human Factors and Ergonomics Society Annual Meeting, 45*(27), 1896–1900. https://doi.org/10.1177/154193120104502709

Jun, H., Miller, M. R., Herrera, F., Reeves, B., & Bailenson, J. N. (2020). Stimulus sampling with 360-videos: Examining head movements, arousal, presence, simulator sickness, and preference on a large sample of participants and videos. *IEEE Transactions on Affective Computing.* https://doi.org/10.1109/TAFFC.2020.3004617

Kalckert, A., & Ehrsson, H. H. (2012). Moving a rubber hand that feels like your own: A dissociation of ownership and agency. *Frontiers in Human Neuroscience, 6*, 40. https://doi.org/10.3389/fnhum.2012.00040

Kamarainen, A. M., Metcalf, S., Grotzer, T., Browne, A., Mazzuca, D., Tutwiler, M. S., & Dede, C. (2013). EcoMOBILE: Integrating augmented reality and probeware with environmental education field trips. *Computers & Education*, *68*, 545–556. https://doi.org/10.1016/j.compedu.2013.02.018

Kaufmann, H., & Schmalstieg, D. (2002, July). Mathematics and geometry education with collaborative augmented reality. In *ACM SIGGRAPH 2002 conference abstracts and applications* (pp. 37–41). https://doi.org/10.1145/1242073.1242086

Kenny, P., Parsons, T. D., Gratch, J., Leuski, A., & Rizzo, A. A. (2007, September). Virtual patients for clinical therapist skills training. In *International Workshop on Intelligent Virtual Agents* (pp. 197–210). Springer, Berlin, Heidelberg.

Ketelhut, D. J., Nelson, B. C., Clarke, J., & Dede, C. (2010). A multi-user virtual environment for building and assessing higher order inquiry skills in science. *British Journal of Educational Technology*, *41*(1), 56–68. https://doi.org/10.1111/j.1467-8535.2009.01036.x

Kilteni, K., Bergstrom, I., & Slater, M. (2013). Drumming in immersive virtual reality: The body shapes the way we play. *IEEE Transactions on Visualization and Computer Graphics*, *19*(4), 597–605. https://doi.org/10.1109/TVCG.2013.29

Kinateder, M., Wirth, T. D., & Warren, W. H. (2018). Crowd dynamics in virtual reality. In L. Gibelli & N. Bellomo (Eds.), *Crowd Dynamics* (Vol. 1, pp. 15–36). Birkhäuser. https://doi.org/10.1007/978-3-030-05129-7_2

Klippel, A., Zhao, J., Jackson, K. L., La Femina, P., Stubbs, C., Wetzel, R., Blair, J., Wallgrün, J. O., & Oprean, D. (2019). Transforming earth science education through immersive experiences: Delivering on a long held promise. *Journal of Educational Computing Research*, *57*(7), 1745–1771. https://doi.org/10.1177/0735633119854025

Klippel, A., Zhao, J., Oprean, D., Wallgrün, J. O., Stubbs, C., La Femina, P., & Jackson, K. L. (2020). The value of being there: toward a science of immersive virtual field trips. *Virtual Reality*, *24*(4), 753–770. https://doi.org/10.1007/s10055-019-00418-5

Klopfer, E., & Squire, K. (2008). Environmental detectives—The development of an augmented reality platform for environmental simulations. *Educational Technology Research and Development*, *56*(2), 203–228. https://doi.org/10.1007/s11423-007-9037-6

Knight, M. M., & Arns, L. L. (2006, July). The relationship among age and other factors on incidence of cybersickness in immersive environment users. In *Proceedings of the 3rd Symposium on Applied Perception in Graphics and Visualization* (p. 162). https://doi.org/10.1145/1140491.1140539

Lee, K. M. (2004). Presence, explicated. *Communication Theory*, *14*(1), 27–50. https://doi.org/10.1111/j.1468-2885.2004.tb00302.x

Lee, K. M., Peng, W., Jin, S. A., & Yan, C. (2006). Can robots manifest personality? An empirical test of personality recognition, social responses, and social presence in human–robot interaction. *Journal of Communication, 56*(4), 754–772. https://doi.org/10.1111/j.1460-2466.2006.00318.x

Lenggenhager, B., Tadi, T., Metzinger, T., & Blanke, O. (2007). Video ergo sum: Manipulating bodily self-consciousness. *Science, 317*(5841), 1096–1099. https://doi.org/10.1126/science.1143439

Li, B. J., & Bailenson, J. N. (2018). Exploring the influence of haptic and olfactory cues of a virtual donut on satiation and eating behavior. *Presence, 26*(3), 337–354. https://doi.org/10.1162/pres_a_00300

Li, B. J., Bailenson, J. N., Pines, A., Greenleaf, W. J., & Williams, L. M. (2017). A public database of immersive VR videos with corresponding ratings of arousal, valence, and correlations between head movements and self report measures. *Frontiers in Psychology, 8*, 2116. https://doi.org/10.3389/fpsyg.2017.02116

Ling, Y., Nefs, H. T., Brinkman, W. P., Qu, C., & Heynderickx, I. (2013). The relationship between individual characteristics and experienced presence. *Computers in Human Behavior, 29*(4), 1519–1530. https://doi.org/10.1016/j.chb.2012.12.010

Lok, B., Ferdig, R. E., Raij, A., Johnsen, K., Dickerson, R., Coutts, J., Stevens, A., & Lind, D. S. (2006). Applying virtual reality in medical communication education: Current findings and potential teaching and learning benefits of immersive virtual patients. *Virtual Reality, 10*(3–4), 185–195. https://doi.org/10.1007/s10055-006-0037-3

Lombard, M., & Ditton, T. (1997). At the heart of it all: The concept of presence. *Journal of Computer-Mediated Communication, 3*(2), JCMC321.

Lombard, M., Ditton, T. B., & Weinstein, L. (2009, November 11–13). Measuring presence: The temple presence inventory. In *Proceedings of the 12th Annual International Workshop on Presence*. Los Angeles, CA, United States.

Loomis, J. M., Blascovich, J. J., & Beall, A. C. (1999). Immersive virtual environment technology as a basic research tool in psychology. *Behavior Research Methods, Instruments, & Computers, 31*(4), 557–564. https://doi.org/10.3758/BF03200735

Ma, X., Cackett, M., Park, L., Chien, E., & Naaman, M. (2018, April 23). Web-based VR experiments powered by the crowd. In *Proceedings of the 2018 World Wide Web Conference*. Geneva, Switzerland. https://doi.org/10.1145/3178876.3186034

Markowitz, D. M., Laha, R., Perone, B. P., Pea, R. D., & Bailenson, J. N. (2018). Immersive virtual reality field trips facilitate learning about climate change. *Frontiers in Psychology, 9*, 2364. https://doi.org/10.3389/fpsyg.2018.02364

Maxhall, M., Backman, A., Holmlund, K., Hedman, L., Sondell, B., & Bucht, G. (2004). Caregiver responses to a stroke training simulator. In *Proceedings of ICDVRAT2004 Conference.* https://doi.org/10.1515/IJDHD.2005.4.3.245

McNamara, A. M. (2011, December). Enhancing art history education through mobile augmented reality. In *Proceedings of the 10th International Conference on Virtual Reality Continuum and Its Applications in Industry* (pp. 507–512). https://doi.org/10.1145/2087756.2087853

Milk, C. (2015, March). *How virtual reality can create the ultimate empathy machine* [Video]. TED Conferences. https://www.ted.com/talks/chris_milk_how_virtual_reality_can_create_the_ultimate_empathy_machine?language=en

Moreno, R., & Mayer, R. E. (2002). Learning science in virtual reality multimedia environments: Role of methods and media. *Journal of Educational Psychology, 94*(3), 598–610. https://doi.org/10.1037/0022-0663.94.3.598

Mujber, T. S., Szecsi, T., & Hashmi, M. S. (2004). Virtual reality applications in manufacturing process simulation. *Journal of Materials Processing Technology, 155–156*, 1834–1838. https://doi.org/10.1016/j.jmatprotec.2004.04.401

Munafo, J., Diedrick, M., & Stoffregen, T. A. (2017). The virtual reality head-mounted display Oculus Rift induces motion sickness and is sexist in its effects. *Experimental Brain Research, 235*(3), 889–901. https://doi.org/10.1007/s00221-016-4846-7

Nakamura, L. (2020). Feeling good about feeling bad: Virtuous virtual reality and the automation of racial empathy. *Journal of Visual Culture, 19*(1), 47–64. https://doi.org/10.1177/1470412920906259

Nicovich, S. G., Boller, G. W., & Cornwell, T. B. (2005). Experienced presence within computer-mediated communications: Initial explorations on the effects of gender with respect to empathy and immersion. *Journal of Computer-Mediated Communication, 10*(2), JCMC1023.

O'Dwyer, N., Young, G. W., Johnson, N., Zerman, E., & Smolic, A. (2020, July). Mixed reality and volumetric video in cultural heritage: Expert opinions on augmented and virtual reality. In *International Conference on Human-Computer Interaction* (pp. 195–214). Springer, Cham.

Oh, S. Y., Bailenson, J. N., Weisz, E., & Zaki, J. (2016). Virtually old: Embodied perspective taking and the reduction of ageism under threat. *Computers in Human Behavior, 60*, 398–410. https://doi.org/10.1016/j.chb.2016.02.007

Oh, S. Y., Bailenson, J. N., & Welch, G. F. (2018). A systematic review of social presence: Definition, antecedents, and implications. *Frontiers in Robotics and AI,* Article 514134. https://doi.org/10.3389/frobt.2018.00114

Park, G. D., Allen, R. W., Fiorentino, D., Rosenthal, T. J., & Cook, M. L. (2006, October). Simulator sickness scores according to symptom susceptibility, age, and gender for an older driver assessment study. *Proceedings of the Human*

Factors and Ergonomics Society Annual Meeting, 50(26), 2702–2706. https://doi.org/10.1177/154193120605002607

Parong, J., & Mayer, R. E. (2018). Learning science in immersive virtual reality. *Journal of Educational Psychology, 110*(6), 785–797. https://doi.org/10.1037/edu0000241

Patel, K., Bailenson, J. N., Hack-Jung, S., Diankov, R., & Bajcsy, R. (2006, August). The effects of fully immersive virtual reality on the learning of physical tasks. In *Proceedings of the 9th Annual International Workshop on Presence, Ohio, USA* (pp. 87–94).

Pausch, R., Crea, T., & Conway, M. (1992). A literature survey for virtual environments: Military flight simulator visual systems and simulator sickness. *Presence, 1*(3), 344–363. https://doi.org/10.1162/pres.1992.1.3.344

Peck, T. C., Seinfeld, S., Aglioti, S. M., & Slater, M. (2013). Putting yourself in the skin of a black avatar reduces implicit racial bias. *Consciousness and Cognition, 22*(3), 779–787. https://doi.org/10.1016/j.concog.2013.04.016

Perry, J., Klopfer, E., Norton, M., Sutch, D., Sandford, R., & Facer, K. (2008). AR gone wild: Two approaches to using augmented reality learning games in Zoos. In *Proceedings of the 8th International Conference of the Learning Sciences, Utrecht, The Netherlands* (pp. 322–329).

Petkova, V. I., & Ehrsson, H. H. (2008). If I were you: Perceptual illusion of body swapping. *PLOS ONE, 3*(12), Article e3832. https://doi.org/10.1371/journal.pone.0003832

Ratan, R., Beyea, D., Li, B. J., & Graciano, L. (2019). Avatar characteristics induce users' behavioral conformity with small-to-medium effect sizes: A meta-analysis of the Proteus effect. *Media Psychology, 23*(5), 651–675. https://doi.org/10.1080/15213269.2019.1623698

Reeves, B., & Nass, C. I. (1996). *The media equation: How people treat computers, television, and new media like real people and places*. Cambridge University Press.

Scavarelli, A., Arya, A., & Teather, R. J. (2021). Virtual reality and augmented reality in social learning spaces: A literature review. *Virtual Reality, 25*, 257–277. https://doi.org/10.1007/s10055-020-00444-8

Schultheis, M. T., & Rizzo, A. A. (2001). The application of virtual reality technology in rehabilitation. *Rehabilitation Psychology, 46*(3), 296–311. https://doi.org/10.1037/0090-5550.46.3.296

Segovia, K. Y., & Bailenson, J. N. (2009). Virtually true: Children's acquisition of false memories in virtual reality. *Media Psychology, 12*(4), 371–393. https://doi.org/10.1080/15213260903287267

Seinfeld, S., Arroyo-Palacios, J., Iruretagoyena, G., Hortensius, R., Zapata, L. E., Borland, D., de Gelder, B., Slater, M., & Sanchez-Vives, M. V. (2018). Offenders become the victim in virtual reality: Impact of changing perspective in domestic violence. *Scientific Reports, 8*(1), 2692. https://doi.org/10.1038/s41598-018-19987-7

Sharar, S. R., Carrougher, G. J., Nakamura, D., Hoffman, H. G., Blough, D. K., & Patterson, D. R. (2007). Factors influencing the efficacy of virtual reality distraction analgesia during postburn physical therapy: Preliminary results from 3 ongoing studies. *Archives of Physical Medicine and Rehabilitation, 88*(12, Suppl. 2), S43–S49. https://doi.org/10.1016/j.apmr.2007.09.004

Slater, M., Pérez-Marcos, D., Ehrsson, H. H., & Sanchez-Vives, M. V. (2009). Inducing illusory ownership of a virtual body. *Frontiers in Neuroscience, 3*(2), 214–220. https://doi.org/10.3389/neuro.01.029.2009

Slater, M., Spanlang, B., Sanchez-Vives, M. V., & Blanke, O. (2010). First person experience of body transfer in virtual reality. *PLOS ONE, 5*(5), Article e10564. https://doi.org/10.1371/journal.pone.0010564

Slater, M., Usoh, M., & Steed, A. (1994). Depth of presence in virtual environments. *Presence, 3*(2), 130–144. https://doi.org/10.1162/pres.1994.3.2.130

Slater, M., & Wilbur, S. (1997). A framework for immersive virtual environments (FIVE): Speculations on the role of presence in virtual environments. *Presence, 6*(6), 603–616. https://doi.org/10.1162/pres.1997.6.6.603

Spitzer, V. M., & Ackerman, M. J. (2008). The Visible Human® at the University of Colorado 15 years later. *Virtual Reality, 12*(4), 191–200. https://doi.org/10.1007/s10055-008-0102-1

Sri Kalyanaraman, S., Penn, D. L., Ivory, J. D., & Judge, A. (2010). The virtual doppelganger: Effects of a virtual reality simulator on perceptions of schizophrenia. *The Journal of Nervous and Mental Disease, 198*(6), 437–443. https://doi.org/10.1097/NMD.0b013e3181e07d66

Stanney, K., Fidopiastis, C., & Foster, L. (2020). Virtual reality is sexist: But it does not have to be. *Frontiers in Robotics and AI, 7*, 4. https://doi.org/10.3389/frobt.2020.00004

Stanney, K. M., Hale, K. S., Nahmens, I., & Kennedy, R. S. (2003). What to expect from immersive virtual environment exposure: Influences of gender, body mass index, and past experience. *Human Factors, 45*(3), 504–520. https://doi.org/10.1518/hfes.45.3.504.27254

Steptoe, W., Steed, A., & Slater, M. (2013). Human tails: Ownership and control of extended humanoid avatars. *IEEE Transactions on Visualization and Computer Graphics, 19*(4), 583–590. https://doi.org/10.1109/TVCG.2013.32

Steuer, J. (1992). Defining virtual reality: Dimensions determining telepresence. *Journal of Communication, 42*(4), 73–93. https://doi.org/10.1111/j.1460-2466.1992.tb00812.x

Stevens, A., Hernandez, J., Johnsen, K., Dickerson, R., Raij, A., Harrison, C., DiPietro, M., Allen, B., Ferdig, R., Foti, S., Jackson, J., Shin, M., Cendan, J., Watson, R., Duerson, M., Lok, B., Cohen, M., Wagner, P., & Lind, D. S. (2006). The use of virtual patients to teach medical students history taking and communication skills. *American Journal of Surgery*, *191*(6), 806–811. https://doi.org/10.1016/j.amjsurg.2006.03.002

Tarr, M. J., & Warren, W. H. (2002). Virtual reality in behavioral neuroscience and beyond. *Nature Neuroscience*, *5*(11), 1089–1092. https://doi.org/10.1038/nn948

van Veen, H. A., Distler, H. K., Braun, S. J., & Bülthoff, H. H. (1998). Navigating through a virtual city: Using virtual reality technology to study human action and perception. *Future Generation Computer Systems*, *14*(3–4), 231–242. https://doi.org/10.1016/S0167-739X(98)00027-2

Vorderer, P., Wirth, W., Gouveia, F. R., Biocca, F., Saari, T., Jäncke, F., Böcking, S., Schramm, H., Gysbers, A., Hartmann, T., Klimmt, C., Laarni, J., Ravaja, N., Sacau, A., Baumgartner, T., & Jäncke, P. (2004). *MEC Spatial Presence Questionnaire (MEC-SPQ): Short documentation and instructions for application* (Report to the European Community, Project Presence: MEC [IST-2001-37661]). https://academic.csuohio.edu/kneuendorf/frames/MECFull.pdf

Walther, J. B. (1992). Interpersonal effects in computer-mediated interaction: A relational perspective. *Communication Research*, *19*(1), 52–90. https://doi.org/10.1177/009365092019001003

Weech, S., Kenny, S., & Barnett-Cowan, M. (2019). Presence and cybersickness in virtual reality are negatively related: A review. *Frontiers in Psychology*, *10*, 158. https://doi.org/10.3389/fpsyg.2019.00158

Wirth, W., Hartmann, T., Böcking, S., Vorderer, P., Klimmt, C., Schramm, H., Saari, T., Laarni, J., Ravaja, N., Ribeiro Gouveia, F., Biocca, F., Sacau, A., Jäncke, L., Baumgartner, T., & Jäncke, P. (2007). A process model of the formation of spatial presence experiences. *Media Psychology*, *9*(3), 493–525. https://doi.org/10.1080/15213260701283079

Witmer, B. G., & Singer, M. J. (1998). Measuring presence in virtual environments: A presence questionnaire. *Presence*, *7*(3), 225–240. https://doi.org/10.1162/105474698565686

Yee, N., & Bailenson, J. (2006). Walk a mile in digital shoes: The impact of embodied perspective-taking on the reduction of negative stereotyping in immersive virtual environments. *Proceedings of PRESENCE*, *24*, 26.

Yee, N., & Bailenson, J. (2007). The Proteus Effect: The effect of transformed self-representation on behavior. *Human Communication Research*, *33*(3), 271–290. https://doi.org/10.1111/j.1468-2958.2007.00299.x

6

Social Media and Psychological Well-Being

Jeffrey T. Hancock, Sunny Xun Liu, Mufan Luo, and Hannah Mieczkowski

Is social media good or bad for us? Our research group at the Stanford Social Media Lab has been fielding this question for years from a wide variety of perspectives: parents concerned about their children's use of social media, policy makers worried about the effects on society, startups excited to spark creativity or new connections, tech companies trying to ensure that their products have a positive impact. The question of how social media is related to well-being has inspired a massive quantity of research by scholars around the globe, with hundreds of studies examining this question since the first paper on the topic came out in 2006. Despite this plethora of empirical evidence, the question remains hotly debated not only in popular discourse but also in the academic literature.

The 2020 American "docudrama" *The Social Dilemma* provides a salient case for one side of that debate, namely, that social media is harmful for people. It describes how using social media is addicting, narcissistic,

https://doi.org/10.1037/0000290-007
The Psychology of Technology: Social Science Research in the Age of Big Data, S. C. Matz (Editor)
Copyright © 2022 by the American Psychological Association. All rights reserved.

and superficial, undermining to our relationships, our cognitive powers, and ultimately increasing our anxiety, loneliness, and depression. It's slickly produced, with powerful metaphors equating social media with other addictive and harmful substances. The portrayal of several fictional stories showing how social media can harm people, complete with dark and gripping music and excellent acting, captures the concerns that many hold about how social media is dangerous and harmful. Indeed, the movie's popularity reflects the widespread concern that parents, teens, educators, and policy makers all have about how social media is affecting our well-being.

The movie leaves many important questions unanswered and highlights many of the difficult conceptual and methodological issues that emerge when asking if social media is good or bad for well-being. We need a better scientific understanding of these questions. For example, what does it mean to "use" social media? The movie unfortunately fails to distinguish between different usages of social media, and this is an important—but often overlooked—component of the methodological challenges researchers face when trying to answer the question. *The Social Dilemma* implicitly assumes that time spent is what matters, with more use leading to more detrimental effects. But there are myriad ways to engage with social media, and an hour spent doing one thing, like creating a TikTok video with close friends, can be radically different from an hour spent doing something else, like stalking an ex-spouse. Similarly, what do we mean by the question "Is social media is good or bad for us?" Typically, this refers to our well-being, which can be about physical, mental, or emotional health and/or social and cognitive function. Without careful conceptualization of well-being, it is difficult if not impossible to make any claims about how social media affects us. Here we focus on psychological well-being and consider both negative indicators, including depression, anxiety, and loneliness, and positive indicators, including eudaimonic (life-satisfaction), hedonic (emotional), and social well-being.

Another open question is the nature of the relationship between social media and well-being. *The Social Dilemma* movie assumes that social media has a causal effect on well-being. Indeed, the movie portrays an extreme version of technological determinism, which assumes that technology has direct and causal effects on human life. In this view, humans have no agency; instead, technology operates on people. Indeed, one of the metaphors from the movie is that people are like drugged patients lying on a surgery bed while social media companies operate on us. This kind of technological determinism is a common ingredient in moral panics that typically arise with new technologies. Orben (2020) documented this form of technological determinism in moral panics regarding radio, comic books, television, and video games. We need to pay careful attention to the kinds of claims that research can make about the direction or causal nature of the relationship between social media and well-being.

Finally, and perhaps most shockingly, given how much research has been conducted on social media and well-being, most debates take place without reference to the vast amount of evidence that has already been collected. News stories tend to focus on a single study, or worse, not include any grounding in science. *The Social Dilemma*, for example, presents almost no scientific evidence. Although the movie includes scientists, there is very little reference to scientific studies, and there is no reference to any research that actually connects social media and well-being.

So, what does science tell us? What is the accumulated evidence from the past decade of social science research into the harms and benefits of social media for well-being? In this chapter, we describe our research group's efforts to tackle the question of social media use and its relationships to psychological well-being. For the past several years, we have reviewed this entire literature from 2006 to 2018 as part of a large meta-analysis (Hancock et al., 2019). Here, we go over some of the key findings from this analysis, along with other insights we gleaned.

The chapter is organized first around this meta-analysis (Hancock et al., 2019) and the empirical findings for key questions concerning social media use and well-being. The second part of the chapter reviews the many conceptual mechanisms that authors have proposed for how social media and well-being may be linked. In the third part, we highlight some of the key methodological issues that became clear in our review. For example, over 75% of all the studies we examined were correlational, cross-sectional studies. In this section, we highlight some of the changes that are required for our field to improve our understanding of social media and well-being. In the last section of the chapter, we describe how the field can move forward focusing on some new methods that we believe will advance the field as well as some new conceptualizations that could be important in rethinking the relationship between social media and well-being, along with more nuanced analyses of different populations.

META-ANALYSIS OVERVIEW

We set out to conduct the most comprehensive meta-analysis to date on social media use and well-being. While other reviews have focused on specific time periods or on specific forms of well-being (Cheng et al., 2019; Domahidi, 2018; D. Liu et al., 2016; Twenge, 2020; Verduyn et al., 2017), we were interested in studying the relationship from the onset of social media research and in considering both positive and negative forms of well-being. Our analysis included all empirical studies examining the relationship between social media use and six types of psychological well-being (depression, anxiety, loneliness, eudaimonic, hedonic, and social) from 2006 to 2018. After reviewing 5,214 articles from the four largest databases in psychology, communication, and human–computer interaction, we applied a careful inclusion and exclusion review, which resulted in a final sample of 226 peer-reviewed papers. Across all the papers, there were a total of $N = 275{,}728$ participants and a total of 1,279 effect sizes calculated. Next, we lay out some of the key questions we were able to ask of this large field of evidence.

Is There an Overall Effect of Social Media Use on Well-Being?

Returning to the main question we posed at the beginning of this chapter—"Is social media use good or bad?"—we first examined the overall effect size of the relationship between social media and well-being. To look at this, we combined the negative and positive forms of well-being into one index. By doing aggregation, we could focus on the relationship regardless of the specific form of well-being measured and include all of the studies from the meta-analysis (i.e., all 1,279 effects). For example, high scores on depression would now be equal to low well-being, while low scores on life-satisfaction would also equal low well-being. Following this approach, we found that the weighted mean effect size across all studies was $r = 0.01$ [–.02, .04]. This effect is not only very small, but it is also a very precise estimate around zero, with the 95% confidence interval indicating that the relationship between social media use and well-being was a correlation somewhere between $r = -.02$ and $r = .04$ and therefore nonsignificant. Thus, when we look across all the studies conducted between 2006 through 2018, including all six types of well-being, social media use is not significantly associated with well-being—it is neither good nor bad.

This finding is consistent with those of several other recent studies that have suggested that any general relationship between social media and well-being is small at best (Heffer et al., 2019; Orben & Przybylski, 2019; Schemer et al., 2020). Orben and Przybylski (2019) found that only 0.4% of the variance of well-being can be accounted for by digital technology use, while Schemer et al. (2020) found that social media use's connections with depression ($b = 0.003$) and life satisfaction ($b = -0.0004$) are close to 0 after controlling for other variables.

Has the Effect of Social Media Use on Well-Being Changed Over Time?

Although we did not observe an overall effect of social media use on well-being, it is possible that looking at this relationship over such a long period is masking effects that change over time. One concern that has been raised recently is that social media may have become more harmful after 2012. While there is no documented causal link between social media and well-being at this time, some authors have noted that mental health for adolescents declined in 2012 and that social media use began to be taken up on mobile phones around this time (Twenge, 2020).

In the top half of Figure 6.1, we show the overall mean effect size by year. In this figure, higher scores represent more positive associations between social media use and well-being. In 2012, there is indeed a decrease from

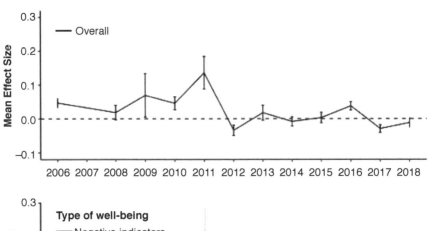
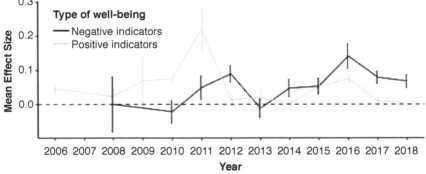

Figure 6.1

Mean effect size for the association between social media use and well-being over time. The black line in the top panel denotes the average effect size aggregating by positive and negative outcomes. The lines in the bottom panel denote the effect sizes for positive (i.e., eudaimonic, hedonic, and social) and negative indicators (i.e., anxiety, depression, loneliness) of well-being. Error bars denote standard errors.

the previous year, although this decrease is driven in part by a paucity of studies in 2011 and 2013 for some forms of well-being (e.g., there was only one study on anxiety in each of 2011 and 2013). Statistically, there is a significant decrease in the effect size over the 12 years. The measured effect sizes between social media use and overall well-being have become more negative over time, although the change represents a very small effect size, going from roughly $r = .05$ in 2006 at the start of research on social media and well-being to $r = -.01$ in 2018.

How Is Social Media Use Related to Different Kinds of Well-Being?

One unique aspect of our meta-analysis (Hancock et al., 2019) is that in addition to looking at well-being generally, we could drill down into how social media is related to specific forms of well-being. This allowed us to see, for example, whether the null effect we observed with overall well-being was replicated in each specific type of well-being or whether there were patterns of effects across the positive and negative forms of well-being that led to an averaging out. When we looked at the negative indicators of well-being, we found that social media use had small but significant associations with anxiety ($r = .13$) and depression ($r = 0.12$) but not loneliness. On the positive side, we found a larger though still small association with social well-being ($r = .20$) but not with eudaimonic or hedonic well-being.

This pattern of results suggests that social media can affect well-being both positively and negatively at the same time. It also revealed that our null effect for overall well-being was, in part, driven by the averaging out of small effects for both negative and positive indicators. These findings suggest that there might be a potential trade-off of social media use between elevated depression and anxiety along with improved social well-being.

Is Active Use Better Than Passive Use?

Prior research has suggested that active (e.g., sending a message, uploading a picture) versus passive use (e.g., reading tweets or looking at Instagram pictures), defined as the extent to which direct information exchange with others is involved, can generate different effects on well-being. This work shows that while passive use undermines people's well-being, active use can have positive effects (Burke & Kraut, 2016; Verduyn et al., 2017). We compared the relationships between passive and active use with well-being and found no significant difference. We then calculated the effect sizes of active and passive social media use and their associations with each well-being type. The findings suggested that active social media use was positively associated with overall well-being, depression, eudaimonic, and social well-being. In contrast, passive social media use was associated only with depression. Therefore, although active use correlated with more positive well-being outcomes, it was also linked with increased depression. Overall, the pattern provides only partial support for the proposition that active use is more positive for well-being than passive use.

Does How We Measure Social Media Use Matter?

When coding the different studies included in our meta-analyses (Hancock et al., 2019), we noticed that the vast majority relied on self-reported social media use, asking participants how much they use social media rather than observing or measuring behavior directly. As a result, we also noticed the different ways in which social media use is framed in these measures. One important framing was whether social media use was addictive or not. We separated studies that focused on addictive social media use and used scales that measured social media addiction from studies that used more neutral framing of use. When we compared these two types of studies, we found that studies that used addiction-related scales produced significantly different relationships from those that did not frame social media use as addictive. Addictive social media use was negatively associated with overall well-being, while neutral social media use was positively associated. This pattern of results might be explained in two ways. First, it is possible that people who experience high levels of addictive social media use indeed experience lower levels of well-being. However, it is also possible that the framing of social media as potentially addictive influenced participants to view their experience with social media more negatively. We explore this possibility in deeper detail in the section on methodological issues.

Do the Effects Change Across Populations?

A major challenge for the field is that studies often differ in the types of populations they use. These populations, in turn, might be highly heterogeneous in their social media use as well as the goals and needs they try to accomplish or satisfy with it. A teenager using social media to develop their identity and friend network is at a very different stage of their developmental trajectory than a grandparent whose identity is well-established and who uses social media to maintain old friendships and family ties. Unfortunately, over half (54%) of the studies we reviewed focused on college students, likely due to their convenience as a sample, but make inferences about the general population. Student populations are important to study because social media use is often higher among younger people. However, it is important to acknowledge that this population is likely different in many aspects from other sociodemographic groups and to be cautious about prematurely generalizing from one population to another.

When we analyzed the effects for specific age populations, we found no statistical differences for any of the participant population types for overall well-being. When we focused on each individual well-being type, we found social media use among college students was associated with higher social well-being, suggesting that for that age group social media use is positive for social connectedness and building social capital. For studies focusing on adolescents, social media use was significantly associated with both higher depression and higher social well-being, again suggesting a trade-off for social media use for young people and their use of social media. Notably, these effects are small ($r = .13$ for depression, $r = .16$ for social well-being) but consistent with a recent high-quality study of adolescents that found little evidence for longitudinal effects of social media and mental health (Jensen et al., 2019).

Another population dimension that we considered was geographical, given that culture and region may be important moderators for well-being effects. We compared studies conducted in North America versus Asia versus Europe. Interestingly, the effects were strikingly different across the three regions. Studies conducted in Asia produced overall positive associations between social media use and well-being, while studies conducted in Europe produced negative associations. Studies from North America were in the middle, though statistically more positive than zero. These data are important, as they suggest that there are notable cultural differences in how social media and well-being are related. If, for example, there is some core psychological effect, then we should not see large differences across these populations. Instead, these results suggest several intriguing possibilities. It could be that different societies are using social media to different effect, or that different societies have varying assumptions about social media, perceptions about privacy, and norms relating to networking. These culturally specific assumptions, perceptions, and norms influence users' expectations. More research is needed here to unpack these cultural differences.

What Do Cross-Sectional Versus Longitudinal Versus Experimental Studies Tell Us?

In our final sample of articles, a majority of studies were cross-sectional in nature: 75% of all articles were cross-sectional studies, 23% were longitudinal, and 2% were experimental. We found that the methodology did not significantly moderate the observed correlations between social media use and overall well-being. However, when breaking the analysis down by individual well-being indicators, we found that cross-sectional studies produced positive associations with anxiety, depression, and social well-being. In contrast, longitudinal studies of social media use produced correlations with only depression and social well-being. There were too few experimental studies to meta-analyze except for eudaimonic and hedonic outcomes, and neither produced a significant association with social media use.

Is There a Causal Link, and if so, What Is the Direction?

The causality of the relationship between social media use and well-being remains speculative and has been debated in a number of contexts. For example, do lonely people use social media more, or does social media use make people lonely? We used longitudinal data in our meta-analysis and used cross-lagged effects to explore the direction of the effect between social media use and well-being. The finding suggested that positive psychological well-being (eudaimonic well-being, hedonic well-being, and social well-being) leads to decreased social media use, not vice versa. For psychological distress (loneliness, depression, and anxiety), neither direction was found to be significant, indicating no causal relationship between social media use and psychological distress.

MAPPING THE CONCEPTUAL SPACE LINKING SOCIAL MEDIA USE AND WELL-BEING

It became clear during our review of the existing literature for our meta-analyses (Hancock et al., 2019) that there is no coherent conceptualization of the relationships between social media use and well-being and no explanation of the mechanisms driving these relationships. There was a striking range of explanations, assumptions, and propositions about how and why researchers believed that using social media would influence well-being or vice versa. Many studies never explicitly articulated the assumed conceptual mechanism linking social media with well-being, whereas others would identify a mechanism but fail to test or measure it.

In an attempt to make sense of the various theoretical connections proposed in the literature and to provide a comprehensive framework for future research, we conducted a qualitative review of the articles. The result is what we believe is one of the first compilations of the various theoretical mechanisms proposed by researchers that connect social media and well-being. In our qualitative analysis, we identified 10 conceptual mechanisms that were assumed, proposed, or tested by study authors in our meta-analysis database. Table 6.1 describes each of these constructs and provides an example. We mapped these 10 constructs into three higher level categories: social structure, psychological processes, and behavioral dynamics. While we acknowledge that this list is not conclusive, we do believe that it provides a useful framework for future research that explores the mechanisms underlying the relationships between social media use and well-being.

Social Structure

We identified two conceptual mechanisms that were based on participants' social structure. *Network metrics* focused on individuals' actual and perceived

Table 6.1
Conceptual Mechanisms Linking Social Media Use and Well-Being

Conceptual mechanisms	Description
Social structure	
Network size	# of online friends, # of likes received, perceived network size, etc.
Social support/Social capital	perceptions of tangible and intangible assistance from one's social network
Psychological processes	
Social comparison	evaluating oneself through comparison with others
Connectedness	perceptions of feeling socially connected with others
Fear of missing out	apprehension that others are having experiences without one
Overload	perceptions of too many social demands
Social compensation	use of social media to compensate for challenges encountered offline
Behavioral dynamics	
Displacement/ enhancement	social media use displaces or reinforces meaningful interpersonal communication and social activities

network size, as well as the expected amount of feedback derived from the audience. These aspects can play important roles in how people engage on social media and how they feel about themselves. For example, the number of received Likes on one's profile pictures can have a positive impact on self-esteem (Burrow & Rainone, 2017). Other research revealed that a higher proportion of actual friends in one's network was associated with lower levels of loneliness (Chang et al., 2015). In contrast, some studies found that a larger network size can exacerbate distress and negative online experience when online friends fail to provide tangible social support (Best et al., 2015).

We defined *social support* as various types of assistance, both tangible and intangible (e.g., informational, emotional, belonging), offered by a social network (Cohen & Hoberman, 1983; Uchino, 2004). Feeling a sense of greater social support has been shown to benefit psychological well-being. Research showed that lower perceptions of friends' support negatively affect people's depressed mood (Frison & Eggermont, 2015), whereas more active use is positively associated with people's perceived social support, which leads to reduced loneliness (Seo et al., 2016). Research has shown that social media use is positively associated with both bridging and bonding social capital, which can lead to higher self-esteem (Brooks et al., 2014; Choi & Kim, 2016).

Psychological Processes

As for psychological processes, *feelings of connectedness* refer to the extent to which individuals feel related to one another in their lives and behavior, which serves an intrinsic need important for well-being (Deci & Ryan, 1985). In contrast, social exclusion and isolation have negative effects on psychological well-being. Research has shown that general Facebook use can facilitate connectedness and reduce social isolation (Ahn & Shin, 2013), which in turn can decrease loneliness (Deters & Mehl, 2013), depression, and anxiety, as well as promote life satisfaction (Grieve et al., 2013). For the same reason, a lack of social network site use can intensify distress due to increased social exclusion (Chiou et al., 2015). As for specific social media use, research suggests that active participation such as chatting and commenting can increase connectedness, thereby promoting positive emotional states (Neubaum & Kramer, 2015).

Social comparison is a prominent and automatic form of self-evaluation (Bandura & Jourden, 1991; Wood, 1989). Festinger (1954) argued that social comparison occurs when there is no objective information on which an individual can base their evaluations. Performing a "downward comparison," or mentally degrading another in some manner, is typically associated with an increase in well-being (Wills, 1991). That is, when someone feels that they are in a better position than others, they tend to feel better about themselves and may also experience a self-esteem boost (Morse & Gergen, 1970). When someone feels they are worse off than others, a decrease in both well-being and self-esteem is common (Wheeler & Miyake, 1992). Evidence to date supports upward comparisons triggered by social media use. For example, more Facebook use corresponds to lower self-esteem and life satisfaction and higher depression because people often perceive themselves as worse off than others (Steers et al., 2014; Vogel et al., 2014). Other mechanisms include *the fear of missing out (FOMO)*—a persuasive apprehension that other people in one's social network may have more rewarding experiences that one is absent from (Przybylski et al., 2013). Research showed that FOMO may mediate how social media is associated with lower self-esteem and higher online vulnerability (Buglass et al., 2017). Authors suggest that individuals with a deep fear of missing out tend to engage in more online self-presentation to compensate for their lack of control over personal lives. Self-disclosure or self-presentation on social media, however, may increase one's exposure to harmful content and harassment and one's overall vulnerability to toxic discourse online (Buglass et al., 2017).

Last, while much work in our meta-analysis examined the direction from use to well-being, a few studies focused on the other direction: investigating how one's well-being status shapes social media use. The category of motivation involves the desire, preference, or need that provides the rationale for why individuals varying in well-being status or characteristics engage in social media use. Prime examples of motivation include social compensation and comparison. *Social compensation* refers to the motive of using social media to compensate for inadequate offline social experiences (J.-E. R. Lee et al., 2012). Articles in our database suggest that lonely people are motivated to compensate for their poor offline network by engaging in more online self-disclosure on social media (Hood et al., 2018). Adolescents feeling left out by friends reported a higher motive to compensate for social skills through Facebook (Teppers et al., 2014).

While *social comparison* serves as a psychological mechanism, as discussed earlier, this tendency to compare oneself with others having a better life can be a consequence of one's well-being status. Using two-wave longitudinal data, Frison and Eggermont (2016) found that people with lower life satisfaction tend to engage in negative comparison on Facebook, whereas negative comparison on Facebook also reduced life satisfaction later. The authors argued that the finding aligns with the selective exposure theory that people dissatisfied with life offline tend to use social media in a similar way: They are motivated to gain information about others to understand themselves and the social world.

Behavioral Dynamics

Finally, social media use can affect well-being through behavioral dynamics. *Displacement/reinforcement* is a set of theoretical perspectives that predict social media use can either reduce or augment actual face-to-face communication (Kraut et al., 1998). Studies in our database generally supported the reinforcement perspective. For example, using two-wave longitudinal data with a 6-month interval, Dienlin et al. (2017) found a positive longitudinal effect of social media use on face-to-face interaction and life satisfaction. Rui et al. (2015), using survey data, showed a positive association between social information seeking on Facebook and offline activities.

METHODOLOGICAL ISSUES

When examining the methodological choices employed in the studies in our meta-analysis, we noticed a number of recurring challenges that limit the conclusions we can draw from existing work. The first is an overreliance on measures of time spent on social media as a measure of social media use compared with other potential measures of interest, such as the type of connections one has in their network and the type of content one is exposed to. The second is a lack of adequately validated scales used to measure social media use, which are often adopted from other media use scales or have significant overlap with the dependent variables researchers are trying to measure. The third is the overwhelming prevalence of cross-sectional studies about social media use and well-being. Last is the insufficient attention paid to questionnaire design in most studies of social media, despite past work emphasizing the importance of survey methodology.

Overreliance on Time Spent as a Measure of Social Media Use

Scholars have been discussing challenges for the measurement of media exposure for several decades. Overreliance on measures that approximate social media use as time spent on the platform, such as the number of times checking a specific social media platform per day, or the duration of time spent on the platform, is not a new problem in media research.

For instance, the growing popularity of television in the 1950s prompted numerous studies on the effects of time spent viewing television on aggressive behavior, sleep disturbances, and mental health (Feshbach & Singer, 1971; Owens et al., 1999; Sirgy et al., 1998). Research on media effects of television, and even older media such as radio, focused on time spent as the predictor variable have had broad and long-lasting impacts on policies like program ratings (Hamilton, 1998/2000) and panicked attitudes toward new technologies (Orben, 2020; Wartella & Reeves, 1985). However, as aptly noted by Junco (2013), "drivers often estimate driving distances in miles and time to destination" but people using media "typically do not estimate frequency and intensity of use in time" (p. 630). In terms of research on digital technologies, Boase and Ling (2013) found that 40% of studies in communication journals about mobile phone use relied on self-report measures of time spent on the phone.

Among others, self-report measurements of media exposure of time spent suffer from only moderate reliability, low predictive power, and low criterion validity (Boase & Ling, 2013). Additionally, social desirability bias and perceptions of norms mean that "heavy" users of a medium are more likely to underestimate their usage, whereas "light" users tend to overestimate, so that a regression to the mean effect occurs (Scharkow, 2016). Some researchers claim that participants may produce estimates of time spent up to 5 times more than their actual time spent (Junco, 2013). These effects may also be compounded by the reliance on single-item measures of social media use, such as "How much time did you spend on social media today?"

Lack of Validated Social Media Use Scales

To gain a more robust understanding of social media use and its consequences, researchers have created dozens of different scales. However, these scales are often modified without validation, rely on very few items, and/or focus on the frequency or duration of social media behaviors (Jenkins-Guarnieri et al., 2013). Even measurements adapted from general smartphone use scales may not be enough to gain a better understanding of social media use because individuals often use a range of devices to access social media content. Compounding this problem is that different devices may be associated with different types of response errors, such as tablet users underreporting their internet use (Araujo et al., 2017). Further, digital media use is well-integrated into daily life, meaning that the activities people engage in are likely to be viewed as more mundane, making attentiveness and accurate self-reports even more of a challenge (Vanden Abeele et al., 2013). Ellis (2019) argued that current scales examining digital media use lack validity because of their inability to predict "comparatively simple behaviors that appear to be stable within participants" (p. 61). Recent work also indicates that many of these scales, even those that are seemingly distinct, are measuring the same construct, and that there is substantial overlap with mental health scales (Davidson et al., 2020). Further, Scharkow (2019) claimed that these issues are aggravated if a researcher is interested in within-person effects because of the "moderate reliability and high stability" (p. 207) of responses, making it difficult to uncover behavioral nuances that may be driving outcomes.

Cross-Sectional Study Designs Dominate the Field

Since researchers often have limited resources to spend on any given study, they often face trade-offs when considering how to best allocate these resources. Cattell's (1952) "data box" heuristic provides an overview of three crucial choices; few researchers can collect data about (a) many variables from (b) many people over (c) many time points. Until recently, most scholars of social media, and scholars of human behavior in general, have focused their resources on *inter*individual analyses—or the "variables" and "persons" dimensions of the data box (Molenaar & Campbell, 2009). Our meta-analysis indicated that over 75% of studies examining social media and well-being used cross-sectional methods. And even though longitudinal studies typically neglect variables for the sake of additional temporal information, they also tend to focus on interindividual analyses rather than *intra*individual ones. Further, there has been a lack of experimental studies in this research area as well, with only 2% of studies in our meta-analysis employing this method. All of these choices can be justified, and all have various advantages and disadvantages, but they necessarily limit the types of questions a researcher can ask and the resulting claims they can make.

Insufficient Attention to Questionnaire Design

Another challenge researchers face is how participants will interpret the questions and response options within the survey. Table 6.2 summarizes

Table 6.2

Summary of Methodological Issues Observed in the Literature

Methodological issue	Examples	Potential solutions
Overreliance on time spent as a proxy for social media use	28% of the effect sizes in our meta-analysis used numerical frequency or duration-based measures of social media use.	Emphasize other aspects of use with more predictive power, like content and network connections, and beliefs about social media
	Most subjective measures also asked about time spent (e.g., "a lot").	
	Reliance on general single-item measures	
Lack of adequately robust social media use scales	Modifying other media use scales without appropriate validation	Focus on aspects and/or affordances exclusive to social media
	Overlap with mental health scales	Gather log data for validation when possible
Cross-sectional study designs dominate the field	75% of studies in our meta-analysis used cross-sectional study designs.	Integrate longitudinal and experimental study designs into research plan to better understand causality and directionality of relationships
Insufficient attention to questionnaire design	Lack of adherence to best practices for question wording and question order	Ask more specific questions (e.g., about different devices)
		Critically consider the impact of effects such as priming that may bias results

methodological issues in previous research. There has been a great deal of work from scholars in the field of psychometrics or survey design discussing the advantages and disadvantages of drafting questionnaires in different ways (e.g., Tourangeau et al., 2000), but best practices are not commonly employed in social media studies or in media studies in general. Junco (2013) contended that a prevalent issue in social media use measurement is the lack of specificity in question wording, which could impact the recall and estimation portions of the response process. Scholars of both traditional and digital media have also emphasized the distinction between passive and active use (Verduyn et al., 2015). Recent work has attempted to parse these distinctions through studying battery usage and screen time separately using log data (Hodes & Thomas, 2021), but this may not be theoretically sufficient, as people may be passive "lurkers" yet deeply engaged in the content they're viewing. Likewise, people may seem like active "clickers" but in reality are indiscriminately engaging with all content (Ellison et al., 2020).

In addition to biases in question wording, biases might also arise from the order in which questions are presented. A common issue in survey design is *priming*, which refers to the implicit memory effects of exposure to prior stimuli on responses to later stimuli (Cesario, 2014). In a survey, this might mean that the order of questions on a survey influences responses (Tourangeau & Rasinski, 1988). Priming effects can alter responses to a wide variety of questions, including those about attitudes, behavior, demographic information, and well-being. For example, Fox and Kahneman (1992) noted that questions about one's dating life affected responses to a later question about life satisfaction, but that answering a question about life satisfaction first did not impact responses to questions about dating life. Similar effects have also been found in studies about current mood and more general subjective well-being (Diener, 1994).

Recent research indicates that this phenomenon applies to the domain of social media use and well-being, as well (Mieczkowski et al., 2020). Over the course of two studies, the authors investigated the relationship between two common social media use scale types (addiction and intensity) and self-reported depression. Results suggested that answering questions about social media use before questions about depression could alter responses to a well-validated, stable depression scale (the Beck Depression Inventory; Beck et al., 1996). However, they did not find evidence for the reverse—participants who answered the depression scale first did not report significant differences in depression, social media addiction, social media intensity, self-reported social media use, or logged screen time. Based on prior work on well-being (e.g., Pavot & Diener, 1993), it is possible that these priming effects could affect responses beyond self-reported depression, especially in the cases of more volatile well-being measures (e.g., state measurements of anxiety; Spielberger, 2010).

Although it is difficult statistically, financially, and cognitively to employ high-quality research methods when studying social media (Kobayashi & Boase, 2012), it is imperative that researchers make use of the resources they do have, so that academics and social media users alike have a better understanding of this technology. In the next section, we propose a number of methodological approaches, as well as potential concepts and populations, that social media and well-being scholars would benefit from integrating into their research.

MOVING THE FIELD FORWARD

Our review makes it clear that there are major challenges facing researchers working on social media and well-being. However, we are optimistic that these challenges can be overcome as the field matures. In particular, we see a need for novel methods for measuring social media use that go beyond time spent or frequency self-report measures. It is always painful to read a carefully designed and executed study that meticulously assessed the well-being dependent variable (e.g., depression, life satisfaction) with a validated and reliable scale but then measured social media use with a self-reported single item about social media (e.g., "On average, how much do you use social media per day?"). A crucial advance for the field will be the development and validation of new methods and measures around social media use. A second area that requires attention is the development of new frameworks and theories outlining how social media might be related to well-being. There are numerous studies that test the impact of passive versus active use or addictive versus nonaddictive use. However, this narrow focus on a handful of mechanisms seems inadequate and undertheorized when considering the broad variety of potential pathways through which social media use might influence our well-being. Next, we lay out some new methodological approaches and novel theoretical concepts that can help researchers in the field to rethink social media's influence on our well-being.

A Need for New Methods

As the previous section makes clear, the field to date faces serious methodological limitations. There is no question that measuring media use is difficult. The habitual nature of digital media and its integration into people's daily lives means that participants in research often have a hard time answering questions about their use of social media (Ellis et al., 2019). We need to reconsider whether these common methods (e.g., cross-sectional, self-report studies) are appropriate for the research questions at hand. Slater (2004) argued that since media effects, by definition, happen over a period of time, it is necessary for researchers to conduct longitudinal and experimental studies.

A great example is the recent work by Allcott et al. (2020) that incorporated both longitudinal and field experimental strategies when they examined the impact of deactivating Facebook on subjective well-being over the course of several weeks. The authors were able to compare behavior and well-being across conditions and over time, providing a more comprehensive picture of the effects of social media on well-being. There are a number of ways that researchers can conduct studies longitudinally and/or experimentally. Simply asking questions via in situ experience sampling methods could reduce recall issues (Cohen & Lemish, 2003), even without the use of log-based measures.

For those interested in investigating social media use over time and with a high level of granularity, a framework like Screenomics (Reeves et al., 2021) is promising. Screenomics is an application that collects screenshots of a participant's device every 5 seconds. Studies using the Screenomics framework have demonstrated that individuals often engage in a wide variety of activities, even over the course of a relatively short time scale (e.g., only a few minutes). An understudied aspect of social media use—the content someone views and interacts with on a platform—is easier to observe and quantify with this research design strategy. C.-j. Lee et al. (2008) noted that examining "exposure" through concepts like screen time "will be of limited use if the content of Internet exposure is poorly understood" (p. 19). Without understanding *what* people are looking at, it is difficult to make theoretical claims regarding the relationships between social media and well-being, or the mechanisms that underlie them.

To gain additional control over these variables, researchers could also consider simulating social media use with tools like Social Media TestDrive (DiFranzo et al., 2019). The tool creates a realistic experience on social media by providing an interface with common features such as a profile and news feed, as well as interactions (e.g., posts, comments) from others. Social Media TestDrive relies on preprogrammed bots to create the simulated experience and not real people, allowing the researchers to manipulate content in which they are interested in a controlled way that can easily be replicated across participants. For example, a researcher could study the effects of positive or negative reactions to a post in real time, under conditions that were already specified and can be held constant.

A Need for Novel Theories

In addition to developing new methods for moving the field forward, we also believe that some new and exciting theoretical directions have the promise of advancing the field beyond the oversimplified question of whether social media use is good or bad for us. Here are some of the concepts that we think have the potential to reshape the field.

Toward a More Functional Approach

In recent years there has been a growing interest in various contexts where social media use is associated with well-being. An active research line is to understand how different types of use and content moderate the association between use and well-being. Verduyn et al. (2015) provided a carefully and explicitly articulated analysis of use, which argues that passive use triggers social comparison and negative well-being outcomes while active use enhances social capital and improves well-being. Examination of use to date also includes but is not limited to addictive versus non—addictive, public versus private (e.g., Frison & Eggermont, 2015), high versus low effort use (e.g., status update vs. one-click; Burke & Kraut, 2016), image based versus text based (Pittman & Reich, 2016), and ephemeral versus persistent (Bayer et al., 2016).

Given a wide range of features and affordances of social media, the conceptualization and operationalization of social media use continue to grow and become more complex. One of our central arguments after reviewing this field is that researchers need to take a more functional approach to understanding people's use of social media that goes beyond time spent or frequency—a perspective of examining what people are trying to accomplish when they use social media, what goals and needs are being met, along with what kind of content they are engaging with and with whom. For example, self-presentation is a major social media use behavior and a core mechanism that links use to well-being, and authenticity is an important dimension of self-presentation. As shown in prior research, people felt more positive (or negative) after sharing positive (or negative) personal events (Choi & Toma, 2014). Authentic self-presentation is shown to correspond to better well-being outcomes, including higher self-esteem (Yang et al., 2017), life satisfaction, positive affect, and mood (Bailey et al., 2020), whereas unauthentic self-presentation can make people feel less socially connected, less satisfied with life (Bailey et al., 2020), and more stressful (Grieve & Watkinson, 2016). Similarly, the psychological implications of social sharing may depend on the valence of the shared content, such that people felt more positive (or negative) after sharing positive (or negative) personal events (Choi & Toma, 2014). We need more of this kind of work that looks at how people use social media and connects that use theoretically to well-being outcomes.

Taking Individual Differences Seriously

Individual-level characteristics also play a significant role in the association between social media use and well-being. A nice example is related to the classic debate, ongoing since the early days of social media research, over who benefits the most from social media use. Specifically, the "rich get richer" hypothesis argues that socially rich people would acquire more social capital by using social media as an additional avenue to leverage existing social resources offline. "Poor get richer" argues that using social media may compensate for some people's deficient offline social capital. Novel findings from a recent meta-analysis (Cheng et al., 2019) suggested that while both extroverts and socially anxious people tend to engage in a higher level of social media, only extroverts can reap the benefits of greater online social capital, bearing out the "rich-get-richer" hypothesis. Along with this rationale, personality traits can not only directly affect use but also moderate the use and well-being outcomes. Future research should continue examining how individual differences affect (a) how much and what types of social media to use, and (b) how such use affects well-being outcomes differently.

Rethinking Social Media and Addiction

One of the most common concerns in the popular discourse about social media is its potentially addictive nature. As the prevalence of social media use has increased worldwide, so have the claims associating it with addiction. A great deal of research has focused on connections between social media addiction and a number of unfortunate outcomes, such as poor academic achievement, cognitive impairments, and mental health problems. According to a study by Statistics Netherlands (2018), an estimated 30% of social media users in that country suffer from social media addiction—a massive increase from the 3% global prevalence of substance use disorders (World Health Organization, 2010). If this claim is accurate and generalizable, then a problem of such magnitude certainly warrants the amount of scientific investigations being conducted, in addition to collaborations with government and healthcare officials alike.

Unfortunately, it is not clear whether social media addiction is as prolific as some scholars argue. In fact, there might not be evidence to suggest that social media addiction—defined using concepts from substance use and behavioral addictions—is even the underlying problem at all. Although the original article search for our meta-analysis showed that there were thousands of scholarly articles examining social media addiction, many lack appropriate theoretical and methodological frameworks to discover whether or not an individual is suffering from social media addiction, much less a robust relationship between social media addiction and any sort of negative outcomes.

If social media addiction is a disorder of the same type as substance use disorders and other behavioral addictions, then it may be symptomatic in similar ways. The symptoms typically do not appear all at once—instead, addiction is often considered a "process" during which an individual has distinct motives to pursue certain effects. Sometimes the individual loses and regains these motives, which is why the process is often thought of as cyclical. Addiction scholars have highlighted a number of well-accepted and common symptoms, including mood modification, salience, tolerance, withdrawal, negative consequences ("conflict") and relapse (Griffiths, 2005). Furthermore, addiction is typically characterized by feelings of "loss of control" (Sussman & Sussman, 2011).

Unlike in clinical psychological work, there are numerous psychometric assessments used to diagnose social media addiction, as well as its aliases (see Mieczkowski et al., 2020, for examples). However, in the same vein as other addiction scholars, social media researchers have focused their efforts on diagnosing social media addiction with most of the aforementioned symptomatic criteria. One of the most common addiction scales is the Bergen Facebook Addiction scale (Andreassen et al., 2012). This scale focuses on salience ("How often during the last year . . . spent a lot of time thinking about Facebook or planned use of Facebook?"), tolerance (". . . felt an urge to use Facebook more and more?"), mood modification (". . . used Facebook in order to forget about personal problems?"), relapse (". . . tried to cut down on the use of Facebook without success?"), withdrawal (". . . become restless or troubled if you have been prohibited from using Facebook?") and conflict (". . . used Facebook so much that it has had a negative impact on your job/studies?").

However, most studies on social media addiction have relied solely on cross-sectional, self-report measures of addiction, so the potential *causal* relationships between variables such as addictive symptoms and negative outcomes cannot be unearthed. Panova and Carbonell (2018) argued that there are six main concerns regarding the current body of research on smartphone addiction, which could be applicable to social media as well: (a) "a lack of longitudinal studies" that would allow researchers to observe the cyclical nature of the addictive process; (b) invalid "screening instruments" for diagnosis; (c) "a large probability of false positives"; (d) "arbitrarily designed" questionnaires; (e) an overreliance on "self-report data, which are collected using convenience samples"; and (f) major inconsistencies in "methodology, definitions, measurement, cut-off scores, and diagnostic criteria across studies." When considering these issues, it is not clear that social media addiction exists in the form that scholars are currently studying it. In order to shed light on social media addiction and remedy the theoretical and methodological problems within this area of research, we argue that scholars need to reconceptualize the current nature of social media addiction.

Perhaps one of the biggest challenges social media addiction researchers face is the current diagnostic criteria. Imagine that a participant responds to a social media addiction questionnaire and indicates that they used social media to lessen feelings of anxiety, attempted to use social media for fewer hours per day but failed, and felt distress when they could not use social media for long periods of time. With the current social media addiction diagnostic criteria, a researcher would likely assume this participant is addicted to social media because they show symptoms of mood modification, loss of control, and withdrawal. Yet these behaviors might be indicative of other issues or even productive functioning in this participant's life. They may feel anxious due to their home environment and use social media to connect with friends. They may have received a recent promotion at work that requires them to spend more time on social media coordinating with colleagues. They may be waiting for an important message from a relative that they don't want to miss.

In 2018, only 55.3% of the adult population in the United States reported drinking in the past month, compared with upwards of 70% of adults who use social media (Perrin & Anderson, 2019; U.S. Department of Health and Human Services, 2019). Unlike alcohol or drug use, social media use is tightly integrated into many aspects of life for a majority of the population. Additionally, due to the private nature of social media use, it is difficult for people to make accurate assumptions about what activities someone might be engaging in. Furthermore, in the United States, there are no legal restrictions on social media use.

As such, measuring social media addiction with the same self-reported diagnostic criteria as other addictions proves erroneous at best and completely misrepresentative at worst. Even the presumably "simple" measurement of time spent on social media as a way to infer tolerance has numerous confounds (King et al., 2018). People suffering from addiction often experience "time distortion," meaning that they could not provide accurate reports of time spent engaging in the addictive behavior even if they were motivated to (Hirschman, 1992; Lin et al., 2015). Even objective measures of time spent on social media are typically not as explanatory as frequency-based measurements, such as how often a user checks their notifications (H. Lee et al., 2014).

Measuring other common symptoms, such as mood modification, may be equally challenging, as social media use has been shown to affect mood both positively and negatively (Mark et al., 2014; Sagioglou & Greitemeyer, 2014). If both "addicted" and "nonaddicted" social media users report mood modification symptoms, the measurement has low discriminant validity, which provides little helpful information for the researcher. Additionally, symptoms of withdrawal in the form of an individual feeling distressed when they cannot use social media may not be indicative of addiction considering the "functional dependence" most of the general public has to digital technology (Parent & Shapka, 2020, p. 183). Without social media, many people would have to "restructure[e] and adapt . . . regular activities" (Panova & Carbonell, 2018, p. 254), an inherently stressful process.

The role of social media in daily life is both qualitatively and quantitatively different from the role of alcohol, drugs, or other behavioral addictions. As such, new theoretical and methodological avenues are necessary to understand the components of social media addiction that do not pathologize everyday behaviors for billions of people.

Considering Mindsets

Quantifying the amount and the types of social media use has been the primary approach to explore the relationship between social media use and well-being. Yet, our meta-analysis indicated that this approach has limitations and that the way people think about social media may matter (e.g., "it's addictive") more than how much they actually use it. We have therefore started to focus on social media mindsets, which refer to the beliefs, expectations, and feelings users have about social media use (A. Y. Lee et al., 2021).

Mindsets are mental frames that selectively organize and encode information, orienting an individual toward a unique way of understanding an experience and guiding one's actions and responses (Dweck, 2008). Mindsets provide mental shortcuts by shaping people's attention, behavior, cognition, and expectations. Previous research has found that people with more adaptive mindsets about intelligence, stress, or even illness have better physiological and mental health than those with maladaptive mindsets (Claro et al., 2016; Crum et al., 2013; Yeager & Dweck, 2012).

Our recent work suggests that there are mindsets for social media, as well. In one preliminary study (A. Y. Lee et al., 2021), we found evidence for two types of mindsets about social media use. In one mindset, people have control of social media use, and they view it positively. People with this kind of "social media as tool" mindset believe that social media is beneficial for them and serves a useful, meaningful purpose in their lives. In the second mindset, people feel that social media controls them, and they have an unsurprisingly negative view of social media. People holding this kind of "social media as addicting" mindset believe that social media is harmful and addictive.

Our initial work suggests that these mindsets are powerful. In one study, we found that social media mindsets can mediate the relationship between well-being outcomes, social media use measures, and depression (Mieczkowski et al., 2020). In another study, people with a tool mindset have positive well-being outcomes regarding social support, depression, anxiety, and stress, while people with an addiction mindset have negative outcomes concerning these well-being dimensions (A. Y. Lee et al., 2021). These studies suggest that perception of use has the potential to be more influential than use itself (Boase & Ling, 2013). We argue here that measuring mindsets around social media perceptions is an important area for future research.

Mechanisms of Blame

After discovering that effect sizes between social media use and well-being tend to be very small, we started wondering why the public discourse is so heavily based on the suggestion that social media can be harmful. One potential explanation is that people may consider the impact of the social media use of *other people* on their own well-being. Put differently, a person's social media use might diminish another person's well-being because high levels of social media use may detract from their relationship or social interaction. Imagine, for example, being ignored by a friend while they check their phone. Indeed, one study found that a partner's phone use was associated with increased depression in married couples (Wang et al., 2017). This indirect association between a person's social media use and another person's well-being is caused by ostracism, defined as excluding and ignoring by individuals or groups (Williams, 2009).

Previous studies have focused on *phubbing*, the behavior of one person ignoring another person due to phone use. Individuals are shut out of social interaction while remaining in the physical presence of other people (Chotpitayasunondh & Douglas, 2018). This type of phone-related social ostracism is associated with a decreased social connection (Kushlev & Heintzelman, 2018), reduced sense of mattering (Kadylak, 2020) and sense of belonging (Hales et al., 2018), and increased distress and depression (Gonzales & Wu, 2016; Wang et al., 2020).

Social media use can induce ostracism, as well. People seem to infer social media use in a phubbing scenario. We conducted a study and asked participants what they thought the other person was doing when using the phone. We gave the five most popular uses of the phone as options and found that almost two thirds of participants believed that the other person was checking social media (Liu & Hancock, 2020). This automatic inference of phone use to social media use in a phubbing scenario can not only undermine people's well-being if they are frequently being ostracized but also magnify people's concerns and resentment of social media.

How people interpret social technology-related ostracism has implications for well-being. Attribution theory (Kelley & Michela, 1980) provides a valuable framework to explore how people perceive these types of social ostracism and how they infer causes based on their self-interest and motivations. Classic attribution theory posits a person (self) versus situation (external) classification of causes. In general, people attribute positive outcomes to themselves while deflecting blame and attributing negative consequences to situations (Kelley & Michela, 1980).

There are multiple possible sources of external attributions for social technology-related ostracism, ranging from blaming the other person, to the context, to the technology. Other work examining attributions regarding technology in social dynamics have applied three types of external attributions: dispositional attribution, situational attribution, and interpersonal attribution. When people attribute actions to the partner's personality and disposition, it is a dispositional attribution. When people believe that context and environmental factors impact the action, it is a situational attribution. When people think that the nature or characteristics of the relationship affect the action, it is an interpersonal disposition (Jiang et al., 2011; Walther et al., 2016).

We developed and measured a new source of external attribution: the technology itself. We applied attribution theory and conducted five experimental studies to examine how people perceive phone-related ostracism and the degree to which people blame the person doing the ostracizing in a phubbing scene or whether they blame the phone (S. X. Liu & Hancock, 2020). We found that across various conditions, people overattribute to the phone, and phone attribution was a significant predictor of beliefs about the negative consequences of phone use, beliefs of the addictive framing of the phone, and support for strict phone use regulations. However, blaming the phone may reduce psychological harms to self-esteem and produce well-being benefits. It hurts less to blame the phone than to blame the self, the partner, or the relationship between the self and the partner. Attribution theory provides a useful guide to explore the mechanisms linking social technology use, attributional dynamics, personal well-being, and societal consequences of social technology use.

Highlighting the Role of Culture and Different Populations

Although culture is likely to have an impact on the relationship between social media and well-being, few studies to date have examined this influence directly. This is especially important given that social media platforms are common across cultures and countries as tech companies such as Facebook have aggressively expanded their international markets. Although social media platforms and tools differ somewhat across cultures, the main affordances and features of social media are similar. Our meta-analysis hints at important cultural differences in our findings across geographic locations. While geographic location is not isomorphic with culture, it can serve as a very rough proxy. Studies conducted in Asia reported an overall positive relationship between social media use and well-being, whereas studies conducted in Europe showed the opposite relationship. This pattern suggests that participants in Asia have a more positive experience and perspective regarding social media and well-being than Europeans, who appear to have a much more negative experience and perspective.

More research into how different cultures experience social media and well-being is needed. People in different cultures domesticate social media platforms in culturally unique ways. Well-being also varies across culture. Psychologists have been debating cultural questions for decades, such as whether well-being is composed of the same or of different components across cultures. Previous research has accumulated abundant evidence that well-being differs across cultures regarding its causes, components, and effects (Diener et al., 2017). Besides the differences in how people use social media and how well-being varies across cultures, the mechanisms that link social media use and well-being may also be distinct across cultures. For example, social capital gathered on social media can be very different in collectivistic cultures than in individualistic cultures. Delineating cultural similarities and cultural differences at both the individual and the cultural level is a key stride in social media and well-being research.

Finally, given how widespread the concerns are about social media, and how widespread its use is, another issue is the lack of diversity represented in addressing the question of social media and well-being. The points of view in *The Social Dilemma*, for example, are limited to mostly White, mostly young, former technology workers providing their insights about how and why social media technologies were developed. This perspective, while valuable for understanding the design of these systems, omits many other perspectives, with very little representation of people of color, people from different socioeconomic classes, or people from different cultures. We need more research with more diversity of perspectives. For example, although seniors, usually defined as adults age 55 and above, are now one of the fastest growing and largest populations using social media, there were too few studies for us to examine the relationship in this particular population. How social media use affects well-being for older adults is hence a pressing and important avenue for future research.

CONCLUSION

In *The Social Dilemma*, all social media use is considered equal, and the more people use it, the more harmful it is. Our review of the science and the substantial field of evidence collected over more than a decade suggests that the relationship between social media use and well-being is far less sinister and much more complicated. Rather than the simplistic "more is worse" heuristic, our review reveals that it matters how we use social media, that the effects are quite small, varied, and most likely not a simple causal effect. People have agency. They use social media to accomplish goals and fulfill needs.

One of our high-level conclusions after this massive undertaking is that *psychology beats technology* when it comes to understanding how social media use is related to well-being. Although media and technology researchers tend to emphasize the effects of technology, it is clear from our review that psychological dynamics matter more. What matters is how we use social media, in which context, and for which purposes. For example, using social media to connect with an old friend is very different from mindlessly scrolling through a social media feed. To accomplish a more nuanced perspective on the relationship between social media and well-being, we must move beyond simple self-report of time spent or frequency of using social media towards a more functional approach to social media use that takes into considerations a person's goals, motivations, and needs. A second implication is that we need new concepts and theories that actually connect the specific use of social media to well-being outcomes. Assuming that social comparison or social compensation are playing a role is no longer sufficient; instead, research needs to articulate these connections, measure them, and test them in order for the field to move forward.

Finally, we need to take seriously the populations we are studying. Different populations have radically different experiences on social media. When we move away from commonly used college student samples or WEIRD (White, educated, industrialized, rich, and democratic) populations, we see nuanced dynamics between social media use and well-being. Even within WEIRD populations, the well-being of teenagers versus seniors are radically different moving through their developmental trajectory (Boyd, 2014; Hargittai & Dobransky, 2017). With different needs, goals, and motivations, the role of social media in a teenager's well-being is likely to be very different from that of a retired person. Our field can gain fundamental theoretical and practical insights by paying attention to underrepresented populations, such as Black or Latino communities, rather than assuming White populations as the default (e.g., Bennett et al., 2012; Brock, 2020; Patton et al., 2019). Although digital literacy and phone use vary significantly across populations, the wide adoption of smartphones across the globe provides an opportunity to study hard-to-reach groups that are underrepresented in traditional media studies. Our field needs to conduct research with these populations.

Finally, social media use is not a dosage, and interactions on social media are not medical procedures. The faster we depart from an oversimplified *Social Dilemma* style of social media addiction, the sooner we can discover interventions and inform public policies that can promote well-being across diverse populations.

REFERENCES

Ahn, D., & Shin, D. H. (2013). Is the social use of media for seeking connectedness or for avoiding social isolation? Mechanisms underlying media use and subjective well-being. *Computers in Human Behavior, 29*(6), 2453–2462. https://doi.org/10.1016/j.chb.2012.12.022

Allcott, H., Braghieri, L., Eichmeyer, S., & Gentzkow, M. (2020). The welfare effects of social media. *American Economic Review, 110*(3), 629–676. https://doi.org/10.1257/aer.20190658

Andreassen, C. S., Torsheim, T., Brunborg, G. S., & Pallesen, S. (2012). Development of a Facebook addiction scale. *Psychological Reports*, *110*(2), 501–517. https://doi.org/10.2466/02.09.18.PR0.110.2.501-517

Araujo, T., Wonneberger, A., Neijens, P., & de Vreese, C. (2017). How much time do you spend online? Understanding and improving the accuracy of self-reported measures of Internet use. *Communication Methods and Measures*, *11*(3), 173–190. https://doi.org/10.1080/19312458.2017.1317337

Bailey, E. R., Matz, S. C., Youyou, W., & Iyengar, S. S. (2020). Authentic self-expression on social media is associated with greater subjective well-being. *Nature Communications*, *11*(1), 4889. https://doi.org/10.1038/s41467-020-18539-w

Bandura, A., & Jourden, F. J. (1991). Self-regulatory mechanisms governing the impact of social comparison on complex decision making. *Journal of Personality and Social Psychology*, *60*(6), 941–951. https://doi.org/10.1037/0022-3514.60.6.941

Bayer, J. B., Ellison, N. B., Schoenebeck, S. Y., & Falk, E. B. (2016). Sharing the small moments: Ephemeral social interaction on Snapchat. *Information Communication and Society*, *19*(7), 956–977. https://doi.org/10.1080/1369118X.2015.1084349

Beck, A. T., Steer, R. A., & Brown, G. K. (1996). *Manual for the Beck Depression Inventory-II*. Psychological Corporation.

Bennett, L., Freelon, D. G., Hussain, M., & Wells, C. (2012). Digital media and youth engagement. In H. A. Semetko & M. Scammell (Eds.), *The SAGE handbook of political communication* (pp. 127–140).

Best, P., Taylor, B., & Manktelow, R. (2015). I've 500 friends, but who are my mates? Investigating the influence of online friend networks on adolescent wellbeing. *Journal of Public Mental Health*, *14*(3), 135–148. https://doi.org/10.1108/JPMH-05-2014-0022

Boase, J., & Ling, R. (2013). Measuring mobile phone use: Self-report versus log data. *Journal of Computer-Mediated Communication*, *18*(4), 508–519. https://doi.org/10.1111/jcc4.12021

Boyd, D. (2014). *It's complicated: The social lives of networked teens*. Yale University Press.

Brock, A., Jr. (2020). *Distributed Blackness: African American cybercultures*. New York University Press. https://doi.org/10.18574/nyu/9781479820375.001.0001

Brooks, B., Hogan, B., Ellison, N., Lampe, C., & Vitak, J. (2014). Assessing structural correlates to social capital in Facebook ego networks. *Social Networks*, *38*, 1–15. https://doi.org/10.1016/j.socnet.2014.01.002

Buglass, S. L., Binder, J. F., Betts, L. R., & Underwood, J. D. (2017). Motivators of online vulnerability: The impact of social network site use and FOMO.

Computers in Human Behavior, 66, 248–255. https://doi.org/10.1016/j.chb.2016.09.055

Burke, M., & Kraut, R. E. (2016). The relationship between Facebook use and well-being depends on communication type and tie strength. *Journal of Computer-Mediated Communication, 21*(4), 265–281. https://doi.org/10.1111/jcc4.12162

Burrow, A. L., & Rainone, N. (2017). How many likes did I get? Purpose moderates links between positive social media feedback and self-esteem. *Journal of Experimental Social Psychology, 69*, 232–236. https://doi.org/10.1016/j.jesp.2016.09.005

Cattell, R. B. (1952). The three basic factor-analytic research designs—their interrelations and derivatives. *Psychological Bulletin, 49*(5), 499–520. https://doi.org/10.1037/h0054245

Cesario, J. (2014). Priming, replication, and the hardest science. *Perspectives on Psychological Science, 9*(1), 40–48. https://doi.org/10.1177/1745691613513470

Chang, P. F., Choi, Y. H., Bazarova, N. N., & Löckenhoff, C. E. (2015). Age differences in online social networking: Extending socioemotional selectivity theory to social network sites. *Journal of Broadcasting & Electronic Media, 59*(2), 221–239. https://doi.org/10.1080/08838151.2015.1029126

Cheng, C., Wang, H.-y., Sigerson, L., & Chau, C.-l. (2019). Do the socially rich get richer? A nuanced perspective on social network site use and online social capital accrual. *Psychological Bulletin, 145*(7), 734–764. https://doi.org/10.1037/bul0000198

Chiou, W. B., Lee, C. C., & Liao, D. C. (2015). Facebook effects on social distress: Priming with online social networking thoughts can alter the perceived distress due to social exclusion. *Computers in Human Behavior, 49*, 230–236. https://doi.org/10.1016/j.chb.2015.02.064

Choi, J., & Kim, H. J. (2016). Influence of SNS user innovativeness and public individuation on SNS usage patterns and social capital development: The case of Facebook. *International Journal of Human–Computer Interaction, 32*(12), 921–930. https://doi.org/10.1080/10447318.2016.1220067

Choi, M., & Toma, C. L. (2014). Social sharing through interpersonal media: Patterns and effects on emotional well-being. *Computers in Human Behavior, 36*, 530–541. https://doi.org/10.1016/j.chb.2014.04.026

Chotpitayasunondh, V., & Douglas, K. M. (2018). The effects of "phubbing" on social interaction. *Journal of Applied Social Psychology, 48*(6), 304–316. https://doi.org/10.1111/jasp.12506

Claro, S., Paunesku, D., & Dweck, C. S. (2016). Growth mindset tempers the effects of poverty on academic achievement. *Proceedings of the National Academy of Sciences, 113*(31), 8664–8668. https://doi.org/10.1073/pnas.1608207113

Cohen, A. A., & Lemish, D. (2003). Real time and recall measures of mobile phone use: Some methodological concerns and empirical applications. *New Media & Society, 5*(2), 167–183. https://doi.org/10.1177/1461444803005002002

Cohen, S., & Hoberman, H. M. (1983). Positive events and social supports as buffers of life change stress. *Journal of Applied Social Psychology, 13*(2), 99–125. https://doi.org/10.1111/j.1559-1816.1983.tb02325.x

Crum, A. J., Salovey, P., & Anchor, S. (2013). Rethinking stress: The role of mindsets in determining the stress response. *Journal of Personality and Social Psychology, 104*(4), 716–733. https://doi.org/10.1037/a0031201

Davidson, B. I., Shaw, H., & Ellis, D. A. (2020). *Fuzzy constructs: The overlap between mental health and technology "use."* PsyArXiv. https://doi.org/10.31234/osf.io/6durk

Deci, E., & Ryan, R. M. (1985). *Intrinsic motivation and self-determination in human behavior.* Springer. https://doi.org/10.1007/978-1-4899-2271-7

Deters, F. G., & Mehl, M. R. (2013). Does posting Facebook status updates increase or decrease loneliness? An online social networking experiment. *Social Psychological and Personality Science, 4*(5), 579–586. https://doi.org/10.1177/1948550612469233

Diener, E. (1994). Assessing subjective well-being: Progress and opportunities. *Social Indicators Research, 31*(2), 103–157. https://doi.org/10.1007/BF01207052

Diener, E., Heintzelman, S. J., Kushlev, K., Tay, L., Wirtz, D., Lutes, L. D., & Oishi, S. (2017). Findings all psychologists should know from the new science on subjective well-being. *Canadian Psychology/Psychologie Canadienne, 58*(2), 87–104. https://doi.org/10.1037/cap0000063

Dienlin, T., Masur, P. K., & Trepte, S. (2017). Reinforcement or displacement? The reciprocity of FTF, IM, and SNS communication and their effects on loneliness and life satisfaction. *Journal of Computer-Mediated Communication, 22*(2), 71–87. https://doi.org/10.1111/jcc4.12183

DiFranzo, D., Choi, Y. H., Purington, A., Taft, J. G., Whitlock, J., & Bazarova, N. N. (2019). Social media testdrive: Real-world social media education for the next generation. In *CHI'19: Proceedings of the 2019 CHI Conference on Human Factors in Computing Systems* (pp. 1–11). https://doi.org/10.1145/3290605.3300533

Domahidi, E. (2018). The associations between online media use and users' perceived social resources: A meta-analysis. *Journal of Computer-Mediated Communication, 23*(4), 181–200. https://doi.org/10.1093/jcmc/zmy007

Dweck, C. (2008). *Mindsets and math/science achievement* [Report]. The Opportunity Equation, Carnegie Corporation of New York.

Ellis, D. A. (2019). Are smartphones really that bad? Improving the psychological measurement of technology-related behaviors. *Computers in Human Behavior, 97*, 60–66. https://doi.org/10.1016/j.chb.2019.03.006

Ellis, D. A., Davidson, B. I., Shaw, H., & Geyer, K. (2019). Do smartphone usage scales predict behavior? *International Journal of Human–Computer Studies, 130*, 86–92. https://doi.org/10.1016/j.ijhcs.2019.05.004

Ellison, N. B., Triệu, P., Schoenebeck, S., Brewer, R., & Israni, A. (2020). Why we don't click: Interrogating the relationship between viewing and clicking in social media contexts by exploring the "non-click." *Journal of Computer-Mediated Communication, 25*(6), 402–426. https://doi.org/10.1093/jcmc/zmaa013

Feshbach, S., & Singer, R. D. (1971). *Television and aggression*. Jossey-Bass.

Festinger, L. (1954). A theory of social comparison processes. *Human Relations, 7*(2), 117–140. https://doi.org/10.1177/001872675400700202

Fox, C. R., & Kahneman, D. (1992). Correlations, causes and heuristics in surveys of life satisfaction. *Social Indicators Research, 27*(3), 221–234. https://doi.org/10.1007/BF00300462

Frison, E., & Eggermont, S. (2015). The impact of daily stress on adolescents' depressed mood: The role of social support seeking through Facebook. *Computers in Human Behavior, 44*, 315–325. https://doi.org/10.1016/j.chb.2014.11.070

Frison, E., & Eggermont, S. (2016). Exploring the relationships between different types of Facebook use, perceived online social support, and adolescents' depressed mood. *Social Science Computer Review, 34*(2), 153–171. https://doi.org/10.1177/0894439314567449

Gonzales, A. L., & Wu, Y. (2016). Public cellphone use does not activate negative responses in others . . . Unless they hate cellphones. *Journal of Computer-Mediated Communication, 21*(5), 384–398. https://doi.org/10.1111/jcc4.12174

Grieve, R., Indian, M., Witteveen, K., Tolan, G. A., & Marrington, J. (2013). Face-to-face or Facebook: Can social connectedness be derived online? *Computers in Human Behavior, 29*(3), 604–609. https://doi.org/10.1016/j.chb.2012.11.017

Grieve, R., & Watkinson, J. (2016). The psychological benefits of being authentic on Facebook. *Cyberpsychology, Behavior, and Social Networking, 19*(7), 420–425. https://doi.org/10.1089/cyber.2016.0010

Griffiths, M. (2005). A "components" model of addiction within a biopsychosocial framework. *Journal of Substance Use, 10*(4), 191–197. https://doi.org/10.1080/14659890500114359

Hales, A. H., Dvir, M., Wesselmann, E. D., Kruger, D. J., & Finkenauer, C. (2018). Cell phone-induced ostracism threatens fundamental needs. *The Journal of Social Psychology, 158*(4), 460–473. https://doi.org/10.1080/00224545.2018.1439877

Hamilton, J. T. (Ed.). (2000). *Television violence and public policy*. University of Michigan Press. https://doi.org/10.3998/mpub.15632 (Original work published 1998)

Hancock, J. T., Liu, S. X., French, M., Luo, M., & Mieczkowski, H. (2019). *Social media use and well-being: A meta-analysis* [Paper presentation]. 69th Annual International Communication Association Conference, Washington, DC, United States.

Hargittai, E., & Dobransky, K. (2017). Old dogs, new clicks: Digital inequality in skills and uses among older adults. *Canadian Journal of Communication*, 42(2), 195–212. https://doi.org/10.22230/cjc.2017v42n2a3176

Heffer, T., Good, M., Daly, O., MacDonell, E., & Willoughby, T. (2019). The longitudinal association between social-media use and depressive symptoms among adolescents and young adults: An empirical reply to Twenge et al. (2018). *Clinical Psychological Science*, 7(3), 462–470. https://doi.org/10.1177/2167702618812727

Hirschman, E. C. (1992). The consciousness of addiction: Toward a general theory of compulsive consumption. *The Journal of Consumer Research*, 19(2), 155–179. https://doi.org/10.1086/209294

Hodes, L. N., & Thomas, K. G. (2021). Smartphone screen time: Inaccuracy of self-reports and influence of psychological and contextual factors. *Computers in Human Behavior*, 115, 106616. https://doi.org/10.1016/j.chb.2020.106616

Hood, M., Creed, P. A., & Mills, B. J. (2018). Loneliness and online friendships in emerging adults. *Personality and Individual Differences*, 133, 96–102. https://doi.org/10.1016/j.paid.2017.03.045

Jenkins-Guarnieri, M. A., Wright, S. L., & Johnson, B. (2013). Development and validation of a social media use integration scale. *Psychology of Popular Media Culture*, 2(1), 38–50. https://doi.org/10.1037/a0030277

Jensen, M., George, M., Russell, M., & Odgers, C. (2019). Young adolescents' digital technology use and mental health symptoms: Little evidence of longitudinal or daily linkages. *Clinical Psychological Science*, 7(6), 1416–1433. https://doi.org/10.1177/2167702619859336

Jiang, L. C., Bazarova, N. N., & Hancock, J. T. (2011). The disclosure–intimacy link in computer-mediated communication: An attributional extension of the hyperpersonal model. *Human Communication Research*, 37(1), 58–77. https://doi.org/10.1111/j.1468-2958.2010.01393.x

Junco, R. (2013). Comparing actual and self-reported measures of Facebook use. *Computers in Human Behavior*, 29(3), 626–631. https://doi.org/10.1016/j.chb.2012.11.007

Kadylak, T. (2020). An investigation of perceived family phubbing expectancy violations and well-being among U.S. older adults. *Mobile Media & Communication, 8*(2), 247–267. https://doi.org/10.1177/2050157919872238

Kelley, H. H., & Michela, J. L. (1980). Attribution theory and research. *Annual Review of Psychology, 31*(1), 457–501. https://doi.org/10.1146/annurev.ps.31.020180.002325

King, D. L., Herd, M. C., & Delfabbro, P. H. (2018). Motivational components of tolerance in Internet gaming disorder. *Computers in Human Behavior, 78*, 133–141. https://doi.org/10.1016/j.chb.2017.09.023

Kobayashi, T., & Boase, J. (2012). No such effect? The implications of measurement error in self-report measures of mobile communication use. *Communication Methods and Measures, 6*(2), 126–143. https://doi.org/10.1080/19312458.2012.679243

Kraut, R., Patterson, M., Lundmark, V., Kiesler, S., Mukopadhyay, T., & Scherlis, W. (1998). Internet paradox. A social technology that reduces social involvement and psychological well-being? *American Psychologist, 53*(9), 1017–1031. https://doi.org/10.1037/0003-066X.53.9.1017

Kushlev, K., & Heintzelman, S. J. (2018). Put the phone down: Testing a complement-interfere model of computer-mediated communication in the context of face-to-face interactions. *Social Psychological & Personality Science, 9*(6), 702–710. https://doi.org/10.1177/1948550617722199

Lee, A. Y., Katz, R., & Hancock, J. (2021). The role of subjective construals on reporting and reasoning about social media use. *Social Media + Society, 7*(3). https://doi.org/10.1177/20563051211035350

Lee, C.-j., Hornik, R., & Hennessy, M. (2008). The reliability and stability of general media exposure measures. *Communication Methods and Measures, 2*(1–2), 6–22. https://doi.org/10.1080/19312450802063024

Lee, H., Ahn, H., Choi, S., & Choi, W. (2014). The SAMS: Smartphone addiction management system and verification. *Journal of Medical Systems, 38*(1), 1–10. https://doi.org/10.1007/s10916-013-0001-1

Lee, J.-E. R., Moore, D. C., Park, E. A., & Park, S. G. (2012). Who wants to be "friend-rich"? Social compensatory friending on Facebook and the moderating role of public self-consciousness. *Computers in Human Behavior, 28*(3), 1036–1043. https://doi.org/10.1016/j.chb.2012.01.006

Lin, Y.-H., Lin, Y. C., Lee, Y. H., Lin, P. H., Lin, S. H., Chang, L. R., Tseng, H. W., Yen, L. Y., Yang, C. C., & Kuo, T. B. (2015). Time distortion associated with smartphone addiction: Identifying smartphone addiction via a mobile application (app). *Journal of Psychiatric Research, 65*, 139–145. https://doi.org/10.1016/j.jpsychires.2015.04.003

Liu, D., Ainsworth, S. E., & Baumeister, R. F. (2016). A meta-analysis of social networking online and social capital. *Review of General Psychology, 20*(4), 369–391. https://doi.org/10.1037/gpr0000091

Liu, S. X., & Hancock, J. (2020). *People don't ostracize people, phones do: Attributional dynamics for ostracism associated with phubbing* [Paper presentation]. International Communication Association (ICA) 70th annual conference. Virtual conference.

Mark, G., Iqbal, S., Czerwinski, M., & Johns, P. (2014). Capturing the mood: Facebook and face-to-face encounters in the workplace. In *CSCW'14: Proceedings of the 17th ACM Conference on Computer Supported Cooperative Work & Social Computing* (pp. 1082–1094). https://doi.org/10.1145/2531602.2531673

Mieczkowski, H., Lee, A. Y., & Hancock, J. T. (2020). Priming effects of social media use scales on well-being outcomes: The influence of intensity and addiction scales on self-reported depression. *Social Media+ Society, 6*(4), 1–15. https://doi.org/10.1177/2056305120961784

Molenaar, P. C., & Campbell, C. G. (2009). The new person-specific paradigm in psychology. *Current Directions in Psychological Science, 18*(2), 112–117. https://doi.org/10.1111/j.1467-8721.2009.01619.x

Morse, S., & Gergen, K. J. (1970). Social comparison, self-consistency, and the concept of self. *Journal of Personality and Social Psychology, 16*(1), 148–156. https://doi.org/10.1037/h0029862

Neubaum, G., & Krämer, N. C. (2015). My friends right next to me: A laboratory investigation on predictors and consequences of experiencing social closeness on social networking sites. *Cyberpsychology, Behavior, and Social Networking, 18*(8), 443–449. https://doi.org/10.1089/cyber.2014.0613

Orben, A. (2020). The Sisyphean cycle of technology panics. *Perspectives on Psychological Science, 15*(5), 1143–1157. https://doi.org/10.1177/1745691620919372

Orben, A., & Przybylski, A. K. (2019). The association between adolescent well-being and digital technology use. *Nature Human Behaviour, 3*(2), 173–182. https://doi.org/10.1038/s41562-018-0506-1

Owens, J., Maxim, R., McGuinn, M., Nobile, C., Msall, M., & Alario, A. (1999). Television-viewing habits and sleep disturbance in school children. *Pediatrics, 104*(3), e27. https://doi.org/10.1542/peds.104.3.e27

Panova, T., & Carbonell, X. (2018). Is smartphone addiction really an addiction? *Journal of Behavioral Addictions, 7*(2), 252–259. https://doi.org/10.1556/2006.7.2018.49

Parent, N., & Shapka, J. (2020). Moving beyond addiction: An attachment theory framework for understanding young adults' relationships with their smartphones. *Human Behavior and Emerging Technologies, 2*(2), 179–185. https://doi.org/10.1002/hbe2.180

Patton, D. U., Leonard, P., Elaesser, C., Eschmann, R. D., Patel, S., & Crosby, S. (2019). What's a threat on social media? How Black and Latino Chicago young men define and navigate threats online. *Youth & Society, 51*(6), 756–772. https://doi.org/10.1177/0044118X17720325

Pavot, W., & Diener, E. (1993). The affective and cognitive context of self-reported measures of subjective well-being. *Social Indicators Research, 28*(1), 1–20. https://doi.org/10.1007/BF01086714

Perrin, A., & Anderson, M. (2019, April 10). *Share of U.S. adults using social media, including Facebook, is mostly unchanged since 2018*. Pew Research Center. https://www.pewresearch.org/fact-tank/2019/04/10/share-of-u-s-adults-using-social-media-including-facebook-is-mostly-unchanged-since-2018/

Pittman, M., & Reich, B. (2016). Social media and loneliness: Why an Instagram picture may be worth more than a thousand Twitter words. *Computers in Human Behavior, 62*, 155–167. https://doi.org/10.1016/j.chb.2016.03.084

Przybylski, A. K., Murayama, K., DeHaan, C. R., & Gladwell, V. (2013). Motivational, emotional, and behavioral correlates of fear of missing out. *Computers in Human Behavior, 29*(4), 1841–1848. https://doi.org/10.1016/j.chb.2013.02.014

Reeves, B., Ram, N., Robinson, T. N., Cummings, J. J., Giles, C. L., Pan, J., Chiatti, A., Cho, M., Roehrick, K., Yang, X., Gagneja, A., Brinberg, M., Muise, D., Lu, Y., Luo, M., Fitzgerald, A., & Leo Yeykelis, L. (2021). Screenomics: A framework to capture and analyze personal life experiences and the ways that technology shapes them. *Human–Computer Interaction, 36*(2) 150–201. https://doi.org/10.1080/07370024.2019.1578652

Rui, J. R., Covert, J. M., Stefanone, M. A., & Mukherjee, T. (2015). A communication multiplexity approach to social capital: On-and offline communication and self-esteem. *Social Science Computer Review, 33*(4), 498–518. https://doi.org/10.1177/0894439314552803

Sagioglou, C., & Greitemeyer, T. (2014). Facebook's emotional consequences: Why Facebook causes a decrease in mood and why people still use it. *Computers in Human Behavior, 35*, 359–363. https://doi.org/10.1016/j.chb.2014.03.003

Scharkow, M. (2016). The accuracy of self-reported internet use—A validation study using client log data. *Communication Methods and Measures, 10*(1), 13–27. https://doi.org/10.1080/19312458.2015.1118446

Scharkow, M. (2019). The reliability and temporal stability of self-reported media exposure: A meta-analysis. *Communication Methods and Measures, 13*(3), 198–211. https://doi.org/10.1080/19312458.2019.1594742

Schemer, C., Masur, P. K., Geiß, S., Müller, P., & Schäfer, S. (2020). The impact of Internet and social media use on well-being: A longitudinal analysis of adolescents across nine years. *Journal of Computer-Mediated Communication, 26*(1), 1–21. https://doi.org/10.1093/jcmc/zmaa014

Seo, M., Kim, J., & Yang, H. (2016). Frequent interaction and fast feedback predict perceived social support: Using crawled and self-reported data of Facebook users. *Journal of Computer-Mediated Communication, 21*(4), 282–297. https://doi.org/10.1111/jcc4.12160

Sirgy, M. J., Lee, D. J., Kosenko, R., Lee Meadow, H., Rahtz, D., Cicic, M., Jin, G. X., Yarsuvat, D., Blenkhorn, D. L., & Wright, N. (1998). Does television viewership play a role in the perception of quality of life? *Journal of Advertising, 27*(1), 125–142. https://doi.org/10.1080/00913367.1998.10673547

Slater, M. D. (2004). Operationalizing and analyzing exposure: The foundation of media effects research. *Journalism & Mass Communication Quarterly, 81*(1), 168–183. https://doi.org/10.1177/107769900408100112

Spielberger, C. D. (2010). State-Trait anxiety inventory. *The Corsini encyclopedia of psychology*. John Wiley & Sons, Inc. https://doi.org/10.1002/9780470479216.corpsy0943

Statistics Netherlands. (2018, May 18). *More and more young adults addicted to social media*. https://www.cbs.nl/en-gb/news/2018/20/more-and-more-young-adults-addicted-to-social-media

Steers, M. L. N., Wickham, R. E., & Acitelli, L. K. (2014). Seeing everyone else's highlight reels: How Facebook usage is linked to depressive symptoms. *Journal of Social and Clinical Psychology, 33*(8), 701–731. https://doi.org/10.1521/jscp.2014.33.8.701

Sussman, S., & Sussman, A. N. (2011). Considering the definition of addiction. *International Journal of Environmental Research and Public Health, 8*(10), 4025–4038. https://doi.org/10.3390/ijerph8104025

Teppers, E., Luyckx, K., Klimstra, T. A., & Goossens, L. (2014). Loneliness and Facebook motives in adolescence: A longitudinal inquiry into directionality of effect. *Journal of Adolescence, 37*(5), 691–699. https://doi.org/10.1016/j.adolescence.2013.11.003

Tourangeau, R., & Rasinski, K. A. (1988). Cognitive processes underlying context effects in attitude measurement. *Psychological Bulletin, 103*(3), 299–314. https://doi.org/10.1037/0033-2909.103.3.299

Tourangeau, R., Rips, L. J., & Rasinski, K. (2000). *The psychology of survey response*. Cambridge University Press. https://doi.org/10.1017/CBO9780511819322

Twenge, J. M. (2020). Increases in depression, self-harm, and suicide among U.S. adolescents after 2012 and links to technology use: Possible mechanisms. *Psychiatric Research & Clinical Practice, 2*(1), 19–25. https://doi.org/10.1176/appi.prcp.20190015

Uchino, B. N. (2004). *Social support and physical health: Understanding the health consequences of relationships*. Yale University Press. https://doi.org/10.12987/yale/9780300102185.001.0001

U.S. Department of Health and Human Services. (2019). *2018 National Survey on Drug Use and Health: Summary of national findings*. Substance Abuse and Mental Health Services Administration. https://www.samhsa.gov/data/release/2018-national-survey-drug-use-and-health-nsduh-releases.

Vanden Abeele, M., Beullens, K., & Roe, K. (2013). Measuring mobile phone use: Gender, age and real usage level in relation to the accuracy and validity of self-reported mobile phone use. *Mobile Media & Communication, 1*(2), 213–236. https://doi.org/10.1177/2050157913477095

Verduyn, P., Lee, D. S., Park, J., Shablack, H., Orvell, A., Bayer, J., Ybarra, O., Jonides, J., & Kross, E. (2015). Passive Facebook usage undermines affective well-being: Experimental and longitudinal evidence. *Journal of Experimental Psychology: General, 144*(2), 480–488. https://doi.org/10.1037/xge0000057

Verduyn, P., Ybarra, O., Résibois, M., Jonides, J., & Kross, E. (2017). Do social network sites enhance or undermine subjective well-being? A critical review. *Social Issues and Policy Review, 11*(1), 274–302. https://doi.org/10.1111/sipr.12033

Vogel, E. A., Rose, J. P., Roberts, L. R., & Eckles, K. (2014). Social comparison, social media, and self-esteem. *Psychology of Popular Media Culture, 3*(4), 206–222. https://doi.org/10.1037/ppm0000047

Walther, J. B., Kashian, N., Jang, J. W., & Shin, S. Y. (2016). Overattribution of liking in computer-mediated communication: Partners infer the results of their own influence as their partners' affection. *Communication Research, 43*(3), 372–390. https://doi.org/10.1177/0093650214565898

Wang, X., Gao, L., Yang, J., Zhao, F., & Wang, P. (2020). Parental phubbing and adolescents' depressive symptoms: Self-esteem and perceived social support as moderators. *Journal of Youth and Adolescence, 49*(2), 427–437. https://doi.org/10.1007/s10964-019-01185-x

Wang, X., Xie, X., Wang, Y., Wang, P., & Lei, L. (2017). Partner phubbing and depression among married Chinese adults: The roles of relationship satisfaction and relationship length. *Personality and Individual Differences, 110*, 12–17. https://doi.org/10.1016/j.paid.2017.01.014

Wartella, E., & Reeves, B. (1985). Historical trends in research on children and the media: 1900–1960. *Journal of Communication, 35*(2), 118–133. https://doi.org/10.1111/j.1460-2466.1985.tb02238.x

Wheeler, L., & Miyake, K. (1992). Social comparison in everyday life. *Journal of Personality and Social Psychology, 62*(5), 760–773. https://doi.org/10.1037/0022-3514.62.5.760

Williams, K. D. (2009). Ostracism: A temporal need-threat model. In M. P. Zanna (Ed.), *Advances in experimental social psychology* (Vol. 41, pp. 275–314). https://doi.org/10.1016/S0065-2601(08)00406-1

Wills, T. A. (1991). Similarity and self-esteem in downward comparison. In J. Suls & T. A. Wills (Eds.), *Social comparison: Contemporary theory and research* (pp. 51–78). Lawrence Erlbaum Associates.

Wood, J. V. (1989). Theory and research concerning social comparisons of personal attributes. *Psychological Bulletin, 106*(2), 231–248. https://doi.org/10.1037/0033-2909.106.2.231

World Health Organization. (2010). *Atlas on substance use (2010): Resources for the prevention and treatment of substance use disorders.* https://www.who.int/publications/i/item/9789241500616

Yang, C. C., Holden, S. M., & Carter, M. D. (2017). Emerging adults' social media self-presentation and identity development at college transition: Mindfulness as a moderator. *Journal of Applied Developmental Psychology, 52,* 212–221. https://doi.org/10.1016/j.appdev.2017.08.006

Yeager, D. S., & Dweck, C. S. (2012). Mindsets that promote resilience: When students believe that personal characteristics can be developed. *Educational Psychologist, 47*(4), 302–314. https://doi.org/10.1080/00461520.2012.722805

7

How Social Media Contexts Affect the Expression of Moral Emotions

William J. Brady and Killian L. McLoughlin

With over 3.8 billion users worldwide (Kemp, 2020), social media has become an integral part of everyday life. One of the most popular areas of social media research is the study of *emotion*: How it can be measured? How frequently do people express it? And what consequences does it have for social interactions online? The number of academic articles published using the term "sentiment analysis" grew from 78 in 2008 to 1,786 in 2018 (https://clarivate.com/webofsciencegroup/solutions/web-of-science/). This research demonstrates that emotion expression on social media is consequential: Emotions drive engagement with various types of content (Brady et al., 2017; Stieglitz & Dang-Xuan, 2013; Valenzuela et al., 2017), are a force behind online political movements (Manikonda et al., 2018; Spring et al., 2018; Yoo et al., 2018), are associated with online hate speech and harassment (Brady & Crockett, 2019; Reichelmann et al., 2020;

https://doi.org/10.1037/0000290-008
The Psychology of Technology: Social Science Research in the Age of Big Data, S. C. Matz (Editor)
Copyright © 2022 by the American Psychological Association. All rights reserved.

Vranjes et al., 2017), and may even provide a measurable signal of clinical symptoms (De Choudhury et al., 2013; Guntuku et al., 2017; Wongkoblap et al., 2017).

While this research made notable discoveries linking emotion expression on social media with important outcomes, there is relatively little research that examines a more basic question: How might the nature of emotion vary in online, social media contexts compared to offline, face-to-face contexts? For instance, questions remain as to whether emotions are expressed more frequently, serve different functions, or are influenced by different factors on social media compared with offline contexts. Most studies assume that theories of emotion developed to explain emotion in offline settings are well-equipped to explain emotion in online settings. Yet psychological research on emotions has shown consistently that many facets of emotion are bound by context (Barrett et al., 2011). This leads to the possibility that online contexts significantly influence and alter the expression and functions of emotions. In this chapter, we provide a systematic analysis of features of the social media context that differ from features of offline social interaction, and we examine their consequences for the expression and function of emotions.

On social media, emotions are expressed through symbolic representation via text and images. As such, it makes sense to define emotion using a psychological constructivist framework (Barrett, 2013). Specifically, we define emotions on social media in terms of expression in language and images: that is, the usage of specific culturally and contextually defined concepts that represent underlying feelings defined by valence and arousal, or "affect" (Lindquist et al., 2015). Inherent in this definition is a key assumption that on social media, it is most relevant to examine emotion *expression* as compared to *experience*, since expressions are the only measurable and perceivable component of emotions as they unfold on social media (Brady et al., 2021).

In this chapter, we focus on emotions tied to morality and politics, often called "moral emotions" (Haidt, 2003b; Hutcherson & Gross, 2011). Building from our definition of emotions, we define *moral emotion expression* on social media as representational expressions of affect that reliably signal, either to others or to the self, that something is relevant to the interests or good of society, as defined by the conceptual knowledge of the expresser (Brady et al., 2020). This definition is inherently a functional account (Keltner & Haidt, 1999), as it specifies what moral emotion expression is and what role it serves in online communication. For instance, moral outrage expression is a prototypical moral emotion expression because it normally indicates that the expresser perceives that some transgression against one's concept of right and wrong has occurred (Rozin et al., 1999; Tetlock et al., 2000).

There is a pressing need to understand the impact of the social media context on moral emotions in particular because their expression and spread—perhaps even more than other emotions—have played a prominent role in social media with offline consequences. For instance, the viral expression of moral outrage has led to the rise of "cancel culture," where outrage "firestorms" spread rapidly through social media platforms with the intention to harm the reputation of an individual or organization (Rost et al., 2016; Stahel & Rost, 2017). Swells of moral emotion expression on social media leveraged with videos of police brutality have also helped keep the messages of the #BlackLivesMatter movement in the spotlight of American media (Hassan & O'Grady, 2020). Moral language on social media regarding protests is even associated with increased offline violence (Mooijman et al., 2018).

On one hand, social media is simply another context in which people express emotions in reaction to morally relevant events. Yet, on the other hand, there seems to be something exaggerated about the frequency and magnitude of moral emotions expressed on social media. For example, recent data from experience sampling suggest that people not only witness more moral transgressions online than offline, but they also experience more moral outrage per transgression (Crockett, 2017). Online and offline environments depart in many meaningful ways (Bayer et al., 2020; Lieberman & Schroeder, 2020) that could impact emotion expression, but social media platforms share a number of specific features that distinguish them from other online and offline contexts. Here, we examine some of the most central features of social media platforms: (a) streamlined social feedback delivery systems, (b) communication via symbolic representation, and (c) organization of users into large social networks. The features we examine here are not intended to serve as an exhaustive list, but we argue that they have key implications for the expression and function of moral emotions.

First, we argue that streamlined social feedback delivery inherent to social media platforms has the potential to amplify moral emotions because they make them socially rewarding, even when the social reward is not necessarily representative of the preferences of all users in the network. Next, we argue that the necessity of expressing emotions via symbolic representation can make moral emotions spread faster and more widely. Third, we contend that the large group dynamics of social media make group identities highly salient and facilitate the expression and contagion of moral emotions. We conclude by examining how the amplification of moral emotion on social media may have notable consequences for intergroup relations.

SOCIAL FEEDBACK DELIVERY SYSTEM

One defining feature of online communication is the limited availability of nonverbal social cues, which are replaced in a large part by technically mediated social cues (e.g., Like or Share buttons). The limited ability to perceive nonverbal social cues is one of the most psychologically consequential differences between off- and online social interactions (Lieberman & Schroeder, 2020). Compared with in-person communication, limited nonverbal social cues can reduce our capacity to assess the veracity of claims (Hall & Schmid Mast, 2007) and worsen estimations of our interlocutors' mental abilities (Schroeder et al., 2017). While some social media platforms—particularly those that rely on video content (e.g., TikTok)—include greater levels of nonverbal social cues than others, they still lack the immediate and manifold nonverbal social cues encountered in typical face-to-face exchanges.

Technically mediated social cues are one feature that platforms deploy to allow users to give and receive a form of nonverbal information. On major social media platforms such as Facebook, Twitter and Instagram, these cues take the form of interactive *social feedback* buttons that are used to express nonverbal social cues between users. In some cases, this social feedback explicitly contains emotional information. For instance, Facebook users can indicate that they are saddened, maddened, in love with, surprised by, or amused at content that their fellow users share or post.

Social Feedback Is Streamlined on Social Media Platforms

Perhaps the most important characteristic of the social feedback system on social media platforms is the method by which feedback is delivered. We refer to the method of delivery as *streamlined*, meaning that the feedback is shown to users in a way that makes the feedback (a) highly salient, (b) directly quantifiable, and (c) occurring in seemingly random bursts from the user perspective. The social feedback is more salient simply by virtue of being displayed alongside any content that a user produces. In essence, there is no way to view social media posts without simultaneously seeing social feedback (or lack a thereof) associated with it. Offline, social feedback is much more heterogeneous in ways that impact its salience—it might be subtle, ambiguous, and come in many forms that vary by personal and cultural context (Lindquist & Gendron, 2013). By making feedback uniform and on display at all times, social media makes feedback one of the most salient features of a user's experience.

The social feedback is directly quantifiable in the sense that users know exactly how much positive feedback they are receiving for any given post. Users receive precise numbers that allow them to directly compare the social value of each post they create. They can compare the social value of their own posts, and they can also compare the value of their posts with that of other people's posts in their social network. Thus, people can use the quantified social value derived from the Likes and/or Shares of each post as information for updating their behavior. In other words, the quantified social feedback delivery makes it likely that social media behavior can be thought of a form of *value-based decision making* (Rangel et al., 2008), where users attempt to maximize the social value of their posts based on evaluative feedback.

Psychological research has shown that receiving social praise is just as rewarding as monetary reward (Ruff & Fehr, 2014); when people receive positive social feedback, reward areas of the brain known to respond to nonsocial reward are highly active (Izuma et al., 2008; Meshi et al., 2013; Sherman et al., 2016). Recent empirical work has confirmed across multiple platforms that social media engagement can be predicted from a computational social reinforcement model (Brady et al., 2021; Lindström et al., 2021), which involves people altering their behavior based on social feedback they use to update their representation of social value.

Finally, social media platforms deliver social feedback in seemingly random "bursts," in which users receive varying quantities of feedback, on a varying schedule depending on the users' activity, as well as when other users see the posts. Understood in the framework of reinforcement learning, this means that social feedback is delivered on a reinforcement schedule that can be described as some combination of "variable ratio" reinforcement and "variable interval" reinforcement, which are likely to produce high frequencies of the reinforced behavior (Dickinson, 1985).

Social Feedback Is Readily Internalized

Whether material or social, reward indicates that a given behavior is valuable. *Social reward*, such as that produced by feedback on social media, signals extra information about what our peers want us to do, that is, what is *normative* (Ho et al., 2017). Humans have an automatic tendency to infer what others intend to communicate when they are given feedback and use this information to guide their behavior (Sage & Baldwin, 2011). As a result of inferring communication intentions, people *internalize* the rewarded behavior such that the behavior is rewarding in itself even in the absence of feedback (Grusec & Goodnow, 1994). Internalization results in expectations that others perform the original rewarded behavior, since the behavior is inferred to be normative (Vredenburgh et al., 2015).

The process of internalization suggests that a view of social media behavior as fully explained by social feedback (akin to a "technological Skinner Box") is oversimplified. Even with streamlined social feedback, people are consistently making inferences about what is normative in their social network and using it to guide their behavior. This process is partially determined by social feedback received, but through the process of internalization people's behavior may become disentangled from social feedback over time because the behavior has become rewarding in itself.

Implications for Moral Emotion Expression

The streamlined social feedback delivery system may *amplify* our propensity to express moral emotions in response to morally relevant content (Brady et al., 2021; Crockett, 2017). Although social feedback delivery can theoretically shape any type of expression, there is reason to believe that moral emotion expression may be particularly susceptible to the rewarding elements of social feedback. A growing body of research has shown that emotion expression on social media is associated with increased social reward (Brady et al., 2017; Hansen et al., 2011; Kramer et al., 2014; Quercia et al., 2011; Stieglitz & Dang-Xuan, 2013), suggesting that reward learning of emotion expression could be amplified merely in virtue of larger reward signals. Indeed, recent work has shown on Twitter and in simulated social media environments that moral outrage expression on social media can be explained in part under the framework of reinforcement learning: Variation in social feedback received for previous outrage posts affects future outrage expression (Brady et al., 2021).

Second, the feedback delivery system has the potential to create a *lower threshold* for the elicitation of moral emotion expression for users who are the most sensitive to social feedback. As described above, the social feedback delivery system delivers feedback on reinforcement schedules that are most likely to increase the frequency of behaviors. Another consequence of these schedules used on social media is that they can lead to habit-like behaviors (Alter, 2017), where people perform behaviors even when the behavior loses its value. Thus, the elicitation of moral emotion expressions can be thought of as an interaction between perception of moral transgressions and social learning that encourages people to express moral emotions to a greater extent. For the users who are most susceptible to the effects of social feedback (e.g., Strickland & Crowne, 1962), moral emotion expression can be triggered more easily even if the frequency or intensity of the transgression is held constant. In other words, people are more motivated to express moral emotions based on previous reinforcement contingencies.

Third, the feedback delivery system creates *technology-mediated social learning* of emotional reactions to moral transgressions. Users on the platform tend to assume that the social feedback they receive (and can learn from) is driven purely by others' preferences for the expressions we post (Hamilton et al., 2014). On the contrary, content algorithms deployed by the companies also impact the social feedback that we receive by down- and up-ranking certain content so that people have a greater chance of providing feedback to some content over others (Rose-Stockwell, 2019). Thus, if people learn to express moral emotions because those posts yield higher likes and shares on average, they are learning from feedback that has been shaped both by people's preferences (people Like content more if they approve or endorse it), *and* content algorithms that are giving some expressions a higher probability to be Liked than others. Technology-mediated social learning of moral emotions has key implications for what types of moral emotional expressions users consider to be normative in their network. For instance, if moral outrage is up-ranked and therefore outrage posts gain higher amounts of social feedback, a user is likely to infer that moral outrage is expected of them based on the preferences of their network. This could lead to situations akin to "pluralistic ignorance" (Prentice & Miller, 1993), where users expect that their network endorses a high frequency of moral outrage expression when in fact individuals privately reject the high frequency of expression. It could also lead to misperception of outgroup hostility (Shelton & Richeson, 2005), if content algorithms amplify outrage that targets one's ingroup.

COMMUNICATION VIA SYMBOLIC REPRESENTATION

On social media, as well as other computer-mediated contexts, users typically communicate via symbolic representation, that is, through language and images. This form of communication is generally associated with a number of socially relevant characteristics including depersonalization, lack of social cues, difficulty with impression management (Kiesler et al., 1984). This form of communication also has specific consequences for the expression of emotion, in particular for how it spreads among individuals ("emotional contagion"; Hatfield et al., 1993). In this section, we outline characteristics of expressing emotions via symbolic representation and examine consequences for moral emotions including amplified contagion.

Emotion Expression in Language and Images

In online media, including social media, emotions are expressed via symbolic representation in text and images. In virtue of this form of expression, many aspects of online emotion depart from offline, face-to-face emotion expression (Brady et al., 2020; Peters & Kashima, 2015). Emotion expressions are more *static* in online environments as compared to dynamic facial expressions. The static nature of emotion expression online makes them available to perceive for longer, increasing the potential for social impact. Interestingly, the static nature of linguistic emotion expressions may also create more uniform perception since language reduces the dimensionality of emotion expression and constrains its perception based on shared cultural concepts of emotion (Allport, 1924; Lindquist & Gendron, 2013).

Related to their static nature, emotion expressions online also maintain *higher fidelity* when they spread to others compared to offline, face-to-face expressions. When someone views an emotion expression on social media, they can share or reproduce the expression as a direct copy so that others can see the expression of the original author as if they had directly viewed it themselves. Although offline emotion expressions spread to some extent via mimicry of fleeting facial or body movements (Hatfield et al., 1993), the emotion is unlikely to maintain the same influence on others many degrees removed from the original expressor because they view filtered versions (i.e., lower fidelity) of the original expression.

Finally, emotion expressions online maintain more direct *object representation* than offline facial expressions. Unlike in offline interactions, online one can directly view the eliciting event via images or video or gain a detailed understanding through verbal descriptions. Direct representation of eliciting events has a greater chance of eliciting emotions in the perceiver, since they can witness an event as if they were there themselves.

Implications for Moral Emotion Expression

Each of the outlined features of emotion expression on social media—its static nature, higher fidelity transmission, and object representation—has the ability to increase the contagion of moral emotions (Brady et al., 2020; Peters & Kashima, 2015). In the context of emotion expression on social media, *contagion* can be defined as the spread of emotional content through online sharing that takes the form of information diffusion. The object being spread is the symbolic representation (language, images) of the emotion expression (information). Exposure to the information can serve as input to one's own evaluation of the object or event in question (Manstead & Fischer, 2001). If emotion expressions are static, by definition they are perceivable for longer periods of time. Perceiving an emotion is a necessary precondition for emotional contagion to occur (Hatfield et al., 1993; Parkinson, 2011), and thus static emotion expressions have a greater chance of spreading to others than fleeting expressions.

Relatedly, emotion expressions that can maintain high fidelity upon each person-to-person transfer will have a greater ability to spread widely than emotion expressions that lose fidelity with each transfer. For example, insofar as emotions as "caught" via mimicry of facial expressions and subsequent feedback (as in the canonical case of emotional contagion), the magnitude of emotion transfer on the perceiver is unclear (Hess & Blairy, 2001). While fleeting facial expressions might lose their intensity and specificity upon each "wave" of transfer, when someone shares a linguistic emotion expression on social media, the information that is shared is an exact replica of the original regardless of how many times it is shared. Thus, an emotion expression with a high chance of spreading due to its object, intensity, or implied group identity (Fischer & Hess, 2017) will continue to have a high chance of spreading regardless of how many times it is shared to others.

That emotion expression on social media has direct object representation allows for users to gain a clearer picture of what the eliciting condition of the emotion was originally. In the case of moral emotions, this entails maintaining a direct representation of a transgression against a moral norm for other users to directly perceive. For instance, users can express outrage and embed a video or image of the transgression, so others can react as if they were there themselves. This ability closes the psychological distance (Liberman et al., 2007) between each user and the transgression that elicited the author's emotion, and has the potential to amplify the perceiver's own emotional responses as well as sharing of the original authors' emotional responses.

The facilitated spread of moral emotions due to the symbolic representation of emotion has a unique implication for moral emotion expression on social media: technology-mediated contagion (Brady et al., 2020; Goldenberg & Gross, 2020). Given that there is a withdrawal effect on social media where users exposed to fewer emotion expressions create fewer posts of their own (Kramer et al., 2014), companies are incentivized to amplify users' exposure to emotional content so that the users remain engaged (Goldenberg & Gross, 2020). Rules to up-rank emotional content are at least implicitly coded in the recommender systems that partially determine the content a user sees when they log in to a platform (Rose-Stockwell, 2019). That is, to keep users engaged, content algorithms may promote content with emotion expressions and may encourage users to express and share emotions themselves. As noted in this section, this effect is likely to be amplified due to the static, high-fidelity, and object-representing nature of emotional expression on social media.

LARGE SOCIAL NETWORKS

Compared with offline interactions, interactions on social media occur within much larger social networks. Rather than maintaining a small number of high-quality relationships, social media relationships are relatively weak but more numerous (Dunbar, 2012). Furthermore, the size of the audience is orders of magnitude larger than the number of social connections our brain has evolved to monitor. Studies have estimated that in offline contexts people are acquainted with 150 people on average (Dunbar, 2004). On social media, estimates of the average social network size range from 350 to 700 depending on the platform (MacCarthy, 2016; A. Smith, 2014). In this section, we examine two characteristics of social interactions that occur within large groups and their implications for moral emotion expression: (a) the locus of social motivation shifts from the interpersonal to the *intergroup*, and (b) communications are less personal due to weak social ties.

From Interpersonal to Intergroup Motives

Compared with more intimate interpersonal interactions, social media communications occur within large groups that are less personal. These larger, less personal social interactions can be described as a context with increased *psychological distance*, that is, the mental distance away from the egocentric here and now (Trope & Liberman, 2010). As psychological distance increases, social cognition is biased toward high-level, abstract judgments (Trope & Liberman, 2010). In such contexts, people's group identities are highly salient as opposed to their individual identity. For instance, people are more likely to conform to group norms under conditions of greater psychological distance (Ledgerwood & Callahan, 2012). Thus, large group sizes create a context in which group identities are highly salient.

Decades of research in social psychology show that in contexts where group identities are highly salient, people tend to represent themselves in terms of their group identity (e.g., as a Democrat), and their cognition, emotions, and attitudes become aligned with the group's well-being (Tajfel & Turner, 2004). For instance, when group identities are salient, people are strongly motivated to differentiate their group from other groups and protect their group's image (intergroup motivations; Tajfel & Turner, 2004) as well as protect their own reputation as a good group member (intragroup motivations; Trivers, 1971). Moreover, social media is a context in which moral and political group identities in particular are highly salient. For instance, survey studies suggest that 90% of people report seeing at least "a little" political content in their social media newsfeeds when they log in (Duggan & Smith, 2016). Due to the size of the social networks, social media is a context in which people are especially likely to behave in ways that are motivated by concerns related to their group, in particular, their moral and political group identities.

Less Personal Communications

Computer-mediated communication is generally associated with greater anonymity and a reduction of personal communication (Matheson & Zanna, 1988). Although people often assume that anonymity increases antisocial behaviors, anonymity more subtly impacts behavior by making group identities more salient and group norm conformity more likely (Postmes et al., 2001). In other words, it is not necessarily the case that any single behavior is entailed by greater anonymity, but instead that group norms of behavior will have greater influence.

Furthermore, when communications are more anonymous and less personal, people report lower levels of self-awareness (Matheson & Zanna, 1988) which leads to a state of deindividuation. *Deindividuation* is a psychological state in which a person experiences reduced self-awareness and often acts in a more disinhibited manner. Again, the specific behavior that occurs in the state of disinhibition is dependent on group norms (Diener et al., 1980; Postmes et al., 1998). More anonymous and less personalized communications result in greater identification with one's group and a greater likelihood to behave in terms of group norms.

Implications for Moral Emotion Expression

If social media platform features make group identities more salient, then there is strong reason to believe that moral emotion expression will be amplified on social media platforms. When people categorize themselves in terms of group identities, the elicitation, experience, and regulation of emotions are heavily influenced by group identity concerns (Goldenberg et al., 2016; E. R. Smith & Mackie, 2015). Moral emotions specifically serve functions related to group identity concerns such as norm enforcement (Fehr & Gächter, 2002; D. S. Wilson & O'Gorman, 2003), reputation management (Jordan & Rand, 2020), and group promotion (Haidt, 2003a). Thus, in contexts like social media where group identities are highly salient, moral emotions are likely to be highly prevalent because they fulfil group motivational needs. Although there is a lack of research specifically investigating the frequency of moral emotion expression online versus offline, recent research has shown that viral swells of online moral emotions like outrage are shaped by group identity concerns, such as norm enforcement and a desire for group recognition (Brady & Van Bavel, 2021; Johnen et al., 2018; Rost et al., 2016).

Although moral emotions serve group functions in both offline and online contexts, questions remain as to whether the more anonymous and less personalized nature of social media can bolster certain communication functions of emotions over others. For instance, in offline contexts with smaller groups, moral outrage functions as a strong deterrent for norm violators to commit the norm violation again in the future by communicating disapproval (Fehr & Gächter, 2002; Gintis et al., 2005). If moral outrage were to serve this function on social media, then it must be the case that targets of outrage actually see the outrage expression (else the disapproval being communicated is not effective). However, recent work suggests that on social media, moral emotion expressions are especially likely to be shared amongst ingroup members rather than outgroup members (Brady et al., 2017; Dehghani et al., 2016), which limits the effectiveness of outrage intended to communicate disapproval to outgroup members.

On the other hand, social media appears to be an environment conducive to enhancing the ingroup reputation function of moral emotions: If each user knows the other less personally than in offline settings, signals of moral emotional expressions may serve as a primary source of information for users to form moral character judgments. This is especially likely because people consistently use moral emotion expressions as signals of moral character (Everett et al., 2016; Jordan et al., 2016). Furthermore, recent work has shown that social media users deliberately use moral emotions like outrage to enhance their group status during political discussions in online platforms (Grubbs et al., 2019). In summary, moral emotions serve various group functions, and the nature of communication on social media is expected to make reputational functions within the ingroup more effective compared to communicating disapproval to outgroups.

CONSEQUENCES OF THE AMPLIFICATION OF MORAL EMOTIONS ON SOCIAL MEDIA

The amplification of moral emotions on social media may contribute to myriad social consequences for both online and offline interactions. Here, we focus on recent phenomena that have garnered worldwide attention including the increase in the speed and reach of disinformation, the rise of cancel culture, the creation and maintenance of diverse social movements, and the growth of affective polarization—an increasing dislike and negative evaluation of political outgroups (Finkel et al., 2020; Iyengar et al., 2019).

Moral Emotions and Disinformation

During the 2016 U.S. presidential election, an estimated 10% to 15% of American internet users interacted with false news websites (Guess et al., 2019). Concerns about the role this content played in disrupting past elections and will play in undermining future elections have been the subject of congressional inquiry as well as inordinate press coverage and political consternation (Isaac, 2020). One cause for concern is the unprecedented speed with which false news is disseminated on social media (Vosoughi et al., 2018). Yet the efforts of tech companies to stem this spread—in particular, their decision to employ third party fact checkers to label news as authentic or false—have been largely unsuccessful (Feldman, 2020). Understanding the role of moral emotions in the spread of disinformation may provide insights for interventions to slow its spread.

Research has shown that false news online is associated with greater emotional content than authentic news (Vosoughi et al., 2018) and that engaging with misinformation online—favoriting it, retweeting it, and so on—is associated with a reliance on emotion versus rational thinking (Martel et al., 2020). These findings suggest that disinformation containing emotional content is the most likely to exert influence on citizens. Given that disinformation typically involves content about political group identities designed to evoke strong emotions, we predict that moral emotion expression in particular may be leveraged by disinformation profiteers to spread fake news.

The association between emotion and viral fake news suggests several potential interventions for detecting fake news with the most potential for influence. While several tools have been developed to automatically detect bots and automated disinformation accounts (Shu et al., 2017), combining these tools with machine learning classifiers that can detect moral emotion expression in social media posts (Brady et al., 2021) might help identify and moderate the most pernicious instances of disinformation that intend to sow discontent among political groups. These tools may act as an early warning system before a viral false news article has time to disseminate. At a minimum, bot-like activity combined with moral emotional expressions could be sent to the top of the line for fact-checker review.

Moral Emotions and Collective Action

From the #MeToo movement to #BlackLivesMatter, we have seen the viral spread of moral emotions help draw attention to social injustice with offline consequences. On one hand, the amplification of moral emotions like outrage appears to be an important catalyst for collective action: By raising awareness of social injustice, the spread of outrage can motivate people to act for a common cause (Spring et al., 2018). On the other hand, it is unclear to what extent moral emotions on social media platforms can help create enduring change (Brady & Crockett, 2019). Furthermore, while the spread of moral emotions on social media is assumed to aid in grassroots activism, recent sociological research demonstrates that it can also lead to a top-down instrumentalization of social media for the maintenance of the status quo (Schradie, 2019).

Furthermore, the spread of moral emotions does not necessarily motivate all groups to engage in collective action in a unified manner. For instance, moral outrage, anger, and fear have been shown historically to motivate the political right more than the political left (Hacker & Pierson, 2020), to benefit men more than women (Brescoll & Uhlmann, 2008), and to mobilize White people more than Black people (Phoenix, 2019). Thus, moral emotions might be effective for catalyzing collective action, but the important question is: "Under what conditions and for which group?" Further research is required to understand the specific impact that engagement with moral emotional content on social media can have in the domain of morality and politics.

Moral Emotions and Affective Polarization

Many countries in the world are now ideologically divided across various aspects of social life (Carothers & O'Donohue, 2019). This division takes many forms, one of which is *affective polarization*: animosity towards members of the opposing political party (Finkel et al., 2020; Iyengar et al., 2019). For instance, in the United States, affective polarization is at the highest point in its measured history (Doherty, 2017; Iyengar et al., 2019). The technically mediated social feedback systems on social media have the potential to exacerbate the growing phenomenon of affective polarization. Earlier, we discussed how technology-mediated social feedback and interactions can lead to situations in which moral–emotional content is seen to an extent that exaggerates the true base rate of moral emotional expression in a network. This situation, which we likened to pluralistic ignorance, has the potential to lead to misperceptions of the intensity of negative feelings that political ingroup and outgroup members have toward one another.

Misperceptions of how extreme political outgroup members' attitudes are can lead to "false polarization" in which both political groups overestimate to the extent to which the other is polarized on various issues (Lees & Cikara, 2020; Levendusky & Malhotra, 2016; A. E. Wilson et al., 2020). This misperception can lead to a self-fulfilling prophecy in which true polarization is created in response to the false assumptions about outgroups (Levendusky & Malhotra, 2013). Similarly, if social media platforms amplify engagement with moral emotional content that derogates political outgroups, it is likely that partisans will come to overestimate the extent to which outgroups feel negative emotions toward their group, creating a "false affective polarization" that could be a mechanism to explain the true rise in affective polarization. Although further research is required to estimate the magnitude of social media's unique impact on polarization (Tucker et al., 2018), it is worth noting that other forms of traditional media also give notable coverage to inflammatory social media content and can jointly amplify inaccurate perceptions of political outgroups.

CONCLUSION

In this chapter, we outlined three key features of the social media context that make it unique from other contexts: (a) the social feedback delivery system, (b) communication via symbolic representation, and (c) organization of users into large social networks. We argued that each of these features has the potential to amplify the expression of moral emotions through mechanisms of social learning, emotional contagion, and group identity motivations. A key theme explored here is the idea that on social media technology can mediate social interaction in a way that shapes the expression and function of emotions. The considerations outlined here are intended to serve as a starting point to help update and extend our understanding of the nature of emotion as social interactions move into digital contexts to a greater extent than ever before.

REFERENCES

Allport, F. H. (1924). *Social psychology.* Houghton Mifflin. https://brocku.ca/MeadProject/Allport/1924/1924_toc.html

Alter, A. (2017). *Irresistible: The rise of addictive technology and the business of keeping us hooked.* Penguin Press.

Barrett, L. F. (2013). Psychological construction: The Darwinian approach to the science of emotion. *Emotion Review, 5*(4), 379–389. https://doi.org/10.1177/1754073913489753

Barrett, L. F., Mesquita, B., & Gendron, M. (2011). Context in emotion perception. *Current Directions in Psychological Science, 20*(5), 286–290. https://doi.org/10.1177/0963721411422522

Bayer, J. B., Triệu, P., & Ellison, N. B. (2020). Social media elements, ecologies, and effects. *Annual Review of Psychology, 71*(1), 471–497. https://doi.org/10.1146/annurev-psych-010419-050944

Brady, W. J., & Crockett, M. J. (2019). How effective is online outrage? *Trends in Cognitive Sciences, 23*(2), 79–80. https://doi.org/10.1016/j.tics.2018.11.004

Brady, W. J., Crockett, M. J., & Van Bavel, J. J. (2020). The MAD model of moral contagion: The role of motivation, attention, and design in the spread of moralized content online. *Perspectives on Psychological Science, 15*(4), 978–1010. https://doi.org/10.1177/1745691620917336

Brady, W. J., McLoughlin, K., Doan, T. N., & Crockett, M. (2021). How social learning amplifies moral outrage expression in online social networks. *Science Advances, 7*, eabe5641. Advance online publication. https://doi.org/10.31234/osf.io/gf7t5

Brady, W. J., & Van Bavel, J. J. (2021). *Social identity shapes antecedents and functional outcomes of moral emotion expression in online networks* [Unpublished manuscript]. OSF Preprints. https://doi.org/10.31219/osf.io/dgt6u

Brady, W. J., Wills, J. A., Jost, J. T., Tucker, J. A., & Van Bavel, J. J. (2017). Emotion shapes the diffusion of moralized content in social networks. *Proceedings of the National Academy of Sciences, 114*(28), 7313–7318. https://doi.org/10.1073/pnas.1618923114

Brescoll, V. L., & Uhlmann, E. L. (2008). Can an angry woman get ahead? Status conferral, gender, and expression of emotion in the workplace. *Psychological Science, 19*(3), 268–275. https://doi.org/10.1111/j.1467-9280.2008.02079.x

Carothers, T., & O'Donohue, A. (2019). *Democracies divided: The global challenge of political polarization*. Brookings Institution Press. https://www.jstor.org/stable/10.7864/j.ctvbd8j2p

Crockett, M. J. (2017). Moral outrage in the digital age. *Nature Human Behaviour, 1*(11), 769–771. https://doi.org/10.1038/s41562-017-0213-3

De Choudhury, M., Counts, S., & Horvitz, E. (2013). Social media as a measurement tool of depression in populations. In *WebSci'13: Proceedings of the 5th Annual ACM Web Science Conference* (pp. 47–56). Association for Computing Machinery. https://doi.org/10.1145/2464464.2464480

Dehghani, M., Johnson, K., Hoover, J., Sagi, E., Garten, J., Parmar, N. J., Vaisey, S., Iliev, R., & Graham, J. (2016). Purity homophily in social networks. *Journal of Experimental Psychology: General, 145*(3), 366–375. https://doi.org/10.1037/xge0000139

Dickinson, A. (1985). Actions and habits: The development of behavioural autonomy. *Philosophical Transactions of the Royal Society of London: Series B. Biological Sciences, 308*(1135), 67–78. https://doi.org/10.1098/rstb.1985.0010

Diener, E., Lusk, R., DeFour, D., & Flax, R. (1980). Deindividuation: Effects of group size, density, number of observers, and group member similarity on self-consciousness and disinhibited behavior. *Journal of Personality and Social Psychology, 39*(3), 449–459. https://doi.org/10.1037/0022-3514.39.3.449

Doherty, C. (2017). *Key takeaways on Americans' growing partisan divide over political values.* Pew Research Center. https://www.pewresearch.org/fact-tank/2017/10/05/takeaways-on-americans-growing-partisan-divide-over-political-values/

Duggan, M., & Smith, A. (2016, October 25). *The political environment on social media.* Pew Research Center. https://www.pewresearch.org/internet/2016/10/25/the-political-environment-on-social-media/

Dunbar, R. I. M. (2004). Gossip in evolutionary perspective. *Review of General Psychology, 8*(2), 100–110. https://doi.org/10.1037/1089-2680.8.2.100

Dunbar, R. I. M. (2012). Social cognition on the Internet: Testing constraints on social network size. *Philosophical Transactions of the Royal Society of London: Series B. Biological Sciences, 367*(1599), 2192–2201. https://doi.org/10.1098/rstb.2012.0121

Everett, J. A. C., Pizarro, D. A., & Crockett, M. J. (2016). Inference of trustworthiness from intuitive moral judgments. *Journal of Experimental Psychology: General, 145*(6), 772–787. https://doi.org/10.1037/xge0000165

Fehr, E., & Gächter, S. (2002). Altruistic punishment in humans. *Nature, 415*(6868), 137–140. https://doi.org/10.1038/415137a

Feldman, B. (2020, March 9). How Facebook fact-checking can backfire. *New York Magazine/Intelligencer.* https://nymag.com/intelligencer/2020/03/study-shows-possible-downside-of-fact-checking-on-facebook.html

Finkel, E. J., Bail, C. A., Cikara, M., Ditto, P. H., Iyengar, S., Klar, S., Mason, L., McGrath, M. C., Nyhan, B., Rand, D. G., Skitka, L. J., Tucker, J. A., Van Bavel, J. J., Wang, C. S., & Druckman, J. N. (2020). Political sectarianism in America. *Science, 370*(6516), 533–536. https://doi.org/10.1126/science.abe1715

Fischer, A., & Hess, U. (2017). Mimicking emotions. *Current Opinion in Psychology, 17,* 151–155. https://doi.org/10.1016/j.copsyc.2017.07.008

Gintis, P. E., Bowles, S., Boyd, R. T., & Fehr, E. (Eds.). (2005). *Moral sentiments and material interests: The foundations of cooperation in economic life.* The MIT Press.

Goldenberg, A., & Gross, J. J. (2020). Digital emotion contagion. *Trends in Cognitive Sciences, 24*(4), 316–328. https://doi.org/10.1016/j.tics.2020.01.009

Goldenberg, A., Halperin, E., van Zomeren, M., & Gross, J. J. (2016). The process model of group-based emotion: Integrating Intergroup emotion and emotion regulation perspectives. *Personality and Social Psychology Review*, *20*(2), 118–141. https://doi.org/10.1177/1088868315581263

Grubbs, J. B., Warmke, B., Tosi, J., James, A. S., & Campbell, W. K. (2019). Moral grandstanding in public discourse: Status-seeking motives as a potential explanatory mechanism in predicting conflict. *PLOS ONE*, *14*(10), e0223749. https://doi.org/10.1371/journal.pone.0223749

Grusec, J. E., & Goodnow, J. J. (1994). Impact of parental discipline methods on the child's internalization of values: A reconceptualization of current points of view. *Developmental Psychology*, *30*(1), 4–19. https://doi.org/10.1037/0012-1649.30.1.4

Guess, A., Nagler, J., & Tucker, J. (2019). Less than you think: Prevalence and predictors of fake news dissemination on Facebook. *Science Advances*, *5*(1), eaau4586. https://doi.org/10.1126/sciadv.aau4586

Guntuku, S. C., Yaden, D. B., Kern, M. L., Ungar, L. H., & Eichstaedt, J. C. (2017). Detecting depression and mental illness on social media: An integrative review. *Current Opinion in Behavioral Sciences*, *18*, 43–49. https://doi.org/10.1016/j.cobeha.2017.07.005

Hacker, J. S., & Pierson, P. (2020). *Let them eat tweets: How the right rules in an age of extreme inequality*. Liveright Publishing/W. W. Norton.

Haidt, J. (2003a). Elevation and the positive psychology of morality. In C. L. M. Keyes & J. Haidt (Eds.), *Flourishing: Positive psychology and the life well-lived* (pp. 275–289). American Psychological Association. https://doi.org/10.1037/10594-012

Haidt, J. (2003b). The moral emotions. In R. J. Davidson, K. R. Scherer, & H. H. Goldsmith (Eds.), *Handbook of affective sciences* (pp. 852–870). Oxford University Press.

Hall, J. A., & Schmid Mast, M. (2007). Sources of accuracy in the empathic accuracy paradigm. *Emotion*, *7*(2), 438–446. https://doi.org/10.1037/1528-3542.7.2.438

Hamilton, K., Karahalios, K., Sandvig, C., & Eslami, M. (2014). A path to understanding the effects of algorithm awareness. In *CHI '14 Extended abstracts on human factors in computing systems* (pp. 631–642). Association for Computing Machinery. https://doi.org/10.1145/2559206.2578883

Hansen, L. K., Arvidsson, A., Nielsen, F. Å., Colleoni, E., & Etter, M. (2011). Good friends, bad news—Affect and virality in Twitter. In J. J. Park, L. T. Yang, & C. Lee (Eds.), *Future information technology* (pp. 34–43). Springer. https://doi.org/10.1007/978-3-642-22309-9_5

Hassan, J., & O'Grady, S. (2020, May 29). Anger over George Floyd's killing ripples far beyond the United States. *The Washington Post*. https://www.washingtonpost.com/world/2020/05/29/world-reacts-george-floyd-minneapolis-protests/

Hatfield, E., Cacioppo, J. T., & Rapson, R. L. (1993). Emotional contagion. *Current Directions in Psychological Science*, *2*(3), 96–100. https://doi.org/10.1111/1467-8721.ep10770953

Hess, U., & Blairy, S. (2001). Facial mimicry and emotional contagion to dynamic emotional facial expressions and their influence on decoding accuracy. *International Journal of Psychophysiology*, *40*(2), 129–141. https://doi.org/10.1016/S0167-8760(00)00161-6

Ho, M. K., MacGlashan, J., Littman, M. L., & Cushman, F. (2017). Social is special: A normative framework for teaching with and learning from evaluative feedback. *Cognition*, *167*, 91–106. https://doi.org/10.1016/j.cognition.2017.03.006

Hutcherson, C. A., & Gross, J. J. (2011). The moral emotions: A social–functionalist account of anger, disgust, and contempt. *Journal of Personality and Social Psychology*, *100*(4), 719–737. https://doi.org/10.1037/a0022408

Isaac, M. (2020, September 3). After being used to spread disinformation, Facebook seeks to limit election chaos. *The New York Times*. https://www.nytimes.com/2020/09/03/us/elections/after-being-used-to-spread-disinformation-facebook-seeks-to-limit-election-chaos.html

Iyengar, S., Lelkes, Y., Levendusky, M., Malhotra, N., & Westwood, S. J. (2019). The origins and consequences of affective polarization in the United States. *Annual Review of Political Science*, *22*(1), 129–146. https://doi.org/10.1146/annurev-polisci-051117-073034

Izuma, K., Saito, D. N., & Sadato, N. (2008). Processing of social and monetary rewards in the human striatum. *Neuron*, *58*(2), 284–294. https://doi.org/10.1016/j.neuron.2008.03.020

Johnen, M., Jungblut, M., & Ziegele, M. (2018). The digital outcry: What incites participation behavior in an online firestorm? *New Media & Society*, *20*(9), 3140–3160. https://doi.org/10.1177/1461444817741883

Jordan, J. J., Hoffman, M., Bloom, P., & Rand, D. G. (2016). Third-party punishment as a costly signal of trustworthiness. *Nature*, *530*(7591), 473–476. https://doi.org/10.1038/nature16981

Jordan, J. J., & Rand, D. G. (2020). Signaling when no one is watching: A reputation heuristics account of outrage and punishment in one-shot anonymous interactions. *Journal of Personality and Social Psychology*, *118*(1), 57–88. https://doi.org/10.1037/pspi0000186

Keltner, D., & Haidt, J. (1999). Social functions of emotions at four levels of analysis. *Cognition and Emotion, 13*(5), 505–521. https://doi.org/10.1080/026999399379168

Kemp, S. (2020, January 30). *Digital 2020: 3.8 billion people use social media*. We Are Social. https://wearesocial.com/uk/blog/2020/01/digital-2020-3-8-billion-people-use-social-media/

Kiesler, S., Siegel, J., & McGuire, T. W. (1984). Social psychological aspects of computer-mediated communication. *American Psychologist, 39*(10), 1123–1134. https://doi.org/10.1037/0003-066X.39.10.1123

Kramer, A. D. I., Guillory, J. E., & Hancock, J. T. (2014). Experimental evidence of massive-scale emotional contagion through social networks. *Proceedings of the National Academy of Sciences, 111*(24), 8788–8790. https://doi.org/10.1073/pnas.1320040111

Ledgerwood, A., & Callahan, S. P. (2012). The social side of abstraction: Psychological distance enhances conformity to group norms. *Psychological Science, 23*(8), 907–913. https://doi.org/10.1177/0956797611435920

Lees, J. M., & Cikara, M. (2020). *Understanding and combating misperceived polarization*. PsyArXiv. https://doi.org/10.31234/osf.io/ncwez

Levendusky, M. S., & Malhotra, N. (2013, October 9). The effect of "false" polarization: Are perceptions of political polarization self-fulling prophecies [Blog post]. *The Washington Post*. https://www.washingtonpost.com/blogs/monkey-cage/files/2014/01/fp_writeup_oct6_for_jop.pdf

Levendusky, M. S., & Malhotra, N. (2016). (Mis)perceptions of partisan polarization in the American public. *Public Opinion Quarterly, 80*(S1), 378–391. https://doi.org/10.1093/poq/nfv045

Liberman, N., Trope, Y., & Stephan, E. (2007). Psychological distance. In A. W. Kruglanski & E. T. Higgins (Eds.), *Social psychology: Handbook of basic principles* (2nd ed., pp. 353–381). The Guilford Press.

Lieberman, A., & Schroeder, J. (2020). Two social lives: How differences between online and offline interaction influence social outcomes. *Current Opinion in Psychology, 31*, 16–21. https://doi.org/10.1016/j.copsyc.2019.06.022

Lindquist, K. A., & Gendron, M. (2013). What's in a word? Language constructs emotion perception. *Emotion Review, 5*(1), 66–71. https://doi.org/10.1177/1754073912451351

Lindquist, K. A., MacCormack, J. K., & Shablack, H. (2015). The role of language in emotion: Predictions from psychological constructionism. *Frontiers in Psychology, 6*, 444. Advance online publication. https://doi.org/10.3389/fpsyg.2015.00444

Lindström, B., Bellander, M., Schultner, D. T., Chang, A., Tobler, P. N., & Amodio, D. M. (2021). A computational reward learning account of social media engagement. *Nature Communications*, *12*(1), 1311. https://doi.org/10.1038/s41467-020-19607-x

MacCarthy, R. (2016, June 23). *The average Twitter user now has 707 followers* [Blog post]. Science of Social Sales. https://kickfactory.com/blog/average-twitter-followers-updated-2016/

Manikonda, L., Beigi, G., Liu, H., & Kambhampati, S. (2018). *Twitter for sparking a movement, Reddit for sharing the moment: #metoo through the lens of social media*. arXiv. http://arxiv.org/abs/1803.08022

Manstead, A., & Fischer, A. (2001). Social appraisal: The social world as object of an influence on appraisal processes. In K. R. Scherer, A. Schorr, & T. Johnstone (Eds.), *Appraisal processes in emotion: Theory, methods, research* (pp. 221–232). Oxford University Press.

Martel, C., Pennycook, G., & Rand, D. G. (2020). Reliance on emotion promotes belief in fake news. *Cognitive Research: Principles and Implications*, *5*(1), 47. https://doi.org/10.1186/s41235-020-00252-3

Matheson, K., & Zanna, M. P. (1988). The impact of computer-mediated communication on self-awareness. *Computers in Human Behavior*, *4*(3), 221–233. https://doi.org/10.1016/0747-5632(88)90015-5

Meshi, D., Morawetz, C., & Heekeren, H. R. (2013). Nucleus accumbens response to gains in reputation for the self relative to gains for others predicts social media use. *Frontiers in Human Neuroscience*, *7*, 439. https://doi.org/10.3389/fnhum.2013.00439

Mooijman, M., Hoover, J., Lin, Y., Ji, H., & Dehghani, M. (2018). Moralization in social networks and the emergence of violence during protests. *Nature Human Behaviour*, *2*(6), 389–396. https://doi.org/10.1038/s41562-018-0353-0

Parkinson, B. (2011). Interpersonal emotion transfer: Contagion and social appraisal. *Social and Personality Psychology Compass*, *5*(7), 428–439. https://doi.org/10.1111/j.1751-9004.2011.00365.x

Peters, K., & Kashima, Y. (2015). A multimodal theory of affect diffusion. *Psychological Bulletin*, *141*(5), 966–992. https://doi.org/10.1037/bul0000020

Phoenix, D. L. (2019). *The anger gap: How race shapes emotion in politics*. Cambridge University Press. https://doi.org/10.1017/9781108641906

Postmes, T., Spears, R., & Lea, M. (1998). Breaching or building social boundaries?: SIDE-effects of computer-mediated communication. *Communication Research*, *25*(6), 689–715. https://doi.org/10.1177/009365098025006006

Postmes, T., Spears, R., Sakhel, K., & de Groot, D. (2001). Social influence in computer-mediated communication: The effects of anonymity on group behavior. *Personality and Social Psychology Bulletin*, *27*(10), 1243–1254. https://doi.org/10.1177/01461672012710001

Prentice, D. A., & Miller, D. T. (1993). Pluralistic ignorance and alcohol use on campus: Some consequences of misperceiving the social norm. *Journal of Personality and Social Psychology*, *64*(2), 243–256. https://doi.org/10.1037/0022-3514.64.2.243

Quercia, D., Ellis, J., Capra, L., & Crowcroft, J. (2011). In the mood for being influential on Twitter. In *2011 IEEE Third International Conference on Privacy, Security, Risk and Trust and 2011 IEEE Third International Conference on Social Computing* (pp. 307–314). IEEE. https://doi.org/10.1109/PASSAT/SocialCom.2011.27

Rangel, A., Camerer, C., & Montague, P. R. (2008). A framework for studying the neurobiology of value-based decision making. *Nature Reviews Neuroscience*, *9*(7), 545–556. https://doi.org/10.1038/nrn2357

Reichelmann, A., Hawdon, J., Costello, M., Ryan, J., Blaya, C., Llorent, V., Oksanen, A., Räsänen, P., & Zych, I. (2020). Hate knows no boundaries: Online hate in six nations. *Deviant Behavior*. Advance online publication. https://doi.org/10.1080/01639625.2020.1722337

Rose-Stockwell, T. (2019, August 12). *This is how your fear and outrage are being sold for profit* [Blog post]. Tobias Rose-Stockwell. https://medium.com/@tobiasrose/the-enemy-in-our-feeds-e86511488de

Rost, K., Stahel, L., & Frey, B. S. (2016). Digital social norm enforcement: Online firestorms in social media. *PLOS ONE*, *11*(6), e0155923. https://doi.org/10.1371/journal.pone.0155923

Rozin, P., Lowery, L., Imada, S., & Haidt, J. (1999). The CAD triad hypothesis: A mapping between three moral emotions (contempt, anger, disgust) and three moral codes (community, autonomy, divinity). *Journal of Personality and Social Psychology*, *76*(4), 574–586. https://doi.org/10.1037/0022-3514.76.4.574

Ruff, C. C., & Fehr, E. (2014). The neurobiology of rewards and values in social decision making. *Nature Reviews Neuroscience*, *15*(8), 549–562. https://doi.org/10.1038/nrn3776

Sage, K. D., & Baldwin, D. (2011). Disentangling the social and the pedagogical in infants' learning about tool-use. *Social Development*, *20*(4), 825–844. https://doi.org/10.1111/j.1467-9507.2011.00624.x

Schradie, J. (2019). *The revolution that wasn't: How digital activism favors conservatives*. Harvard University Press. https://doi.org/10.4159/9780674240438

Schroeder, J., Kardas, M., & Epley, N. (2017). The humanizing voice: Speech reveals, and text conceals, a more thoughtful mind in the midst of disagreement. *Psychological Science*, *28*(12), 1745–1762. https://doi.org/10.1177/0956797617713798

Shelton, J. N., & Richeson, J. A. (2005). Intergroup contact and pluralistic ignorance. *Journal of Personality and Social Psychology, 88*(1), 91–107. https://doi.org/10.1037/0022-3514.88.1.91

Sherman, L. E., Payton, A. A., Hernandez, L. M., Greenfield, P. M., & Dapretto, M. (2016). The power of the *like* in adolescence: Effects of peer influence on neural and behavioral responses to social media. *Psychological Science, 27*(7), 1027–1035. https://doi.org/10.1177/0956797616645673

Shu, K., Sliva, A., Wang, S., Tang, J., & Liu, H. (2017). Fake news detection on social media: A data mining perspective. *ACM SIGKDD Explorations Newsletter, 19*(1), 22–36. https://doi.org/10.1145/3137597.3137600

Smith, A. (2014). *What people like and dislike about Facebook*. Pew Research Center. https://www.pewresearch.org/fact-tank/2014/02/03/what-people-like-dislike-about-facebook/

Smith, E. R., & Mackie, D. M. (2015). Dynamics of group-based emotions: Insights from intergroup emotions theory. *Emotion Review, 7*(4), 349–354. https://doi.org/10.1177/1754073915590614

Spring, V. L., Cameron, C. D., & Cikara, M. (2018). The upside of outrage. *Trends in Cognitive Sciences, 22*(12), 1067–1069. https://doi.org/10.1016/j.tics.2018.09.006

Stahel, L., & Rost, K. (2017). Angels and devils of digital social norm enforcement: A theory about aggressive versus civilized online comments. In *#SMSociety17: Proceedings of the 8th International Conference on Social Media & Society* (pp. 1–9). Association for Computing Machinery. https://doi.org/10.1145/3097286.3097304

Stieglitz, S., & Dang-Xuan, L. (2013). Emotions and information diffusion in social media—Sentiment of microblogs and sharing behavior. *Journal of Management Information Systems, 29*(4), 217–248. https://doi.org/10.2753/MIS0742-1222290408

Strickland, B. R., & Crowne, D. R. (1962). Conformity under conditions of simulated group pressure as a function of the need for social approval. *Journal of Social Psychology, 58*(1), 171–181. https://doi.org/10.1080/00224545.1962.9712366

Tajfel, H., & Turner, J. (2004). An integrative theory of intergroup conflict. In M. J. Hatch & M. Schultz (Eds.), *Organizational identity: A reader* (pp. 56–65). Oxford University Press.

Tetlock, P. E., Kristel, O. V., Elson, S. B., Green, M. C., & Lerner, J. S. (2000). The psychology of the unthinkable: Taboo trade-offs, forbidden base rates, and heretical counterfactuals. *Journal of Personality and Social Psychology, 78*(5), 853–870. https://doi.org/10.1037/0022-3514.78.5.853

Trivers, R. L. (1971). The evolution of reciprocal altruism. *The Quarterly Review of Biology*, *46*(1), 35–57. https://doi.org/10.1086/406755

Trope, Y., & Liberman, N. (2010). Construal-level theory of psychological distance. *Psychological Review*, *117*(2), 440–463. https://doi.org/10.1037/a0018963

Tucker, J. A., Guess, A., Barbera, P., Vaccari, C., Siegel, A., Sanovich, S., Stukal, D., & Nyhan, B. (2018). Social media, political polarization, and political disinformation: A review of the scientific literature [SSRN Scholarly Paper ID 3144139]. *Social Science Research Network*. https://doi.org/10.2139/ssrn.3144139

Valenzuela, S., Piña, M., & Ramírez, J. (2017). Behavioral effects of framing on social media users: How conflict, economic, human interest, and morality frames drive news sharing. *Journal of Communication*, *67*(5), 803–826. https://doi.org/10.1111/jcom.12325

Vosoughi, S., Roy, D., & Aral, S. (2018). The spread of true and false news online. *Science*, *359*(6380), 1146–1151. https://doi.org/10.1126/science.aap9559

Vranjes, I., Baillien, E., Vandebosch, H., Erreygers, S., & De Witte, H. (2017). The dark side of working online: Towards a definition and an emotion reaction model of workplace cyberbullying. *Computers in Human Behavior*, *69*, 324–334. https://doi.org/10.1016/j.chb.2016.12.055

Vredenburgh, C., Kushnir, T., & Casasola, M. (2015). Pedagogical cues encourage toddlers' transmission of recently demonstrated functions to unfamiliar adults. *Developmental Science*, *18*(4), 645–654. https://doi.org/10.1111/desc.12233

Wilson, A. E., Parker, V. A., & Feinberg, M. (2020). Polarization in the contemporary political and media landscape. *Current Opinion in Behavioral Sciences*, *34*, 223–228. https://doi.org/10.1016/j.cobeha.2020.07.005

Wilson, D. S., & O'Gorman, R. (2003). Emotions and actions associated with norm-breaking events. *Human Nature*, *14*(3), 277–304. https://doi.org/10.1007/s12110-003-1007-z

Wongkoblap, A., Vadillo, M. A., & Curcin, V. (2017). Researching mental health disorders in the era of social media: Systematic review. *Journal of Medical Internet Research*, *19*(6), e228. https://doi.org/10.2196/jmir.7215

Yoo, J., Brown, J., & Chung, A. (2018). Collaborative touchdown with #Kaepernick and #BLM: Sentiment analysis of tweets expressing Colin Kaepernick's refusal to stand during the national anthem and its association with #BLM. *Journal of Sports Media*, *13*(2), 39–60. https://doi.org/10.1353/jsm.2018.0008

8

Big Data in the Workplace

Peter J. Mancarella and Tara S. Behrend

Technological advances related to data collection and storage have created an explosion in the amount of work- and worker-related data that can be easily accessed and used by researchers and organizations, often incidentally and automatically. As a result, both our analytic frameworks and underlying philosophical assumptions need to adapt. Psychologists have long relied on data to answer questions about behavior in the workplace, emphasizing ethics, reliability, and validity. In an environment that permits nearly infinite data collection, these are still essential factors in determining the ways that Big Data should be used in workplace contexts, though the ways these issues manifest have changed dramatically.

This chapter focuses on the science and practice of Big Data in the workplace. First, we introduce Big Data and describe the regulatory environment unique to workplace Big Data. Next, we review sources and

The authors acknowledge and thank Jerod C. White for his comments and editing on this chapter. The authors also thank Bradley D. Pitcher, Daniel M. Ravid, and the other WAVE lab members for their assistance with drafts of this chapter.

https://doi.org/10.1037/0000290-009
The Psychology of Technology: Social Science Research in the Age of Big Data, S. C. Matz (Editor)
Copyright © 2022 by the American Psychological Association. All rights reserved.

applications of Big Data to the human resources functions of selection, training, and performance management. We examine employee reactions to the collection and use of Big Data, as well as considerations of using Big Data in workplace diversity-and-inclusion efforts. We close by discussing the future of workplace Big Data and psychology.

CHARACTERISTICS OF BIG DATA

Big Data sets have different characteristics from traditional data sets. These characteristics have been described as the "Vs": volume, variety, velocity, and veracity, among others (McAfee & Brynjolfsson, 2012; Oswald et al., 2020; Shafer, 2017). *Volume* refers to the number of observations. *Variety* refers to the different types of data in the set—for example, unstructured text, videos or images, or movement data. *Velocity* refers to the speed or continuity of change in the data; a traditional data set may be a static snapshot or updated occasionally, whereas a Big Data set may have new data flowing in continuously or at regular intervals. *Veracity* refers to the validity or quality of the data. Quality is an important concern with all data, but ensuring quality in data with a large volume, high variety, and high velocity is particularly difficult.

In attempting to quantify these Vs, it is apparent that what is "big" for one researcher is typical for another. In this chapter, we use "big" to mean one order of complexity higher than what would ordinarily be used for a particular research question or application. One of the most critical features of what makes a data set big can be summarized in simple logic: It has outstripped the capacity for researchers to store, analyze, and handle it using their traditional tools and methods (Braun et al., 2018). In other words, data that fall under the appellation of Big Data are just data for which inspecting each observation would be so cumbersome that it is practically impossible; thus, different statistics, tools, and methods must be used.

Most organizations own vast amounts of data, and these data can benefit organizations when used appropriately. Data may describe internal actors and groups of actors (e.g., employees, teams) as well as external ones (e.g., clients, customers, partners, competitors). As organizations seek to gain or maintain a competitive edge in the market, all incentives lead to the increasing use of these data to inform decisions, define problems, identify areas for improvement, and predict future outcomes. Big Data analytics allow firms to create actionable intelligence from vast and continuously updated data, such as those from emails or smartphones (Gandomi & Haider, 2015). Importantly, a large proportion of these data can be understood to be *perishable*, meaning they lose value to the organization as time passes. Therefore, organizational data often must be acted upon as quickly as possible.

BIG DATA AND PRIVACY

The privacy of individuals is an issue of great importance in the era of Big Data. Chapter 12 of this volume provides an excellent overview of privacy and ethics in the context of Big Data generally, so in this section we focus on how these topics apply in the context of work. People spend a lot of time at work, and they generate a large amount of behavioral data in doing so; emails, physical movements, computer activity, and even physiological activity (e.g., body temperature; Mendelson, 2020) are routinely captured and owned by employers. The breadth and depth of information contained in these data pose threats to workers' personal boundaries and privacy perceptions. These behavioral data carry the same privacy concerns as consumer data owned by organizations, but they also have an additional layer of complexity and sensitivity due to the employment relationship. In the context of organization research, data about everyday behaviors hold great value, especially when combined with other sources of workplace and individual data (e.g., Chaffin et al., 2017; Olguín Olguín et al., 2009).

Just as organizations have to deal with privacy and data protection concerns, researchers working with Big Data created in organizational contexts have to carefully consider the privacy risks of using such data for research purposes. Employees are not specified as a vulnerable population in the Common Rule (standards and processes researchers must follow to protect human subjects, as codified by the U.S. Department of Health and Human Services; Protection of Human Subjects, 2017); however, they are, in effect, an at-risk population due to "paycheck vulnerability." Employers or unions may explicitly or implicitly encourage employees to participate (or not participate) in research, and there may be consequences for not complying. Thus, the ability of employees to give informed consent may be diminished or eliminated. Researchers working with employee populations must therefore build additional safeguards and protections into their research and data stewardship plans.

Researchers are developing strategies to identify and protect against threats to individual privacy. Consider "anonymized" data, which may appear to protect individuals' privacy but can be de-anonymized with relative ease. For example, algorithms have been developed that can match anonymous social network profiles with their users' real identities (Halimi & Ayday, 2020), and it has been widely cited that a majority of Americans can be identified using the combination of their zip code, gender, and birth date (Sweeney, 2000). These three pieces of information do not pose any privacy threat individually, but they create a vulnerability if they are present in a data set along with sensitive information (e.g., financial information, health data). Although the computer sciences have started to develop analytic techniques to protect privacy (e.g., differential privacy technologies[1] or federated learning), these techniques are not yet widely used and are not applied consistently to most workplace data.

Data Privacy Regulations

In this section, we discuss regulations pertaining to data protection and privacy. We preface this discussion by noting that although regulations are an important aspect of data protection, they are not sufficient on their own; organizational culture and leadership play a critical role in protecting employees. The existence of these regulations alone does not guarantee that employees will be aware of their rights regarding their personal data, nor that they will exercise those rights. Employees may not read their contracts carefully, may not feel comfortable addressing issues in those contracts, or may not feel comfortable withdrawing consent post hoc. These issues

---

[1] See https://gking.harvard.edu/category/research-interests/methods/missing-data for differential privacy procedure tutorials; these models introduce random noise into data sets to prevent re-identification of de-identified data.

may lead to unjustified use of employee data, but they are not explicitly addressed in the following regulations.

General Data Protection Regulation

The General Data Protection Regulation (GDPR) is a data privacy regulation passed by the European Union (EU) Parliament in April 2016 (GDPR, 2016). The regulation took effect in all EU member states in May 2018 and is focused on organizations that collect and use the personal data of individuals in the EU. Notably, this regulation includes organizations that process the personal data of, or provide goods and services to, individuals in the EU, even if the organizations are not physically located in the EU. *Personal data* in the GDPR is broadly defined as "any information relating to an identified or identifiable natural person" (GDPR, 2016). There is an emphasis that the responsibility for fair, safe, and transparent data processing falls on the data processors (i.e., the organizations using the data), and the regulation is predicated on individuals' right to privacy. Organizations must get consent from the data subjects, and organizations that fail to comply with the regulation face steep fines (for a high-level summary of the GDPR, see *What Is GDPR, the EU's New Data Protection Law?*, 2018).

Under the GDPR, requirements for transparency around data collection and privacy apply to employee data much as they do to site-visitor data. Organizations must inform employees what data are being collected about them and for what purpose and must obtain explicit consent for these practices. Notably, this consent must be freely given separately from any other contract and must be able to be withdrawn at any time, and the employee must be made aware of these practices. Per the regulation, the only type of consent for data collection and processing that can be required as part of an employment contract is for data that is necessary for the performance of the job (GDPR, 2016).

California Consumer Privacy Act

In the United States, data privacy regulations exist in some states, but there is no single federal regulation similar to the GDPR. California implemented the California Consumer Privacy Act (CCPA) in 2018, which took effect on January 1, 2020. The act, modeled after the GDPR, seeks to protect the personal information of individuals located in California. It specifies that organizations must disclose any information gathered from an individual, as well as the uses of that data, upon request from that individual. Also, organizations must delete any personal information upon request from that individual. The CCPA is less comprehensive and strict than the GDPR, mainly because it is based on individuals opting out rather than opting in (except for minors, who must opt in).

A recent amendment to the CCPA clarified that CCPA includes employees, applicants, businesses owners, and contractors, among others, in the definition of "consumer," affording them the same protections as customers under the law (AB-25 California Consumer Privacy Act, 2019). Notably, this amendment exempted employee data from being protected in the same way as other consumer data until January 1, 2021, which may have increased the likelihood that employees did not realize CCPA applies to data collected from them as *employees* as well as data collected from them as *private citizens*.

SOURCES OF BIG DATA

Having discussed the organizational incentives to use Big Data, and issues related to data privacy and protection in the workplace, we now review how workplace data are generated and acquired.

Personnel Data

One common source of Big Data is applicant information during the organizational recruitment-and-selection process. Data collected from applicants may include text data from résumés and applications, behavioral data from video and in-person interviews and assessments, and structured quantitative data from tests and assessments. These data can be used to make inferences about a broad variety of employee characteristics, including demographics, professional background, and psychological characteristics such as personality, intelligence, knowledge, skills, and abilities. These data can be collected without the knowledge or consent of applicants or even incidentally without explicit intent from the organization. Such tools may generate rich data about facial movements, voice, language, and other behavioral characteristics.

Workplace data also come from current employees, traditionally collected in the form of performance reviews but more recently with continuous technology-assisted measurements. Performance data have historically come from managerial performance reviews, which are often conducted quarterly, annually, or at other regular intervals. Traditional performance data consist of both objective measures (e.g., sales measured in dollars or units) and subjective measures (e.g., supervisor ratings of effort).

Technological advances in the workplace have created an opportunity to collect behavioral performance data that are continuous, more comprehensive, and more multifaceted via electronic performance monitoring (EPM). EPM refers to the use of technology to observe, record, and analyze information that relates to a worker's job performance (Ravid et al., 2020). Examples of EPM data include computer data (e.g., keystrokes and clicks, history of the applications or web pages accessed) and camera data. Cameras can monitor the movements and behaviors of employees, both during work time and on breaks; they are also increasingly being placed on employees like police officers (Adams & Mastracci, 2017, 2019). Organizations may also collect data from employee cell phones (Harari et al., 2020; Wang et al., 2018). See Harari et al. (2017) for a review of smartphone sensing methods in the behavioral sciences. Wearable devices like Fitbits may be provided to employees by the organization, which then can give the organization access to highly personal data such as heart rate, steps, activity, sleep patterns, and mental health information (O'Hara, 2019). Other wearable devices may be designed to track workers in physically demanding jobs; for example, wearable GPS badges can detect when employees step into hazardous terrain (Kanan et al., 2018), and ultrasonic wristbands can track hand movements for tasks requiring manual dexterity (e.g., Ong, 2018).

Public Data

It may not be feasible or desirable in certain situations to collect and monitor Big Data about individual employees; in these cases, existing archives of public Big Data shared by researchers or government entities may be valuable. The practice of making one's data publicly available has been advanced by the open science movement. The Open Science Framework (https://osf.io) hosts many data sets and supports researchers in the design and dissemination of projects per the open science philosophy. The Occupational Information Network (https://onetonline.org), maintained by the U.S. Department of Labor, is a source of rich occupation-level data, providing information on the knowledge, skills, and abilities; education requirements; and work activities for over 900 occupations. Both Data.gov (https://www.data.gov) and Google's data set search (https://datasetsearch.research.google.com) can be used to locate publicly available data sets. The American Psychological Association publishes *Archives of Scientific Psychology*, an open-access and open-data scientific journal. Public-data sources support open science and serve as valuable resources for researchers and practitioners; independently or in combination with other sources, they can be used to answer many questions relevant to the workplace.

Validity and Reliability

Using Big Data in organizational contexts is predicated on the assumption that the insights drawn from it are both reliable and valid. *Reliability* is the overall consistency of a measure (i.e., scores are stable across setting and time), and *validity* is the extent to which evidence supports inferences drawn from test scores (i.e., scores represent what they are intended to and show the expected relationships with other scores; American Educational Research Association [AERA] et al., 2014; Campbell & Fiske, 1959).

Validity represents a key consideration for organizations using new measures to select employees. Research has shown that behavioral trace data can be used to accurately infer job-relevant characteristics like personality (Chorley et al., 2015; de Montjoye et al., 2013). Still, a critical question to answer is whether the inferences one can draw from new, nontraditional measures show the same levels of validity and reliability as expected of traditional assessments of the same constructs. Validating new predictive models against validated assessments is critical. However, new behavioral data sources may also provide *incremental validity* (defined as the additional outcome variance explained by including a new predictor in an analysis) and may do so at a low cost to the organization. Furthermore, these methods do not require an applicant to invest additional time and effort in the application process, which can reduce fatigue and attrition of applicants.

Some researchers have noted that nontraditional and traditional selection measures are similar in that they both aim to assess the same constructs. Chamorro-Premuzic et al. (2016) argued that nontraditional measures of work-related potential are largely technological modifications of old methods and that these new modifications are simply new ways to assess known constructs that predict job performance. For example, they noted that synchronous, in-person interviews can be replaced by digital, asynchronous video interviews with voice profiling, and résumés can be replaced by LinkedIn scraping. While the analysis process is different, the constructs being assessed are the same: expertise, social skills, past experiences, and so on. This is a particularly poignant justification for the use of Big Data and related analytics. These technologies are not assessing anything inherently new; rather, they are assessing known constructs in new and efficient ways.

To strike a balance between efficiency and validity, more research is needed to ensure novel methods are valid. This validation step is critically important for organizations and the quality of their decision-making processes. For example, existing work shows that for some constructs (e.g., neuroticism), technology-mediated measures fail to converge with traditional measures (Hickman et al., 2019). Using invalid methods might not just undermine the outcome of hiring or promotion decisions but could also lead to legal ramifications if job candidates or employees are able to show that they were discriminated against by algorithms that were insufficiently tested and validated.

Relatedly, it is an open question whether faking is an issue with new, Big Data-based assessments. When presented with predictive assessments, employees and applicants will eventually alter their behavior to optimize it for the desired outcome (i.e., Goodhart's law; Goodhart, 1984, Chapter 3). However, faking may be less of an issue for game-based assessments than for self-assessment questionnaires because it may be more difficult to figure out the desired behavior and then behave consistently in that way over a period of time. Also, research has shown that, for the most part, faking on personality tests does not pose a serious threat to the validity of assessments in the context of selection (Hogan et al., 2007). The ability to "fake good" (i.e., answer inauthentically in the desired direction) may demonstrate knowledge of the desired behavior and an ability to apply that knowledge. While behavioral change in response to assessments is not necessarily a bad thing, losing predictive power because of range restriction on a predictor *is* (Salkind, 2010). In sum, more research is needed to determine if faking is an issue in the context of Big Data-based assessments, and it is important to regularly test and update predictive algorithms to ensure they remain reliable and valid.

BIG DATA APPLICATIONS IN THE WORKPLACE

Having reviewed Big Data sources and considerations for the legal and appropriate use of Big Data, we now consider the ways organizations can use Big Data to solve work-related problems. It is important to preface this discussion with an acknowledgment of some of the effects of the COVID-19 pandemic on the workplace. As a result of the pandemic, organizations have drastically altered their business practices by moving employees out of the office, using virtual interviews to select new employees, and shifting to virtual training. These alterations were enabled by technology—from videoconferencing services to virtual-interview software to activity-tracking applications. For example, organizations (a) made heavy use of software like Zoom and Microsoft Teams to hold meetings and communicate with remote employees and applicants, (b) used asynchronous video-interview services that automatically score applicant responses, and (c) had employees install activity- and productivity-tracking software on their devices to ensure the employees are working. The increased use of these technologies gave organizations the capability to collect much more data on employees and applicants than before and may have facilitated the collection and use of novel types of data. It is likely that many of these changes will be permanent as organizations and employees realize the benefits (e.g., efficiency, cost) of using these technologies. Fundamentally, COVID-19 has likely accelerated the collection and use of Big Data in the workplace, which increases both the potential benefits and drawbacks associated with using these data.

Big Data in Selection

Big Data can be useful in selecting applicants. A specific example of using Big Data and automation in selection processes is the automatic analysis of text application materials (e.g., résumés, cover letters). Hiring managers can use programs to read résumés, cover letters, short responses, and other texts generated by applicants. These programs can analyze such texts much more quickly and consistently than human readers, which may reduce human bias (e.g., order, halo effects) and increase the speed and accuracy of applicant vetting. Text analysis can be expanded beyond résumés and cover letters to other texts generated by the applicant, such as posts on professional and social media sites.

There are different methods of automated text analysis, each with strengths and weaknesses. The automated reading, coding, and analysis of text (computer-aided text analysis [CATA]) can be conducted using either a top-down or bottom-up design. A *top-down* design typically involves the use of a prespecified dictionary and analyzes the frequency and co-occurrences of these dictionary words in a specific text. This approach is broadly used in psychology (Oswald et al., 2020). A commonly used software package is Linguistic Inquiry and Word Count (LIWC), which counts words in psychologically meaningful categories (Tausczik & Pennebaker, 2010). Organizational researchers have made use of certain dictionaries in LIWC—in particular, the dictionaries regarding positive and negative behavior (e.g., Friedman et al., 2004) and need for affiliation. LIWC and NVivo, another CATA software, account for over half (56%) of CATA software used by organizational researchers (as assessed by Short et al., 2018, who also provided recommendations for future research involving CATA). LIWC-based text analysis has also been used to study differences in the language used to describe male and female applicants in their letters of recommendation (Madera et al., 2009). In their study, Madera et al. (2009) determined that women were typically described using fewer agentic (e.g., "independent," "earn," "gain") and more communal (e.g., "helpful," "caring") words in letters of recommendation and that communal characteristics had a negative relationship with hiring decisions based on letters of recommendation in academia. These results can be applied in many ways, from informing letter writers about effective writing strategies to identifying barriers that certain individuals and groups face in the hiring process.

In contrast to top-down designs, *bottom-up* text analysis does not rely on predefined dictionaries. Instead, it involves analyzing word and phrase co-occurrence and frequency. This type of analysis requires a large amount of text data but is not constrained by the contents of a predefined dictionary. There are various methods for bottom-up text analysis, including feature extraction and word co-occurrence analysis. *Feature extraction* is used to identify particular words that differentiate texts from each other (Chen & Wojcik, 2016); Diermeier et al. (2012) used feature extraction to identify the words from U.S. Senate speeches that were most indicative of liberal and conservative ideologies. A similar approach could be used to identify words that best differentiate between qualified and underqualified applicants or between applicants who tend to perform well on the job and those who do not. *Word co-occurrence* analyzes how words are used together in texts; a common word co-occurrence approach is latent semantic analysis. *Latent semantic analysis* is the theory and method for extracting and representing the contextual usage and meaning of words and phrases using statistics applied to text (Landauer & Dumais, 1997). Marksberry et al. (2014) studied Toyota's employee suggestion programs using latent semantic analysis.

In organizational contexts, both top-down and bottom-up approaches to text analysis can be useful. Using a dictionary (as in LIWC) allows organizations to connect text data to validated psychological constructs. This supports generalizability to other contexts and integration with other assessments that may measure the same constructs. Bottom-up analysis will likely describe an organization's data more closely and may extract patterns from words that are organization- or industry-specific and are thus not included in a dictionary. Organizational researchers should assess the likelihood that important insights are being missed by a dictionary-based analysis in their decision on which approach to use. A combination of the two may be ideal, allowing organizational researchers to generalize the results to other settings and data as well as examine organization-specific idiosyncrasies. As an example, Campion et al. (2016) used semi-structured natural language processing to evaluate achievement records in the context of selection. Such an approach takes advantage of previously constructed resources like dictionaries but also increases flexibility by giving the user the option to make additional manipulations to the knowledge structures extracted.

Another important data source during the personnel selection process is publicly available data from social media. Organizations regularly scrutinize social media pages of applicants during the selection process (Segal, 2018). Much of the time, it is an employee involved in the hiring process who is looking at and drawing inferences from these sites; however, this process can also be automated using machine learning. Research is conflicted regarding whether data from social media are valid. Some research suggests that social media data may not be effective at predicting job performance or turnover (Van Iddekinge et al., 2016), while other research has found that these data may be fairly good at predicting self-reported personality scores. Park et al. (2015) used linguistic analysis of Facebook posts to assess the Big Five personality traits. Their results were that Facebook-based personality measurements converged with self-reports, discriminated between facets, showed stability over time, and correlated with other constructs in similar patterns to self-reported personality. Other researchers have had similar results with other sources of social media data, including tweets and other Twitter data (Golbeck et al., 2011) and Facebook likes (Youyou et al., 2015), though further validation research is needed (Tay et al., 2020).

Social media data offer exciting opportunities for organizational assessment and prediction. Traditionally, organizations have had to administer personality tests to accurately assess personality in applicants. However, if personality could be accurately assessed without the administration of a test, organizations could streamline selection and possibly reduce applicant attrition due to fatigue. If validated by organizational researchers, this type of assessment could also be used on internal data (e.g., emails) to assess and continuously update personality profiles of employees. Such assessments could inform professional development and training opportunities or inform team building to maximize productivity and minimize conflict.

Big Data in Recruitment

Less is known about how Big Data may help during recruitment efforts prior to selection. *Recruitment* is the process of attracting human capital and involves identifying potential employees, informing them about job and organizational characteristics, and convincing them to join the organization (Barber, 1998). Big Data may help organizations improve job–organization attractiveness (i.e., an applicant's overall evaluation of the attractiveness of a job and/or organization; Chapman et al., 2005). For example, an organization could collect behavioral trace data from applicants as they apply online and combine those data with their application materials (e.g., Hernandez et al., 2020) and use those data to match the candidate to an open position ("SAP Resume Matching, Powered by NVIDIA: Finding the Best Talent," 2018). These data could also be used to generate profiles of successful candidates that can then be used to target future candidates (many firms offer such services). Recruiters can scrape candidate information from sites like LinkedIn or ZipRecruiter and invite qualified individuals to apply, thereby actively reaching potential candidates and possibly improving the quality of the applicant pool.

Big Data in Training

Big Data is also informative during training. *Training* refers to any planned effort by a company to facilitate learning of job-related competencies, knowledge, skills, and behaviors by employees (Noe, 2020). Training programs vary greatly in practice, ranging from a single, short session meant to address a specific technical skill (e.g., familiarization with a point-of-sale system for a new server in a restaurant) to an extensive, long-term training program (e.g., boot camp for new military service members). Here we consider primarily formal training programs, which are preplanned, have specific learning objectives, and involve post hoc evaluation of the training for the sake of making improvements.

Needs Analysis

Big Data can be used to identify training needs easily and accurately. *Training needs analysis* is the systematic review of both job requirements (i.e., job analysis; Sanchez & Levine, 2012) and the current knowledge and skills of the population to be trained, with the goal of identifying gaps between job requirements and skills to determine if training is needed (Goldstein, 1993). This step is crucial in identifying learning outcomes of the training and defining how successful training would look. Algorithms can improve the needs analysis process by automatically combining multiple data sources like job descriptions, performance assessments, and behavioral data. For example, artificial intelligence can report that high performers in a particular role use spreadsheets for around 2 hours a day and that the most frequently used functions are pivot tables, basic descriptive statistics, and generating graphical reports. These artificial-intelligence reports can be created quickly and precisely without relying on incumbent or supervisor reports, reducing error associated with human subjectivity and freeing personnel from such tasks. The results from these reports can then be automatically compared to behavioral data from other employees in the same role to identify discrepancies. This automated needs analysis can be done continuously and without much human oversight and can direct training on specific skills in a timely and targeted manner.

Technology-Enhanced Training

Technology-enhanced training opens up the opportunity for the collection of vast and multifaceted data from learners. These data can be used for functions like tracking learner growth and monitoring training effectiveness. Gamification and virtual reality (VR) training are examples of opportunities supported by computer-based training that can improve the learning process and provide large quantities of multifaceted data about learners.

Gamification is the process of using game elements in nongame settings (Deterding et al., 2011). A gamified training might use game elements like leaderboards, earned badges, or points to motivate learners throughout the training. Research has shown that game elements can improve persistence and task performance (Groening & Binnewies, 2019), as well as engagement with the content (da Rocha Seixas et al., 2016; Hamari, 2017). In addition, learners tend to expect greater value from gamified instruction (Landers & Armstrong, 2017). Gamified trainings are often delivered online, and they provide the opportunity for the collection of large amounts of data throughout the training process. These data can be used to model learner progression at a granular level throughout the training, as well as assess the extent to which trainees have a firm grasp of the topics and can transfer that knowledge to various work situations.

VR is another training delivery method that can generate a lot of real-time Big Data and remove barriers associated with learning. Many VR trainings use gamification to improve learner experiences and provide real-time performance feedback as the learner progresses. VR training has many advantages that make it a good option in certain situations, including its ability to support the training of skills that are expensive, dangerous, or difficult to teach in "real life." VR training can be used to teach welding (Pitcher et al., 2019) or surgery (Buckley et al., 2012; Chellali et al., 2016) in an immersive learning experience that is cheaper, safer, and easier to observe than traditional methods. For example, the traditional apprenticeship learning model used in surgery of "see one, do one, teach one" may not be appropriate for learning certain surgical techniques, such as laparoscopy (Chellali et al., 2016), and VR can be used to allow trainees to gain the skills they need with minimal risk. VR simulators can also incorporate real-time feedback and guidance in the form of indicators of performance elements. These feedback mechanisms serve to guide performance in real time and represent data about specific performance metrics that can be used to model, with great detail, trainee learning and performance over time. For more on VR in the context of Big Data, see Chapter 8 of this volume.

Big Data collected from training methods like those described previously can be used to develop in-depth feedback for trainees on specific elements of their performance. This feedback can be generated automatically based on the trainee performance data and built into a dashboard for easy interpretation (organizations like CrossKnowledge and Domo create such products). Such a dashboard, automatically generated, can support the learning of individual trainees with individualized feedback that is specific to their strengths, weaknesses, and progression through the training. This is beneficial because it can (a) reduce the demand on instructors to provide individualized feedback to each trainee, (b) be scaled easily and inexpensively, (c) integrate data from wearable sensors or physiological tracking devices (e.g., eye trackers) to precisely measure factors like engagement, and (d) extend beyond training into the job itself. For example, algorithms could monitor work performance for application of learned skills or materials and provide feedback to the employee as they work, make recommendations for retraining if needed, and generate a longitudinal report of training effectiveness that could be used to improve the training program. All of these applications are enabled by automated integration, processing, and analysis of data from a wide variety of longitudinal sources around the training and the employee.

Big Data in Performance Management

Organizations often collect vast amounts of performance data about employees in real time, which can help optimize performance, reduce counterproductive work behaviors, and ensure the safety of the employees and the organization. Employee reactions to these monitoring techniques vary based on characteristics of the monitoring tool, characteristics of the individual being monitored, and the presentation of the monitoring by the organization. While the monitoring of employee performance is not a new phenomenon, organizations are increasingly using advanced technologies to do so (Ravid et al., 2020)—a trend that accelerated in 2020 and 2021, due to the COVID-19 pandemic, as organizations sought to monitor now-remote employees.

As discussed in the Sources of Big Data section of this chapter, using technology to gather performance information (i.e., EPM) is multifaceted and complex. Ravid and colleagues (2020) provided a framework with four broad elements to characterize the various forms of EPM. First, EPM can differ in the explicit or perceived *purpose* of the monitoring—for example, performance management, loss prevention, safety and security, development and training, or even simply authoritarian surveillance (i.e., having no clear, justifiable purpose). Second, EPM can differ in the *invasiveness* of the monitoring. For example, the target of the monitoring can vary from the behaviors of the employees—such as hand movements (Ong, 2018)—to their thoughts or feelings; employees may have some degree of control over the monitoring, such as the ability to turn off website tracking while on a break or to take off a location-tracking badge when on lunch. Third, EPM can differ in the *synchronicity* of the monitoring: Traditional, in-person performance management is largely asynchronous (or noncontinuous); however, many EPM technologies have the capability to constantly monitor employees, often in the background and without the conscious awareness of the employees. Fourth, EPM can differ in the *transparency* of the monitoring—that is, the degree to which monitoring procedures are clearly communicated to employees. Collectively, these four elements (purpose, invasiveness, synchronicity, and transparency) help describe differences across various forms of EPM.

Employees react differently to different types of EPM. For example, research has shown that employees are less likely to perceive EPM as invasive when they have greater control over monitoring practices (Douthitt & Aiello, 2001; Ravid et al., 2020). Likewise, employees tend to react more negatively to person-focused than task-focused monitoring (Jeske & Santuzzi, 2015; Ravid et al., 2020). Further, employees tend to react more positively to EPM that is high in transparency (Ravid et al., 2020). Recent research in the area of EPM reactions suggested that the rate with which employees accepted certain EPM practices depended heavily on the data sources being used. As discussed by Tomczak, Zarsky, et al. (2020), the three categories of monitoring that were able to be distinguished with regard to their acceptance rates were (a) physiological (e.g., heart rate, body temperature), (b) personal information (e.g., web browsing, personal cell phones), and (c) company property (break rooms, work cell phones). Across a variety of occupations, employees tended to be most likely to accept monitoring of company property and least likely to accept monitoring of physiology (Tomczak, Zarsky, et al., 2020), though it remains to be seen if these perceptions have changed as a result of COVID-19. It may be that employees have become more accepting of physiological monitoring, due to the regularity of temperature checks and other pandemic-related practices.

An example of EPM use comes from Amazon, which uses an app called Mentor to track its delivery drivers (Peterson, 2019). Mentor scores drivers on their driving, monitoring behaviors like hard brakes, swerving, phone use, and seatbelt habits. The app provides the drivers with daily feedback on these behaviors and shares driving reports with their managers. Employees have reported mixed reactions to the app: Some have said it makes them safer by deterring phone use and using bonuses to incentivize safe driving, while others claim it is inaccurate and may even increase hazardous behavior; for example, according to Peterson (2019), one driver said that they are always concerned with getting reported for a hard brake, but when driving their truck down a hill and seeing a light turn yellow ahead, they may have to decide between being safe (i.e., braking hard) and protecting their driving metrics (i.e., running the light). This illustrates how employees' concerns over their Mentor scores might actually induce unsafe behavior in certain situations. Research has shown that when organizations monitored workers on one aspect of the job, those workers tended to prioritize their performance on that monitored aspect at the cost of their performance on unmonitored ones (Stanton & Julian, 2002). The granularity of the behavioral monitoring of the Mentor app reflects a Tayloristic emphasis on behavioral optimization. For example, the app directs drivers to turn the vehicle off at every stop to save gas, and to avoid reversing when possible; according to Amazon and eDriving, the company behind the Mentor app, the arguments for this level of control are safety and efficiency.

In extreme cases, algorithms can be deployed with real-time feedback to fully replace human managers in the delivery of performance management. At Uber, for example, each driver's performance is continuously tracked and reported back in the form of immediate and weekly reports of their passengers' ratings. Data within these reports are often used to inform performance management decisions, such as terminating drivers whose past 50 ratings fall below 4.6 out of 5. Although algorithms present organizations with a convenient means of performance management, employee reaction to the practice remains a concern, as protests among Uber drivers demonstrate (O'Connor, 2016).

An argument for the use of behavioral monitoring like EPM in organizations is that, in theory, EPM provides data related to the work process of the employee. These data allow for the improvement and optimization of employee processes, which is necessary to improve the output of work. One must recognize, however, that behavioral data that are often captured using EPM may not be a strong indicator of an employee's work process in all situations. It is likely that the type of job moderates this relationship, with behavioral data from physical or repetitive jobs having more relevance than those data from knowledge work. This is simply because the optimal way to produce outcomes associated with repetitive or physical jobs is often more easily defined and measured than that of knowledge jobs. Research has shown that reactions to EPM vary based on the characteristics of the job, the monitoring, the individual, and the implementation of the monitoring (White et al., 2020; Yost et al., 2019) and may include feeling that the psychological contract (explained in greater detail later in this chapter) has been breached. Consequently, organizations need to be intentional in their monitoring decisions and implementation.

Organizations should consider the relevance of Big Data gathered via EPM. As discussed, EPM clearly represents an organizational use of Big Data: It generates a large amount of data (volume); can do so in a continuously updating manner (velocity); can be of many different types, such as video, keystrokes, and location data (variety); and may not be perfectly accurate or of high quality (veracity). Importantly, these characteristics do not imply that EPM data are relevant or useful. Simply because an organization can collect biometric data from employees does not mean that collection is necessary. In high-stress, dangerous jobs, the relevance of temperature and heart-rate data, for example, would be much clearer than the relevance of those data for a desk job in an office. Industrial–organizational psychology professionals can guide organizations on the relevance of such data based on the situation, and such guidance can protect the privacy of employees, protect the organization from liability, and support a positive perception of the monitoring.

Big Data in Culture Assessments

Big Data is also useful in understanding an organization's culture. For example, organizations can use social network mapping to understand the patterns of communication in their organization. Social network analysis has been, for this purpose, in organizations for a long time, with social data collected using interviews, observations, and surveys (Tichy et al., 1979). More recently, social networks have been mapped using wearable devices such as sociometric badges, which track proximity to other badges and may record patterns of communication (Fischbach et al., 2010; Kim et al., 2012), and by collecting email or Slack messages. Social network analysis can reveal which individuals may be more central to the network (i.e., star employees; Tichy et al., 1979) and can provide insights into relationships between employees. For example, researchers used sociometric badges and social network analysis to map employee social groups and then coordinated breaks so those groups could be together—which strengthened them and increased employee productivity (Waber et al., 2010).

Other forms of Big Data analysis can support culture assessment. For example, sentiment analysis conducted on tweets or Glassdoor posts about a company, or emails exchanged within a company, can reveal potential issues. The language employees use on a daily basis can reflect their underlying values, and this language may or may not be aligned with established communication norms within the organization. For example, Srivastava et al. (2018) used over 10 million internal emails to analyze communication patterns of employees within a midsized technology firm. Using a language-based model they built, they identified patterns of communication that indicated problems related to culture fit. The model demonstrated that the speed with which an employee adapted an organization's cultural language was related to their likelihood of being fired, and employees who had adapted to the culture but later showed declines in cultural fit were more likely to voluntarily leave the organization. Such data provide organizations with useful information to identify the antecedents and consequences of cultural fit among their employees.

Big Data in Turnover

Data are important for predicting and preventing *turnover* (i.e., the act of employees leaving an organization). *Voluntary turnover* is initiated by the employee (e.g., quitting), whereas *involuntary turnover* is initiated by the organization (e.g., firing). Turnover can be related to external factors like unemployment rate (i.e., how many job opportunities are available at that time), work-related factors like job performance and satisfaction, and personal characteristics like age (Cotton & Tuttle, 1986). While some turnover can benefit an organization, it is generally considered to be negative for a number of reasons. First, there is a loss of productivity associated with turnover. There is a period of time during which a position is vacant, which likely means that the value normally created by that role is not being created. When turnover is high, employees are typically less satisfied and engage in more counterproductive behaviors (Heavey et al., 2013). In addition, it is costly to recruit and select a new employee to fill a vacant role and to train new employees such that they learn their way around the organization to achieve their full productivity potential. Overall, higher rates of turnover tend to result in worse immediate outcomes like production efficiency and customer satisfaction, as well as longer term outcomes like sales efficiency and profit margin (Heavey et al., 2013).

Building on decades of academic research (Cotton & Tuttle, 1986), researchers have recently turned to Big Data to investigate ways of reducing turnover. The Society for Industrial and Organizational Psychology (Division 14 of the American Psychological Association) hosted a competition, before its 2018 annual conference, in which 12 teams of researchers from 14 different institutions and organizations competed to develop the best predictive algorithm for voluntary turnover of employees of Eli Lilly, a pharmaceutical company (Putka, 2018). The data used in this competition were drawn from 32,296 Eli Lilly employees active in December 2009.

The data included 162 predictor variables, ranging from demographics to location to job performance, and was split into a training set ($n = 24{,}205$) and a testing data set ($n = 8{,}091$). The winning team's model could predict the outcome of employees in the test data with an accuracy of about 83%. That means the model was 33% better than a coin flip in determining whether the employee was still active with the company or had left voluntarily. This level of accuracy is comparable to those of other models based on Big Data (e.g., 74% predictive accuracy; Rombaut & Guerry, 2018). Models of turnover based on traditional data tend to look at specific individual predictors, each of which typically can explain no more than 19% of the variance in turnover ($R^2 < .185$; see Heavey et al., 2013, for a meta-analytic review). A model such as the one developed in the SIOP challenge could be used as an early warning system to identify employees who are likely to voluntarily leave the organization in the future. Using these insights proactively, the organization could implement preemptive strategies to retain those employees.

Big Data and the Psychological Contract

In addition to opportunities for Big Data use in organizations and organizational research, it is important to also consider how employees may react to Big Data use by employers. Reactions may be positive if the data are used to benefit the employee (e.g., to help them develop; Wells et al., 2007); however, organizations may not always be transparent or justified in their use of Big Data, and data leaks or hacks can endanger the privacy of employees. In this section, we review consequences that may follow from such issues.

Misuse and mismanagement of employee data can harm employees, may cause negative reactions, and could constitute a breach of the psychological contract. The *psychological contract* is an employee's implicit belief system regarding their exchange relationship with their employer (Rousseau, 2011). Breach of the psychological contract is related to negative outcomes like lower trust, higher turnover intentions, feelings of violation, lower job satisfaction, and decreased in-role performance (Morrison & Robinson, 1997; Zhao et al., 2007). As EPM facilitates the collection of a greater volume of increasingly personal employee data, the chance that these data are misused or mismanaged by the organization increases as well. For example, health-indicative data such as heart rate and blood pressure, if collected on the job, may be used to fire those who have a higher risk of health problems or perhaps to raise the insurance costs of those employees; a range of psychological and legal consequences could be expected to follow. Hacks and data leaks could also lead to breaches, and employees would lose their trust in the organization to protect and responsibly steward their sensitive data. EPM itself may also lead to feelings of psychological contract breach, especially in jobs with high autonomy (Tomczak, Behrend, et al., 2020). Therefore, responsible data stewardship is a precondition for organizational data collection and use.

The case of Castlight Health Inc. constitutes a recent example of how irresponsible use of personal employee data can quickly result in a psychological contract violation with lasting consequences. Castlight partnered with organizations (e.g., Walmart, Time Warner) to manage their employee health data (Silverman, 2016). Among other services, Castlight offered to predict when an employee was pregnant, or trying to become pregnant, using variables like the woman's age, information about her birth control prescription from her health insurance, and her search history on the company health app. Castlight reported the results of these predictions to the employers in the form of an aggregated and anonymized report.

While these kinds of practices may provide benefits (e.g., targeting health programs to the needs of the workforce, addressing general areas of weakness), the employee may feel that this type of monitoring invades their privacy, constituting a breach of the psychological contract. This could cause the employee to be less satisfied with their job or to leave the company. When made public, cases like this can also damage the image of the organizations involved. There are also obvious concerns regarding privacy and discrimination: One especially salient concern is that employers may try to use this information to discriminate against women who are believed to be trying to get pregnant, which is both unethical and illegal under Title VII of the Civil Rights Act (1964).

IMPLICATIONS OF BIG DATA FOR DIVERSITY AND INCLUSION

The intersection of Big Data with diversity-and-inclusion efforts in organizations is complex and multifaceted. The bottom line is that Big Data is information, and the decisions people make—either intentionally or unintentionally—will determine whether that information helps or harms organizational diversity-and-inclusion efforts. Big Data can be used to support diversity and inclusion but may also be used to inadvertently harm such efforts through the reproduction of biased data and systems.

Big Data can inform organizational leaders on struggles faced by employees from disadvantaged groups and can improve recruitment of such employees. Collecting data about social networks in organizations can reveal patterns of connection among employee groups and individuals and can provide an early warning if employees from disadvantaged groups seem to have difficulty connecting to organizational systems or individuals. These data would also make it possible to monitor the results of any changes or interventions in real time, which supports dynamic, fast responses to these issues. In terms of selection, collecting thorough data about applicants' engagement and experience throughout the recruitment process can help identify if any points in the process are causing a disproportionate loss of minority candidates. In addition, Big Data can make it easier to identify sources of minority candidates (e.g., particular job sites or recruitment events that tend to attract minority candidates), which can then be emphasized in active recruiting efforts.

Algorithms (trained using Big Data sets) are effective only at replicating the data on which they are trained. Algorithms have the potential to help detect and thus prevent discrimination (Kleinberg et al., 2018, 2020); however, if biased data (i.e., nonrepresentative or inaccurate) are used to train an algorithm, the algorithm will replicate that bias in its decisions. Consequently, failing to identify bias in training data is an action that can lead to biased algorithms and decision aids; for example, if the traditional hiring processes have favored men or certain majority groups and discriminated against women and minority groups, this bias will be reflected in the algorithm and will affect its output.

Bias in the context of predictive algorithms is defined as the extent to which an algorithm or model systematically overpredicts or underpredicts the outcomes of a specific group of people (AERA et al., 2014). In other words, the algorithm makes systematic mistakes in predicting whether a candidate, for example, is suited for the job or not; that is, a biased algorithm might overestimate the chances of a White man performing well on the job, while it might underestimate the likelihood of a Black woman doing so. While no algorithm perfectly predicts employment outcomes, a biased algorithm produces systematically skewed outcomes (AERA et al., 2014) such that false positives (e.g., recommending an unsuited candidate) and false negatives (e.g., failing to recommend a suited candidate) are unevenly distributed across groups.

From a purely statistical prediction perspective, one could argue that all variables that increase predictive accuracy should be included, even if those variables relate to protected class membership. Despite the logic of this approach, the Civil Rights Act (1991) explicitly prohibits, in the United States, the altering of scores on employment-related tests in any way based on protected class membership (sex, race, color, religion, or national origin). This effectively makes it illegal to adjust scores based on protected class membership in employment-related algorithms as can be done in other contexts (see Giorgi et al., 2020, for an example of using restratification to reduce bias from undersampling). Researchers are developing methods to increase predictive validity while reducing adverse impact—using Pareto optimization, for example. The Pareto-optimization method involves determining the weighting scheme for a given set of predictors that optimizes for two different outcomes (De Corte et al., 2008). This method shows promise in reducing adverse impact while maintaining validity in selection decisions (De Corte et al., 2008; Song et al., 2017).

Organizations must consider more than bias in algorithmic outputs. It may be the case that an algorithm is unbiased (i.e., it does not systematically discriminate against a certain group) yet still predicts differences in performance for different groups. Although it may not be biased in technical terms, organizations still may not wish to use an algorithm that results in differential outcomes, and they must make the decision to use it or not based on their goals and norms. It is important not to blame an algorithm that produces a biased outcome if biased data were used to train it (National Council on Measurement in Education, 2019); rather, one must tackle the more difficult, but ultimately more critical, task of identifying and understanding the causes of biased data.

An increasing number of companies that sell HR algorithms (e.g., selection-decision aids) practice "algorithm debiasing." They debias algorithms by removing predictor variables that show mean differences between protected classes. This practice, while well-intentioned, can be counterproductive, as it reduces the accuracy of the algorithm, introducing error and increasing the randomness of the prediction. The more random a decision-making process, the less biased it will be—but the less diagnostic it will be as well. For example, a coin flip is perfectly random, making it both perfectly unbiased and perfectly useless for making decisions. Importantly, debiasing an algorithm can make leaders feel that they have solved the problem of bias, when it in fact has just blinded them to the true causes of inequities and fooled them into thinking they have taken measures to combat it. In sum, by misidentifying the causes of inequities, reducing the accuracy of predictive information, and allowing leaders to believe they have sufficiently accounted for bias, the debiasing of algorithms can, in practice, severely harm diversity efforts rather than support them. Researchers and leaders must use Big Data as a way to see clearly how inequities manifest in the workplace and use that information to direct their efforts at effective remedies. Ultimately, organizations need to ensure that the decisions they make, whether based on algorithms or not, are free from bias and do not result in discrimination.

THE FUTURE OF WORKPLACE BIG DATA

We have more data at our disposal now than ever before. At the same time, we have less of a monopoly on those data; computer scientists now own more data about people than psychologists do (Griffiths, 2015).

In this environment, making a case for the value of psychological science is essential. Psychologists have worked for over a century with data representing human behavior, thought, and affect. Throughout this period, psychologists have generated a vast amount of knowledge about the human condition and have developed theories to synthesize and generalize those findings. A purely empirical approach, in which analysts identify patterns without any theorizing, can lead to spurious findings and misunderstandings (as seen in the dustbowl empiricism of the 1950s). Big Data leads to true insight only when social scientists (a) apply their expertise to the interpretation of models, (b) actively engage in the process of generating meaningful features for training predictive models, and (c) raise questions regarding the causality of relationships and the reliability and validity of new technological tools.

Psychologists can ensure that Big Data is collected and used ethically. Knowledge of professional guidelines like the Common Rule (Protection of Human Subjects, 2017), *Ethical Principles of Psychologists and Code of Conduct* (American Psychological Association, 2017), and participation in human-subjects research training (e.g., NIH "Protecting Human Subjects Training," CITI Program) is a necessary component of psychological-research training and supports ethical human-subjects research regardless of the scale of the data. In organizations, ethical training is particularly important due to the vulnerability of employees, workplace power dynamics, issues of data ownership, and legal hurdles. Psychologists can ensure that research and products that rely on Big Data are designed responsibly and with well-being as an outcome, which may help reduce algorithmic pitfalls that contribute to conflict, the spread of false information, and maladaptive behaviors.

Ultimately, using Big Data in the workplace offers tremendous opportunities but also poses considerable challenges. As with any tool, Big Data may not be well suited to all needs. Psychologists often have extensive experience with small convenience samples collected from a specific population of interest. Such samples will not cease to be useful in the era of Big Data, as particular research questions will always be best answered with an intentionally collected and well-defined data set. Most important, the combination of Big Data and traditionally sized data sets might hold the greatest promise for psychologists in navigating the goals of (a) external versus internal validity and (b) explanation versus prediction.

CONCLUSION

Big Data can be used to understand and change the way people work. This chapter has reviewed the science and practice of Big Data in the workplace, including sources of workplace Big Data and applications of Big Data analytics for organizations and individual workers. Big Data provides an opportunity for researchers and practitioners to improve the quality of HR functions like recruitment, selection, retention, and performance management. At the same time, Big Data carries ethical, practical, and legal challenges. These challenges include issues with (a) conflicts of interest (e.g., employees analyzing the data of their coworkers), (b) identifiable and linkable data (data that is anonymous on its own but can be linked with other data to identify the owner), and (c) informed consent (see Chapter 17 of this volume for a more detailed treatment of such issues). Psychologists are well equipped to handle these challenges, use Big Data when appropriate to answer interesting and meaningful questions about the nature of people at work, and implement scientifically informed practices to improve the well-being of employees and organizations.

REFERENCES

AB-25 California Consumer Privacy Act of 2018. (2019). https://leginfo.legislature.ca.gov/faces/billTextClient.xhtml?bill_id=201920200AB25

Adams, I., & Mastracci, S. (2017). Visibility is a trap: The ethics of police body-worn cameras and control. *Administrative Theory & Praxis*, *39*(4), 313–328. https://doi.org/10.1080/10841806.2017.1381482

Adams, I., & Mastracci, S. (2019). Police body-worn cameras: Effects on officers' burnout and perceived organizational support. *Police Quarterly*, *22*(1), 5–30. https://doi.org/10.1177/1098611118783987

American Educational Research Association, American Psychological Association, & National Council on Measurement in Education. (2014). *Standards for educational and psychological testing*. American Educational Research Association.

American Psychological Association. (2017). *Ethical principles of psychologists and code of conduct* (2002, Amended June 1, 2010, and January 1, 2017). http://www.apa.org/ethics/code/index.aspx

Barber, A. E. (1998). *Recruiting employees: Individual and organizational perspectives*. Sage.

Braun, M. T., Kuljanin, G., & DeShon, R. P. (2018). Special considerations for the acquisition and wrangling of big data. *Organizational Research Methods, 21*(3), 633–659. https://doi.org/10.1177/1094428117690235

Buckley, C. E., Nugent, E., Ryan, D., & Neary, P. C. (2012). Virtual reality—A new era in surgical training. In C. Eichenberg (Ed.), *Virtual reality in psychological, medical and pedagogical applications* (pp. 141–166). IntechOpen. https://doi.org/10.5772/46415

California Consumer Privacy Act of 2018, no. AB-375, California State Legislature. (2018). https://leginfo.legislature.ca.gov/faces/billTextClient.xhtml?bill_id=201720180AB375

Campbell, D. T., & Fiske, D. W. (1959). Convergent and discriminant validation by the multitrait-multimethod matrix. *Psychological Bulletin, 56*(2), 81–105. https://doi.org/10.1037/h0046016

Campion, M. C., Campion, M. A., Campion, E. D., & Reider, M. H. (2016). Initial investigation into computer scoring of candidate essays for personnel selection. *Journal of Applied Psychology, 101*(7), 958–975. https://doi.org/10.1037/apl0000108

Chaffin, D., Heidl, R., Hollenbeck, J. R., Howe, M., Yu, A., Voorhees, C., & Calantone, R. (2017). The promise and perils of wearable sensors in organizational research. *Organizational Research Methods, 20*(1), 3–31. https://doi.org/10.1177/1094428115617004

Chamorro-Premuzic, T., Winsborough, D., Sherman, R. A., & Hogan, R. (2016). New talent signals: Shiny new objects or a brave new world? *Industrial and Organizational Psychology: Perspectives on Science and Practice, 9*(3), 621–640. https://doi.org/10.1017/iop.2016.6

Chapman, D. S., Uggerslev, K. L., Carroll, S. A., Piasentin, K. A., & Jones, D. A. (2005). Applicant attraction to organizations and job choice: A meta-analytic review of the correlates of recruiting outcomes. *Journal of Applied Psychology, 90*(5), 928–944. https://doi.org/10.1037/0021-9010.90.5.928

Chellali, A., Mentis, H., Miller, A., Ahn, W., Arikatla, V. S., Sankaranarayanan, G., De, S., Schwaitzberg, S. D., & Cao, C. G. L. (2016). Achieving interface and environment fidelity in the Virtual Basic Laparoscopic Surgical Trainer. *International Journal of Human-Computer Studies, 96*, 22–37. https://doi.org/10.1016/j.ijhcs.2016.07.005

Chen, E. E., & Wojcik, S. P. (2016). A practical guide to big data research in psychology. *Psychological Methods*, *21*(4), 458–474. https://doi.org/10.1037/met0000111

Chorley, M. J., Whitaker, R. M., & Allen, S. M. (2015). Personality and location-based social networks. *Computers in Human Behavior*, *46*, 45–56. https://doi.org/10.1016/j.chb.2014.12.038

Civil Rights Act of 1964 § 7, 42 U.S.C. § 2000e *et seq.* (1964). https://www.eeoc.gov/statutes/title-vii-ciil-rights-act-1964

Civil Rights Act of 1991 § 109, 42 U.S.C. § 2000e *et seq.* (1991). https://www.eeoc.gov/civil-rights-act-1991-original-text

Cotton, J. L., & Tuttle, J. M. (1986). Employee turnover: A meta-analysis and review with implications for research. *Academy of Management Review*, *11*(1), 55–70. https://doi.org/10.5465/amr.1986.4282625

da Rocha Seixas, L., Gomes, A. S., & de Melo Filho, I. J. (2016). Effectiveness of gamification in the engagement of students. *Computers in Human Behavior*, *58*, 48–63. https://doi.org/10.1016/j.chb.2015.11.021

De Corte, W., Lievens, F., & Sackett, P. R. (2008). Validity and adverse impact potential of predictor composite formation. *International Journal of Selection and Assessment*, *16*(3), 183–194. https://doi.org/10.1111/j.1468-2389.2008.00423.x

de Montjoye, Y.-A., Quoidbach, J., Robic, F., & Pentland, A. (2013). Predicting personality using novel mobile phone-based metrics. In A. M. Greenberg, W. G. Kennedy, & N. D. Bos (Eds.), *Social computing, behavioral-cultural modeling and prediction* (pp. 48–55). Springer Berlin Heidelberg. https://doi.org/10.1007/978-3-642-37210-0_6

Deterding, S., Dixon, D., Khaled, R., & Nacke, L. (2011). From game design elements to gamefulness: Defining "gamification." In *Proceedings of the 15th International Academic MindTrek Conference: Envisioning future media environments* (pp. 9–15). https://doi.org/10.1145/2181037.2181040

Diermeier, D., Godbout, J. F., Yu, B., & Kaufmann, S. (2012). Language and ideology in Congress. *British Journal of Political Science*, *42*(1), 31–55. https://doi.org/10.1017/S0007123411000160

Douthitt, E. A., & Aiello, J. R. (2001). The role of participation and control in the effects of computer monitoring on fairness perceptions, task satisfaction, and performance. *Journal of Applied Psychology*, *86*(5), 867–874. https://doi.org/10.1037/0021-9010.86.5.867

Fischbach, K., Gloor, P. A., Lassenius, C., Olguín Olguín, D., Pentland, A., Putzke, J., & Schoder, D. (2010). Analyzing the flow of knowledge with sociometric badges. *Procedia: Social and Behavioral Sciences*, *2*(4), 6389–6397. https://doi.org/10.1016/j.sbspro.2010.04.048

Friedman, R., Anderson, C., Brett, J., Olekalns, M., Goates, N., & Lisco, C. C. (2004). The positive and negative effects of anger on dispute resolution: Evidence from electronically mediated disputes. *Journal of Applied Psychology*, 89(2), 369–376. https://doi.org/10.1037/0021-9010.89.2.369

Gandomi, A., & Haider, M. (2015). Beyond the hype: Big data concepts, methods, and analytics. *International Journal of Information Management*, 35(2), 137–144. https://doi.org/10.1016/j.ijinfomgt.2014.10.007

General Data Protection Regulation, no. (EU) 2016/679, European Parliament, 88. (2016).

Giorgi, S., Lynn, V., Matz, S., Ungar, L., & Schwartz, H. A. (2020). *Correcting sociodemographic selection biases for population prediction from social media*. arXiv. https://arxiv.org/abs/1911.03855

Golbeck, J., Robles, C., Edmondson, M., & Turner, K. (2011). Predicting personality from Twitter. *2011 IEEE international conference on privacy, security, risk and trust and IEEE international conference on social computing* (pp. 149–156). Institute of Electrical and Electronics Engineers. https://doi.org/10.1109/PASSAT/SocialCom.2011.33

Goldstein, I. L. (1993). *Training in organizations: Needs assessment, development, and evaluation* (3rd ed.). Brooks/Cole.

Goodhart, C. A. E. (1984). *Monetary theory and practice: The UK experience*. Macmillan Publishers Limited. https://doi.org/10.1007/978-1-349-17295-5

Griffiths, T. L. (2015). Manifesto for a new (computational) cognitive revolution. *Cognition*, 135, 21–23. https://doi.org/10.1016/j.cognition.2014.11.026

Groening, C., & Binnewies, C. (2019). "Achievement unlocked!"—The impact of digital achievements as a gamification element on motivation and performance. *Computers in Human Behavior*, 97, 151–166. https://doi.org/10.1016/j.chb.2019.02.026

Halimi, A., & Ayday, E. (2020). *Profile matching across online social networks*. arXiv. https://arxiv.org/abs/2008.09608

Hamari, J. (2017). Do badges increase user activity? A field experiment on the effects of gamification. *Computers in Human Behavior*, 71, 469–478. https://doi.org/10.1016/j.chb.2015.03.036

Harari, G. M., Müller, S. R., Aung, M. S., & Rentfrow, P. J. (2017). Smartphone sensing methods for studying behavior in everyday life. *Current Opinion in Behavioral Sciences*, 18, 83–90. https://doi.org/10.1016/j.cobeha.2017.07.018

Harari, G. M., Müller, S. R., Stachl, C., Wang, R., Wang, W., Bühner, M., Rentfrow, P. J., Campbell, A. T., & Gosling, S. D. (2020). Sensing sociability: Individual differences in young adults' conversation, calling, texting, and app use behaviors in daily life. *Journal of Personality and Social Psychology, 119*(1), 204–228. https://doi.org/10.1037/pspp0000245

Heavey, A. L., Holwerda, J. A., & Hausknecht, J. P. (2013). Causes and consequences of collective turnover: A meta-analytic review. *Journal of Applied Psychology, 98*(3), 412–453. https://doi.org/10.1037/a0032380

Hernandez, I., Kim, S., Sanders, A., & Towe, S. (2020, June). *Deep selection: Inferring employee's traits from resume style using neural networks*. Symposium presented at the annual meeting of the Society for Industrial and Organizational Psychology, Austin, TX, United States. https://whova.com/xems/whova_backend/get_event_s3_file_api/?eventkey=82ce78f543e19389a027669dd00d8a9a95596313c560929ad80a22c51d163164&event_id=siop_202004&file_url=https://whova.com/xems/whova_backend/get_event_s3_file_api/?event_id=siop_202004&eventkey=82ce78f543e19389a027669dd00d8a9a95596313c560929ad80a22c51d163164&file_url=https://d1keuthy5s86c8.cloudfront.net/static/ems/upload/files/eayun_99251_Liu.pdf

Hickman, L., Tay, L., & Woo, S. E. (2019). Validity evidence for off the shelf language-based personality assessment using video interviews: Convergent and discriminant relationships with self and observer ratings. *Personnel Assessment and Decisions, 5*(3), 12–20. https://doi.org/10.25035/pad.2019.03.003

Hogan, J., Barrett, P., & Hogan, R. (2007). Personality measurement, faking, and employment selection. *Journal of Applied Psychology, 92*(5), 1270–1285. https://doi.org/10.1037/0021-9010.92.5.1270

Jeske, D., & Santuzzi, A. M. (2015). Monitoring what and how: Psychological implications of electronic performance monitoring. *New Technology, Work and Employment, 30*(1), 62–78. https://doi.org/10.1111/ntwe.12039

Kanan, R., Elhassan, O., & Bensalem, R. (2018). An IoT-based autonomous system for workers' safety in construction sites with real-time alarming, monitoring, and positioning strategies. *Automation in Construction, 88*, 73–86. https://doi.org/10.1016/j.autcon.2017.12.033

Kim, T., McFee, E., Olguín Olguín, D., Waber, B., & Pentland, A. (2012). Sociometric badges: Using sensor technology to capture new forms of collaboration. *Journal of Organizational Behavior, 33*(3), 412–427. https://doi.org/10.1002/job.1776

Kleinberg, J., Ludwig, J., Mullainathan, S., & Sunstein, C. R. (2018). Discrimination in the age of algorithms. *The Journal of Legal Analysis, 10*, 113–174. https://doi.org/10.1093/jla/laz001

Kleinberg, J., Ludwig, J., Mullainathan, S., & Sunstein, C. R. (2020). Algorithms as discrimination detectors. *Proceedings of the National Academy of Sciences, 117*(48). https://doi.org/10.1073/pnas.1912790117

Landauer, T. K., & Dumais, S. T. (1997). A solution to Plato's problem: The latent semantic analysis theory of acquisition, induction, and representation of knowledge. *Psychological Review, 104*(2), 211–240. https://doi.org/10.1037/0033-295X.104.2.211

Landers, R. N., & Armstrong, M. B. (2017). Enhancing instructional outcomes with gamification: An empirical test of the Technology-Enhanced Training Effectiveness Model. *Computers in Human Behavior, 71*, 499–507. https://doi.org/10.1016/j.chb.2015.07.031

Madera, J. M., Hebl, M. R., & Martin, R. C. (2009). Gender and letters of recommendation for academia: Agentic and communal differences. *Journal of Applied Psychology, 94*(6), 1591–1599. https://doi.org/10.1037/a0016539

Marksberry, P., Church, J., & Schmidt, M. (2014). The employee suggestion system: A new approach using latent semantic analysis. *Human Factors and Ergonomics in Manufacturing & Service Industries, 24*(1), 29–39. https://doi.org/10.1002/hfm.20351

McAfee, A., & Brynjolfsson, E. (2012). Big data: The management revolution. *Harvard Business Review, 90*(10), 60–66, 68, 128. https://hbr.org/2012/10/big-data-the-management-revolution

Mendelson, L. (2020, November 11). This won't hurt a bit: Employee temperature and health screenings—A list of statewide orders. *Littler Mendelson P.C.* https://www.littler.com/publication-press/publication/wont-hurt-bit-employee-temperature-and-health-screenings-list

Morrison, E. W., & Robinson, S. L. (1997). When employees feel betrayed: A model of how psychological contract violation develops. *The Academy of Management Review, 22*(1), 226–256. https://doi.org/10.2307/259230

National Council on Measurement in Education. (2019, November 22). *Misconceptions about group differences in average test scores.* https://www.ncme.org/publications/statements/new-item2

Noe, R. A. (2020). *Employee training and development* (8th ed.). McGraw Hill Education.

O'Connor, S. (2016, September 6). When your boss is an algorithm. *Financial Times.* https://www.ft.com/content/88fdc58e-754f-11e6-b60a-de4532d5ea35

O'Hara, D. (2019, June 6). *Wearable technology for mental health.* American Psychological Association. https://www.apa.org/members/content/wearable-technology

Olguín Olguín, D., Waber, B. N., Kim, T., Mohan, A., Ara, K., & Pentland, A. (2009). Sensible organizations: Technology and methodology for automatically measuring organizational behavior. *IEEE Transactions on Systems, Man, and Cybernetics. Part B (Cybernetics), 39*(1), 43–55. https://doi.org/10.1109/TSMCB.2008.2006638

Ong, T. (2018, February 1). Amazon patents wristbands that track warehouse employees' hands in real time. *The Verge.* https://www.theverge.com/2018/2/1/16958918/amazon-patents-trackable-wristband-warehouse-employees

Oswald, F. L., Behrend, T. S., Putka, D. J., & Sinar, E. (2020). Big data in industrial-organizational psychology and human resource management: Forward progress for organizational research and practice. *Annual Review of Organizational Psychology and Organizational Behavior, 7*(1), 505–533. https://doi.org/10.1146/annurev-orgpsych-032117-104553

Park, G., Schwartz, H. A., Eichstaedt, J. C., Kern, M. L., Kosinski, M., Stillwell, D. J., Ungar, L. H., & Seligman, M. E. P. (2015). Automatic personality assessment through social media language. *Journal of Personality and Social Psychology, 108*(6), 934–952. https://doi.org/10.1037/pspp0000020

Peterson, H. (2019, December 18). Amazon is tracking delivery workers' every move with an app that assigns them scores based on their driving. *Business Insider.* https://www.businessinsider.com/amazon-scores-delivery-workers-driving-skills-using-tracking-app-2019-12

Pitcher, B. D., Ravid, D. M., Shephard, C. L., & Behrend, T. S. (2019, October 3–5). *Improving student attitudes and performance in STEM through virtual reality and constructive feedback.* Symposium presented at the annual Technology, Mind and Society Conference, Washington, DC.

Protection of Human Subjects, 45 C.F.R. § 46. (2017).

Putka, D. J. (2018, April 20). *A SIOP machine learning competition.* Society for Industrial and Organizational Psychology 2018 Annual Conference, Chicago, IL, United States.

Ravid, D. M., Tomczak, D. L., White, J. C., & Behrend, T. S. (2020). EPM 20/20: A review, framework, and research agenda for electronic performance monitoring. *Journal of Management, 46*(1), 100–126. https://doi.org/10.1177/0149206319869435

Rombaut, E., & Guerry, M.-A. (2018). Predicting voluntary turnover through human resources database analysis. *Management Research Review, 41*(1), 96–112. https://doi.org/10.1108/MRR-04-2017-0098

Rousseau, D. M. (2011). The individual–organization relationship: The psychological contract. In S. Zedeck (Ed.), *APA handbook of industrial and organizational psychology. Vol. 3. Maintaining, expanding, and contracting the organization* (pp. 191–220). American Psychological Association. https://doi.org/10.1037/12171-005

Salkind, N. J. (Ed.). (2010). *Encyclopedia of research design* (Vols. 1–3). SAGE Publications, Inc. https://doi.org/10.4135/9781412961288

Sanchez, J. I., & Levine, E. L. (2012). The rise and fall of job analysis and the future of work analysis. *Annual Review of Psychology, 63*(1), 397–425. https://doi.org/10.1146/annurev-psych-120710-100401

SAP resume matching, powered by NVIDIA: Finding the best talent. (2018). http://images.nvidia.com/content/pdf/infographic/analytics-print-nvidia-and-sap-resume-matching-joint-solution-brief-final-web.pdf

Segal, J. A. (2018, April 11). Legal trends social media use in hiring: Assessing the risks. *HR Magazine.* https://www.shrm.org/hr-today/news/hr-magazine/pages/0914-social-media-hiring.aspx

Shafer, T. (2017). *The 42 V's of big data and data science.* KDnuggets. https://www.kdnuggets.com/the-42-vs-of-big-data-and-data-science.html/

Short, J. C., McKenny, A. F., & Reid, S. W. (2018). More than words? Computer-aided text analysis in behavior and psychology research. *Annual Review of Organizational Psychology and Organizational Behavior, 5*(1), 415–435. https://doi.org/10.1146/annurev-orgpsych-032117-104622

Silverman, R. E. (2016, February 18). Bosses tap outside firms to predict which workers might get sick. *The Wall Street Journal.* https://www.wsj.com/articles/bosses-harness-big-data-to-predict-which-workers-might-get-sick-1455664940

Song, Q. C., Wee, S., & Newman, D. A. (2017). Diversity shrinkage: Cross-validating pareto-optimal weights to enhance diversity via hiring practices. *Journal of Applied Psychology, 102*(12), 1636–1657. https://doi.org/10.1037/apl0000240

Srivastava, S. B., Goldberg, A., Manian, V. G., & Potts, C. (2018). Enculturation trajectories: Language, cultural adaptation, and individual outcomes in organizations. *Management Science, 64*(3), 1348–1364. https://doi.org/10.1287/mnsc.2016.2671

Stanton, J. M., & Julian, A. L. (2002). The impact of electronic monitoring on quality and quantity of performance. *Computers in Human Behavior, 18*(1), 85–101. https://doi.org/10.1016/S0747-5632(01)00029-2

Sweeney, L. (2000). *Simple demographics often identify people uniquely* (Data Privacy Working Paper No. 3). http://ggs685.pbworks.com/w/file/fetch/94376315/Latanya.pdf

Tauszik, Y. R., & Pennebaker, J. W. (2010). The psychological meaning of words: LIWC and computerized text analysis methods. *Journal of Language and Social Psychology, 29*(1), 24–54. https://doi.org/10.1177/0261927X09351676

Tay, L., Woo, S. E., Hickman, L., & Saef, R. (2020). Psychometric and validity issues in machine learning approaches to personality assessment: A focus on social media text mining. *European Journal of Personality, 34*(5), 826–844. https://doi.org/10.1002/per.2290

Tichy, N. M., Tushman, M. L., & Fombrun, C. (1979). Social network analysis for organizations. *Academy of Management Review, 4*(4), 507–519. https://doi.org/10.5465/amr.1979.4498309

Tomczak, D. L., Behrend, T. S., Willford, J. C., & Jimenez, W. P. (2020). *"I didn't agree to these terms": Electronic performance monitoring violates the psychological contract.* PsyArXiv. https://doi.org/10.31234/osf.io/qax9u

Tomczak, D. L., Zarsky, S., Mancarella, P. J., & Behrend, T. S. (2020, June 15–17). *An instrument for measuring electronic performance monitoring* [Poster presentation]. Society for Industrial and Organizational Psychology 35th Annual Conference [Virtual].

Van Iddekinge, C. H., Lanivich, S. E., Roth, P. L., & Junco, E. (2016). Social media for selection? Validity and adverse impact potential of a Facebook-based assessment. *Journal of Management, 42*(7), 1811–1835. https://doi.org/10.1177/0149206313515524

Waber, B. N., Olguín Olguín, D., Kim, T., & Pentland, A. (2010). Productivity through coffee breaks: Changing social networks by changing break structure. *SSRN Electronic Journal.* https://doi.org/10.2139/ssrn.1586375

Wang, W., Harari, G. M., Wang, R., Müller, S. R., Mirjafari, S., Masaba, K., & Campbell, A. T. (2018). Sensing behavioral change over time: Using within-person variability features from mobile sensing to predict personality traits. *Proceedings of the ACM on Interactive, Mobile, Wearable, and Ubiquitous Technologies, 2*(3), 1–21. https://doi.org/10.1145/3264951

Wells, D. L., Moorman, R. H., & Werner, J. M. (2007). The impact of the perceived purpose of electronic performance monitoring on an array of attitudinal variables. *Human Resource Development Quarterly, 18*(1), 121–138. https://doi.org/10.1002/hrdq.1194

What is GDPR, the EU's new data protection law? GDPR.eu. (2018, November 7). https://gdpr.eu/what-is-gdpr/

White, J. C., Ravid, D. M., & Behrend, T. S. (2020). Moderating effects of person and job characteristics on digital monitoring outcomes. *Current Opinion in Psychology, 31*, 55–60. https://doi.org/10.1016/j.copsyc.2019.07.042

Yost, A. B., Behrend, T. S., Howardson, G., Badger Darrow, J., & Jensen, J. M. (2019). Reactance to electronic surveillance: A test of antecedents and outcomes. *Journal of Business and Psychology, 34*(1), 71–86. https://doi.org/10.1007/s10869-018-9532-2

Youyou, W., Kosinski, M., & Stillwell, D. (2015). Computer-based personality judgments are more accurate than those made by humans. *Proceedings of the National Academy of Sciences, 112*(4), 1036–1040. https://doi.org/10.1073/pnas.1418680112

Zhao, H., Wayne, S. J., Glibkowski, B. C., & Bravo, J. (2007). The impact of psychological contract breach on work-related outcomes: A meta-analysis. *Personnel Psychology, 60*(3), 647–680. https://doi.org/10.1111/j.1744-6570.2007.00087.x

9

Human–Robot Interaction Challenges in the Workplace

Guy Hoffman, Alap Kshirsagar, and Matthew V. Law

In this chapter, we discuss challenges for designing robots that interact with people in the workplace. We do so from the perspective of the research field called *human–robot interaction* (HRI), which studies the engineering methods, design aspects, social issues, and psychological outcomes arising from interacting with robots. The "workplace" in this chapter is defined broadly and can include any of a number of contexts in which robots could work alongside humans, such as factories, supermarkets, offices, care facilities, and construction sites. We do, however, exclude some specialized work environments, which already have humans interacting closely with robots, specifically military, space exploration, and surgery contexts. These settings have been covered at length in the literature. Instead, we discuss workplace settings in which robots are currently rare, but which are considered by the research community as likely future contexts for HRI.

---

https://doi.org/10.1037/0000290-010
The Psychology of Technology: Social Science Research in the Age of Big Data, S. C. Matz (Editor)
Copyright © 2022 by the American Psychological Association. All rights reserved.

While there is no commonly agreed-upon definition of "robot," we accept a vague boundary delineation and consider a robot to be a programmable machine that does mechanical work that is not contained fully in an enclosure. Most commonly, a robot would be in the shape of a robotic arm akin to those found on factory floors, a mobile robot in the form of a computer-controlled vehicle, or a human- or animal-like robot designed for interaction.

We present human–robot interaction in the workplace against the background of two better-studied application contexts in which robots are envisioned to operate around humans: at home and in public spaces. Commercially available robots in the home today include mostly cleaning or lawn-mowing robots, but researchers and commercial companies have proposed home robots for companionship, elder care, child care, teaching, therapy, and entertainment purposes. In public spaces, we see sporadic experimental deployments of robots, mainly in transit centers, such as airports and train stations, as well as in museums and shopping malls. Still, robots in public spaces are not yet widespread and are an active area of HRI research. Both the home and public spaces have received the majority of attention in the HRI research literature.

This chapter's main focus is a third deployment context: everyday workplaces where robots interacting with humans are virtually nonexistent. We particularly consider nonspecialized workplaces, such as offices, retail stores, warehouses, workshops, care facilities, and factories. Some of these workplaces have already started to introduce robots. For example, in many manufacturing areas, preprogrammed robots are in operation, but their interaction with humans is usually minimal. That said, large and small factories are experimenting with collaborative robots, which interact more closely with humans. While this is not yet widespread, researchers expect this type of deployment to grow further, along with other, currently nonexistent, uses of robots in everyday workplaces, such as offices, retail stores, and care facilities.

Each of the three application categories described above (home, public spaces, and work) poses both *physical* and *psychological* challenges for developers of human-interactive robots (see Figures 9.1 and 9.2).

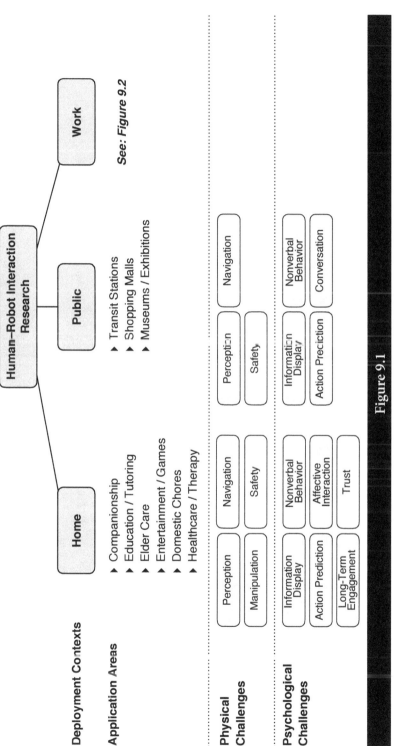

Figure 9.1

Human–robot interactions across different life domains. Application areas for human–robot interaction at home and in public spaces, along with research challenges related to physical and psychological aspects of the interaction. Most physical challenges are common to both contexts, whereas home interaction with robots presents more complex psychological challenges, such as affective interaction, long-term engagement, and trust.

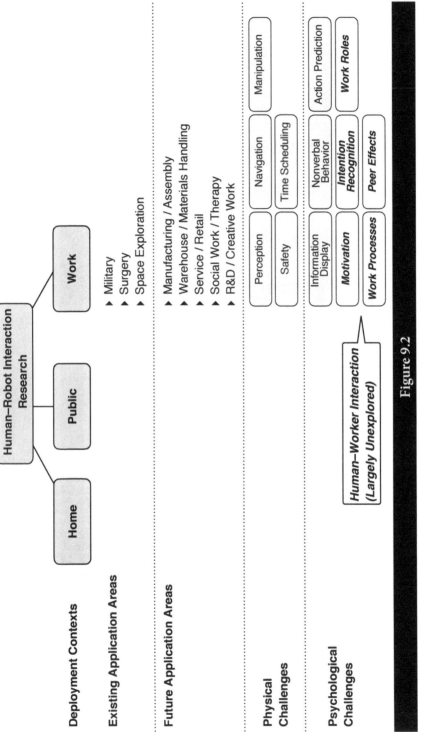

Figure 9.2

Applications for research on human–robot interaction. Established and future application areas for human–robot interaction in the workplace, along with research challenges related to physical and psychological aspects of the interaction. Most physical challenges are similar to those of the home and public context. There are important psychological challenges related to robot–worker interaction that are currently unexplored (clear boxes on bottom right). These include (a) worker motivation, (b) intention recognition in complex creative work, (c) negotiation of roles between humans and robots, (d) understanding work

Some challenges cut across application contexts, while others are unique to the setting in which the robot is placed. For example, two physical HRI challenges, common to all contexts, are the perception of human activity (Martínez-Villaseñor & Ponce, 2019) and safe navigation around humans (Kruse et al., 2013). Some psychological challenges are also common to all three contexts. These include conveying relevant, accurate, and timely information (Robinette et al., 2014), understanding and using nonverbal behaviors (Cha et al., 2018; Saunderson & Nejat, 2019), and predicting human actions (Kong & Fu, 2018). That said, each context also comes with its unique set of psychological challenges. A therapy robot in the home may have to take into account compliance and the emotional well-being of a patient (Cabibihan et al., 2013). An entertainment robot might be concerned with narrative and engagement (Ligthart et al., 2020). In contrast, a collaborative robot in a repair workshop would be more focused on skill adaptation and understanding workplace hierarchies (Parker & Draper, 1999).

On the following pages, we briefly survey the physical and psychological challenges posed by home and public deployment of robots, to provide a sense of what questions the HRI literature has been concerned with over the past decade and a half. Where available, we will refer the reader to other excellent surveys on the topic for more detail. This chapter then moves to discuss interaction with robots in the workplace. The discussion begins with an overview of workplace deployment contexts and a review of some of the associated challenges currently studied in the HRI literature, which are mostly physical in nature. We then turn to the less-discussed psychological challenges of working with robots, detailing three studies conducted in our laboratory, which are relevant to the use of robots in the future workplace. The first study relates to the effects of competing with a robot for monetary reward and the resulting demotivation and loss of self-esteem that we found. The second study tackles challenges that occur when a robot attempts to help a worker make complex decisions. The third study also concerns human decision making and explores whether an artificial intelligence (AI) agent can infer a human's intentions by tracking the outcomes of the human's decision making.

ROBOTS IN THE HOME

A large segment of research in HRI studies robots in the home context. Some of the envisioned applications include companionship, education, elder care, entertainment, domestic chores, and personal health. One of the most studied application areas of home robots is in support of elder care (Broekens et al., 2009; Kachouie et al., 2014). Other therapeutic applications include physical therapy (Fasola & Mataric, 2012, 2013), cognitive therapy (Schroeter et al., 2013), feeding (Gallenberger et al., 2019), and companionship (Heerink et al., 2008; Robinson et al., 2013; Wada & Shibata, 2007). Robot companions have also been studied as devices to help mitigate feelings of loneliness in populations beyond elder care (K. M. Lee et al., 2006; Odekerken-Schröder et al., 2020).

Outside of the therapy context, researchers have proposed to use robots for entertainment and more casual purposes. Robots have been studied in the role of music listening companions (Hoffman & Vanunu, 2013), game playing companions (Marti & Giusti, 2010; Volkhardt et al., 2011), and exercise companions (Graether & Mueller, 2012; Schneider & Kummert, 2016). In contrast to this vision of home robotic companions, research has also shown that not all people accept this role for robots (Deutsch et al., 2019) and that preferred roles may be those of a butler or an assistant rather than that of a friend (Dautenhahn et al., 2005). An additional major application area of robots in the home is domestic chores (Cakmak & Takayama, 2013), such as floor cleaning (Fiorini & Prassler, 2000; Prassler et al., 2000), organizing (Pantofaru et al., 2012), laundry (Estevez et al., 2020), and kitchen assistance (Pham et al., 2017).

Finally, a large portion of HRI research addresses the educational use of social robots at home. This includes robot-assisted learning of languages (Gordon et al., 2016) and mathematics (Kennedy et al., 2016) and improvement of reading skills (Gordon & Breazeal, 2015), among others. Despite the interest in learning from robots, their effectiveness as compared with human tutors has been questioned (Kennedy et al., 2016), and the robot's social behavior has been shown to negatively affect learning in some cases (Kennedy et al., 2015).

The physical human–robot interaction challenges are minimal for most home robots. This is because robots for assisted learning, entertainment, and companionship are most often designed as stationary, desktop devices that do not manipulate any objects. However, if robots are to perform domestic chores and physical care tasks for humans, such as feeding, they need to physically interact with humans and objects (Bhattacharjee et al., 2019). These applications pose challenges related to accurate sensing and perception (Yan et al., 2014), dexterous manipulation (Ozawa & Tahara, 2017), home navigation (Bacciu et al., 2014), fault handling (Khalastchi & Kalech, 2018), and human-safe controllers (Duchaine & Gosselin, 2009).

Beyond these physical challenges, there are a host of psychological challenges that developers of home robots have to address. One is the understanding and generation of nonverbal behavior. Much work has been done toward recognizing different modes of human nonverbal behavior and generating robot nonverbal behavior in human–robot interaction, and there are several excellent surveys. Some of these surveys focus on individual modes of nonverbal behavior, such as arm gestures (Nehaniv et al., 2005), body movements (Bethel & Murphy, 2008), eye gaze (Admoni & Scassellati, 2017), and proxemics, the use of physical space to communicate (Rios-Martinez et al., 2015). Others take a broader perspective with multiple modes and interactions between them (Cha et al., 2018; Mandal, 2014; Saunderson & Nejat, 2019). Although researchers have long tried to categorize nonverbal behavior (Ekman & Friesen, 1969), the complexity and variety of its components make it difficult to devise predictive and generative models.

Emotional appropriateness is another psychological challenge for the developers of home robots. Generating appropriate emotional robot expressions in human–robot interactions requires detection of human emotions and modeling of emotional interaction. Furthermore, the coupling between nonverbal behavior and emotional expression aggravates the challenge. A large body of work has investigated recognition of human emotion through speech (El Ayadi et al., 2011), body gesture (Noroozi et al., 2019), and facial expressions (Ko, 2018). Several researchers have proposed emotional models (Savery & Weinberg, 2020), with some of the most prominent models being the PAD (Pleasure, Arousal, Dominance) emotional model (Mehrabian, 1980), Ekman's (1999) categorizations, Plutchik's (2001) wheel of emotions, and the circumplex model of valence and arousal (Posner et al., 2005).

While research in HRI has made some strides in the nonverbal and affective aspects of interaction, one of the most significant outstanding challenges in home robotics is maintaining long-term engagement with users. There is not enough research to understand the influence of novelty effects on many of the phenomena studied in HRI (Baxter et al., 2016; Smedegaard, 2019). The majority of the existing work studies human–robot interaction in a laboratory setting for a short duration. In contrast, there is a paucity of literature investigating long-term human–robot interaction in the real-world home environment (Aylett et al., 2011; Leite et al., 2013). Some of the impediments to long-term engagement are repetitiveness in interaction, ill-defined use cases, and ethical concerns. Researchers have explored solutions such as personalization (Clabaugh & Matarić, 2018), making robots "imperfect" through human-like cognitive biases (Biswas & Murray, 2015), using narratives (Goodrich et al., 2018), adaptive behavior coordination (Bajones, 2016), and social media integration (Mavridis et al., 2009), among others. Still, this area of research presents a large opportunity for new knowledge generation.

Another open area of research is evaluating and building trust between humans and home robots. If long-term interaction is the goal, a robot would have to develop some model of the human's trust in the robot, to help it mitigate breakdowns. Such a model would also have to take into account that people might have a baseline lack of trust in robots resulting from fear of technology (Hancock, Billings, & Schaefer, 2011; Liang & Lee, 2017; Szollosy, 2017). Mistakes committed by the robot also affect its trustworthiness and reliability (Salem et al., 2015). Trust could be measured in the context of trust-in-technology (Heerink et al., 2009), but also using methods of interpersonal trust (J. J. Lee et al., 2013). Quantitatively measuring trust is challenging, and many of the measures of trust used in HRI research have been qualitative and descriptive (Hancock, Billings, Schaefer, Chen, et al., 2011).

Finally, developers of home robots have to navigate design challenges resulting from preconceived notions and metaphors that people have about robots, people, animals, and objects. Robot design can span from completely functional (e.g., floor cleaning robots), through animal-shaped (like the robot seal Paro or the robotic dog AIBO), to extremely human-like (e.g., the android robot Geminoid). The design of the robot could by itself have such a strong effect on people's interaction with it, that it may overshadow any behavioral or cognitive aspect of the robot's software.

In summary, noncleaning robots in the home are still rare outside of the research laboratory, but many HRI researchers envision desktop robotic companions for therapy, tutelage, and entertainment. Such robots will have to address a number of psychological challenges, many of which are studied in the HRI research community, including reading and producing nonverbal behaviors and emotions, evoking trust, and maintaining long-term engagement.

ROBOTS IN PUBLIC SPACES

A second context that has attracted the attention of researchers in HRI is that of public spaces. Unlike home robots, which tend to be personal, long-term, and focused on a few users, robots in public spaces engage in short-term customer service interactions with many people. The majority of robots considered for this context are designed to provide information or guide people through the space, leading research to center on communication and navigation in crowds. There is also an emphasis on field research and long-term deployments, examining expectations and acceptance in these public spaces, as well as challenges around robustness over time.

One such public context where robots have been studied is in transit stations, including airports and train terminals. Mobile robots can be used to guide customers through a busy airport (Triebel et al., 2016), answer questions (Shiomi et al., 2008), communicate with large groups (Sakamoto et al., 2009), and help travelers check in (Tonkin et al., 2018). Other projects have studied robots as a way to provide round-the-clock security in transit stations (Capezio et al., 2007).

Another public application context is in malls and commercial spaces, where robots can be used to serve, entertain, or attract customers. Kanda et al. (2009) deployed an affective robot in a shopping mall to give shoppers verbal directions and shopping recommendations. However, the robot was also designed to build rapport with customers over time, through self-disclosure, remembering details about customers, and increasing friendliness. In this field study, the researchers connected the relationships that customers perceived with the robot to the effectiveness of its word-of-mouth advertising. Both Foster et al. (2016) and Aaltonen et al. (2017) studied how to create engaging interactions with humans and robots in commercial settings, under the premise that entertaining robots will invoke greater acceptance and better customer experiences. Aaltonen et al. in particular found a need for a robot in a mall to balance entertainment and practical behavior, with potential overlap between the two.

Studies in commercial spaces have also explored expectations and acceptance from stakeholders who more permanently inhabit the space. Niemelä et al. (2017) found that, while consumers expected a robot to play practical roles (e.g., guidance, information), retailers and management hoped a robot could contribute to a warm and entertaining atmosphere. Shi et al. (2016) found that managers were interested in using robots to attract visitors to stop and visit a store, a responsibility they considered stressful for a human to perform. Along these lines, Foster et al. (2016) designed a robot with socially acceptable verbal, nonverbal, and navigation behaviors in a large shopping center with both consumer and business stakeholders.

A third popular public space in which robots have been studied is in museums or expositions. Thrun et al. (1999) deployed a mobile robot in the Smithsonian Museum for 2 weeks as a tour guide designed to educate and entertain. The primary challenges this project tackled included safe navigation in packed and dynamic environments and short-term interactions with humans using familiar human-like behaviors. Kuno et al. (2007) modeled a robot museum tour guide on human tour guide behavior, having found that imitating human guides' head movements while talking could increase engagement with museum visitors.

In terms of physical challenges, robots in public contexts perform roles that require them to navigate through a busy space and communicate information to humans in that space. As a result, "social navigation" is a major focus of HRI research. Social navigation includes not only how to navigate a crowded space in a way that respects social norms and makes humans feel safe (Bera et al., 2017; Chen et al., 2017) but also how to plan shared trajectories alongside a human (Campos et al., 2019; Ferrer et al., 2017; Murakami et al., 2014). For a recent survey of robots in public spaces through the lens of human-aware navigation, see Kruse et al. (2013), who identified comfort, naturalness, and sociability as key challenges in the field.

In addition to navigation, there is also a research emphasis on social communication, including nonverbal behavior and conversational fillers (Kanda et al., 2009), head movements (Kuno et al., 2007), and bowing (Hayashi et al., 2007). This work can go beyond direct communication with humans: Sakamoto et al. (2009) drew on human "passive social" communication (i.e., eavesdropping) to design robot–robot interactions that broadcast information in public spaces. Blending these two areas of research is the work on proxemics, which studies how to use physical space and trajectories to communicate with humans (Mead & Matarić, 2017).

The demands of operating in a public space and interacting with groups and individuals accentuates the blurring of the boundary between the social and the functional for robot design. Taking a broader perspective, Mussakhojayeva and Sandygulova (2017) studied how robots should adapt their behavior to the social and cultural contexts of the space. For example, should a robot focus on the needs of a child or their parent in a shopping mall? Such directions may be complicated by cross-cultural studies that have found different perceptions and norms around interaction with robots in public spaces (Fraune et al., 2015).

ROBOTS IN THE WORKPLACE

We now shift our focus to the third application context: the human–robot shared workplace. While there has been much research on psychological challenges in the context of the home and public spaces, and there are several excellent surveys on human–robot interaction in general (Breazeal et al., 2016; Fong et al., 2003; Goodrich & Schultz, 2007; Thomaz et al., 2016), the specific psychological issues arising from humans sharing their workplace with robots have not received as much focused attention. The workplace environment poses unique research perspectives related to psychological outcomes, such as challenges associated with self-determination, inspiration, and role-related social factors (Froman, 2010). The workplace also imposes different behavioral structures, work processes, societal rules, and expectations than the home or public places (e.g., Mann, 2007). This merits a separate discussion from the perspective of human–robot interaction. Clearly, some of the contexts discussed above, especially those in public spaces, function as workplaces for the people employed in them. Airports, train stations, and malls have employees who may be interacting with robots as part of their workday. However, most of the research in public-space HRI, as evident from the previous discussion, focuses on customers and members of the public and not on workers. In a similar vein, therapy, elder care, and health care robots, including in the home, also interact with care workers, and some of these work relationships have been studied in the past (Gombolay et al., 2018; Mutlu & Forlizzi, 2008; Turja et al., 2019). Still, the bulk of HRI research in these domains investigates end-user–robot interaction.

This chapter highlights questions that arise around human–robot interaction in common workplaces, such as offices, workshops, factories, retail, health care facilities, and points of service. The workplace should be of particular interest to HRI research, as much of the public concern, both economic and societal, surrounds the relationship between humans and robots at work (Moniz & Krings, 2016). Yet most of the research published in the HRI community does not specifically address the psychological challenges that arise specifically from working with robots. The exception to this would be research on HRI in the realm of military, surgery, and outer space deployment, where scholars have looked at relationships between robots and military personnel, surgeons, and astronauts (Ambrose et al., 2000; Barnes & Evans, 2010; Wasen, 2010). These workspaces are extremely specialized, and robots in them interact with highly trained individuals. In addition, these workplaces are unique, in that they already see a significant use of robots and can be considered *current* human–robot workplaces.

The remainder of this chapter instead tackles HRI in everyday workplaces, inspired by potential *future* application contexts for interactive robots that are under development or merely imagined. We begin with an overview of where robots are considered to enter the workplace and some of the challenges currently being studied in these contexts. These challenges are mostly physical in nature, and where they touch on psychological challenges, they deal with rudimentary problems, such as nonverbal behavior and action prediction. We then follow up with research we conducted, which touches on some more complex psychological challenges, such as motivation and creative intention recognition.

Application Areas and Challenges

Robots are imagined to interact with people in many lines of work, which we broadly divide into five categories: (a) manufacturing and assembly; (b) warehouse logistics and materials handling; (c) service and retail; (d) social work and therapy; and (e) creative work, including research and development (R&D).

Manufacturing and Assembly

Production factories are one of the most commonly imagined application areas for robots. Traditional industrial robots are large and dangerous machines that operate at a safe distance or physically separated from humans and are programmed to do a single operation (Sanneman et al., 2020). Recently, a new type of lightweight, reprogrammable collaborative robot ("cobot") has been introduced to the factory floor (Djuric et al., 2016). These robots, due to their smaller size and lower cost, can also be used in smaller fabrication and craft workshops. The main research questions studied in this context relate to the manipulation of objects, factory processes, task flexibility, and safety (e.g., Billard & Kragic, 2019; Tsarouchi et al., 2016; Zanchettin et al., 2016). There are also a number of studies concerned with the timing and coordination of operations between a human and a factory robot (e.g., Wilcox et al., 2013).

Warehouse Logistics and Materials Handling

A second, somewhat related, application of robots is in the context of logistics and materials handling. Improvements in mobile robot navigation and swarm-inspired control of distributed robot systems allowed for commercial deployment of robotic systems in warehouses (D'Andrea, 2012). While robotic warehouse systems include interaction challenges with humans, these are primarily in the form of scheduling, and there is little one-to-one interaction between the human worker and the robot. In most cases, the robots in question are mobile bases for shelving units and are nearly invisible to the workers in the warehouse, who experience the robot's activity more as "moving shelves." There is a strong research thrust to also develop robotic arms for sorting and packing operations (Eppner et al., 2016). In this case, the level of interaction is likely to rise and raise similar challenges as those in the manufacturing and assembly context presented above.

Services and Retail

While much of the robotics literature has traditionally focused on manufacturing and logistics, by far the largest sector in the U.S. economy (both in terms of domestic product and in terms of the percentage of the labor force) is the service sector (U.S. Bureau of Economic Analysis, 2019, 2021). Within this sector, the largest industries (by labor force) are retail, business services, health care, leisure, and government. Despite its relative importance in the economy of the United States and other industrial nations, the service industry has not seen much robotics in real-world deployment. In recent years, there have been forays in retail and health care robotics, mainly in the context of stock monitoring (Bogue, 2019). Two examples include the Moxi robot for hospital wards (Thomaz, 2021) and the Badger robot for grocery stores (Gibson, 2021). In contrast, the service context has captured the interest of HRI researchers for years, including the potential for robots in hospitality applications (Gockley et al., 2005) and retail (Kamei et al., 2010; Paolanti et al., 2019) Many research problems studied in this context are similar to those in transit centers and museums, tackling primarily navigation and information display (Lichtenthäler et al., 2013). Moreover, research in service and retail HRI focuses mainly on the customer and not the robot's human coworkers.

Social Work and Therapy

At the intersection between the personal space and the workplace, robots could be a promising technology for social work and therapy (Wada et al., 2005). Robots have shown to be successful in improving people's sense of perceived responsiveness in a personal disclosure scenario (Birnbaum et al., 2016). An advantage of using robots in this context is that they can be programmed to preserve a speaker's privacy and are perceived as less judgmental than people. This could make robots a promising technology for counseling or taking testimony after traumatic events, such as natural disasters, expulsion, or violence. A robot can provide responsive behavior, a promise of privacy, and some psychological support. In terms of HRI research, the focus has once more been mostly on the relationship between the robot and the patient, and less on that between the health care or service worker and the robotic assistant (with some exceptions, as listed in the previous section on robots in the home).

Creative Work, Design, and R&D

Finally, a large part of the service economy, and a large percentage of the product development workforce in the United States and other countries, is in the design, research, and development sector (Boroush, 2020). While there has been a large body of research around AI-based assistance to design and other creative labor (e.g., Kim & Cho, 2000), there has been little work exploring the potential use of robotics in this sector. Given the relative significance of this work, we present the results from a study focusing on robot assistance in creative work in a subsequent section of this chapter.

Summary

Examining the literature that studies robotics in the workplace context, we identify several trends. Research on robots in manufacturing and logistics focuses on the physical challenges of the system. These challenges include navigating in space, sensing objects and environments, and manipulating objects. In some cases these physical challenges involve a human worker in the robot's environment (Huang et al., 2015; Kemp et al., 2007; Sisbot et al., 2007). However, the psychological aspects of these studies are restricted to sensing and predicting human actions; the goal of this research is to use these physical–psychological cues to better solve the physical challenge.

The second group of studies, concerning robots in the service, retail, and health care industries mostly focuses on the end-user, not the worker in these spaces. Lastly, the area of research, design, and creative work has seen almost no research in human–robot interaction.

In those cases where psychological challenges are raised in human–robot collaborative work, these challenges are often related to rudimentary nonverbal behaviors, such as gestures and gaze behavior. Gaze, for instance, has been taken into account in robot navigation and handover tasks (Kshirsagar et al., 2020; Wiltshire et al., 2013). However, the literature has not sufficiently addressed some of the more subtle psychological issues that come with introducing robots into the workplace, which are separate from the physical, sensory, and body language issues.

The following two sections focus on two such issues that we have recently studied in our laboratory: robots as competitors to humans and robots that aid in decision making. We present three studies in these contexts. In the first study, we examined whether competing with a robot on a mundane task can affect a person's motivation to perform well on the task. We drew upon economics theory to predict that a faster robot competing with a human worker will cause reduced motivation in the human worker.

In the second and third studies, we turned to human–robot collaborative decision making. We developed a robot that assists a human designer in a complex design task. First, we present qualitative evidence from video and interview analysis that uncovered considerations for robots in creative contexts, including the negotiation of the physical workspace, roles in the decision-making process, and the alignment of goals. Then, we present a study that aimed to determine if an AI algorithm could infer a human's intentions and preferences from their decision-making actions. Our results showed that "simulated human decisions" can help artificial agents infer what a real human is trying to achieve when searching for solutions for complex cognitive problems.

Humans and Robots Competing in Mundane Tasks

The first study was a controlled experiment studying the effort of people competing for a monetary prize with a robot on a mundane and repetitive task, and how the robot's performance affected people's effort and their attitude toward the robot (Kshirsagar et al., 2019). We manipulated the robot's performance and the monetary incentive level and measured people's effort level and attitudes toward the robot and toward themselves. These attitudes included people's liking of the robot, and their perception of the ability of the robot and of themselves on the task.

It is worth noting here that much of the existing HRI literature is concerned with collaboration between a human and a robot, where both agents share the same goal (Bauer et al., 2008). In contrast, the public perception and the economic reality is that robots and AI may compete for work with humans. These scenarios are almost never studied in HRI. The small number of studies that have investigated competition between humans and robots (Fraune et al., 2019; Mutlu et al., 2006; Short et al., 2010) looked at game-playing scenarios. Game playing is often a reasonable model for competition, but it is also inherently enjoyable, whereas most of the real-world application contexts for robots will involve mundane and repetitive jobs. Also, in most of the HRI research, participants are paid a fixed show-up fee for participating in the study. This method assumes that people are motivated to do well in the experimental task, and as a result these studies do not measure the effect of incentives on people's effort. Such monetary incentives are a key feature of real-world behavior. Our study, thus, makes two contributions: studying human–robot competition on mundane tasks and introducing monetary reward as a factor in human–robot interaction.

When people make decisions on whether and how much to work on a task, they are motivated by a number of factors. These include the direct value of the reward for the work, but also other psychological and cognitive biases, which result in a deviation from rational behavior. For example, behavioral economics researchers have shown that people perceive the disappointment resulting from loss of a reward more strongly than the satisfaction resulting from gaining an equivalent reward (Kahneman &

Tversky, 2013). In competitive scenarios, the gains and losses perceived by people are relative to a reference point based on the expected outcome of the competition, which depends on their own performance and the performance of the competitor. These factors suggest that the competency of a competitor could affect a person's motivation to do well on a task.

Economists almost always think of human competitors in this kind of analysis. But, as robots become ubiquitous in the workforce, we can expect situations to arise where this competitor is a robotic coworker. People might compete (or feel like they are competing) to win resources or demonstrate capabilities. Therefore, we were interested to investigate the effect of a competitor robot's performance and monetary rewards on people's effort on a mundane and repetitive task. We were also interested to study the effect of a competitor robot's performance on people's attitudes towards the robot and towards themselves.

We designed an experimental scenario to study human–robot competition in the workplace. Participants were rewarded cash prizes with some uncertainty depending on the difference between their performance and the robot's performance. The task consisted of counting a specific letter in strings of random characters and placing a block in a bin corresponding to the count. The participants were allowed to do as many such tasks as they wanted in a time limit of 2 minutes. For each correct block placement, they received one point. The robot counted characters in a different set of strings and moved a block to the corresponding bins in its workspace, which was adjacent to the human's workspace. The robot also received one point for each correct placement. Each 2-minute competition round had a cash prize associated with it, and the participant's chances of winning the cash prize were proportional to the difference between their score and the robot's score. For example, if the participant's and the robot's scores were equal, the participant had a 50% chance of winning the cash prize. For every point scored by the participant, their chance of winning the prize increased by 1%, and for every point scored by the robot, the participant's chance decreased by 1%. Our experiment setup consisted of a robot arm with an overhead camera, a screen with the user interface, and the robot's and the human's workspaces. The screen displayed the current and the predicted scores of the robot and the human, remaining time in the round, prize amount of the round, and the human's chances of winning the prize.

Each participant competed against the robot in 10 competition rounds, but they were not told about the number of rounds ahead of time. In each round, we manipulated the robot's speed and the cash prize. The robot's speed was kept constant throughout a round and was randomly chosen such that the robot's score was in the range of 5 to 45. The prize for each round was also randomly drawn, from $0.10 to $3.80. Our experimental design was inspired by a previous study conducted by Gill and Prowse (2012), in which they examined a competitive scenario between two humans.

We used the theoretical model of expectations-based reference dependent (EBRD) loss aversion (Kőszcgi & Rabin, 2006, 2007, 2009) to base our hypotheses about the effect of the robot's performance and the prize value on the human's performance. This model predicted a discouragement effect of the robot's score on the human's performance, and the discouragement effect would increase with increases in the prize value. For more details about the theoretical predictions and our hypotheses, please refer to Kshirsagar et al. (2019).

We analyzed the data obtained from 60 participants. A fixed-effects multivariate regression controlling for round number and participant number revealed a small discouragement effect of the robot's performance on the participants' performance ($p = .002$), that is, the participant's performance decreased if the robot performed better. Increasing the robot's score from 5 to 45 decreased the human's score by an average of 1.72, which was 8.4% of the human's average score (20.42). In contrast to Gill and Prowse (2012), we did not find an interaction effect of the prize on this discouragement effect. Even more surprisingly, the effect of the prize value on the human's performance was negligible.

We also elicited participants' attitudes towards the robot and toward themselves after each competition round. Specifically, we asked participants to rate the robot's competence and the robot's likability. We also asked participants to report their perceived ability of doing this task. We found that participants liked a low-performing robot competitor more than a high-performing one ($p < .0001$), even though they found a low-performing robot to be less competent ($p < .0001$). Also, they perceived themselves to be more competent at performing the task when the robot did not perform well ($p < .0001$). This suggests that people might assess their ability to perform a task relative to that of their robotic competitor.

In summary, the better the robot performed, the less effort the participants exerted. Moreover, a highly performing robot competitor was rated as less likable and led to lower self-assessment by the human competitors.

Our findings could have implications for the incentive schemes in workplaces where robots are being introduced to work alongside human workers. While it may be tempting to design such robots for optimal productivity, engineers and managers need to take into consideration how the robots' effort may affect the human workers' effort and attitudes toward the robot and even toward themselves.

Human–Robot Collaborative Decision Making

Whereas the first study was concerned with the understudied challenge of humans and robots competing, the second and third studies returned to collaboration in the workplace. However, in contrast to most research in HRI, here we explored how a robot might assist a human in a creative task including complex decision making, rather than in a mundane and repetitive task. We also are interested in a robot as a *cognitive* assistant, rather than as a tool for physical labor.

Most research on collaborative robotics in the product development process focuses almost exclusively on the late stages of the process, such as fabrication (Gleeson et al., 2013), assembly (Hayes & Scassellati, 2013), and warehouse operation (Rosenfeld et al., 2016). However, as mentioned previously, much of the U.S. product development economy is in research and development, and it has been shown that a large portion of the value added stems from this portion of the process (Boroush, 2020; Gemser & Leenders, 2001; Roy & Potter, 1993). We therefore wanted to investigate whether robots can help in workplace contexts where the main task is related to creative work, design, and decision making. We first discuss a qualitative study examining how humans interact with a decision-making assistance robot, and how they negotiate the physical and creative space that occurs between the human and the robot (Law et al., 2019). Our study revealed psychological challenges that could be applicable to a large number of creative tasks in the workplace. We then discuss a study that tests whether an AI assistant can infer a human designer's intentions in a complex decision-making task (Law et al., 2020).

Robots are well-suited for collaboration with humans on complex unstructured tasks like complex decision making due to their complementary skills. Humans are excellent at contextualizing and prioritizing problems, generalizing ideas across contexts, and reasoning abductively (Dorst, 2011; Egan & Cagan, 2016; Gonzalez & Haselager, 2005; Kolko, 2010). Machines are able to generate and evaluate potential solutions with superhuman scale and precision. The potential in blending these capacities has inspired many AI decision-making support tools (Babbar-Sebens & Minsker, 2012; Banerjee et al., 2008; Cho, 2002; Karimi et al., 2018; Smith et al., 2010). Furthermore, based on research that draws on embodied and enactive cognition (Davis et al., 2015), we argue that creative work can benefit from physical and social interaction between a human and an embodied AI partner (i.e., a robot).

We illustrate this premise with two studies that examine human–robot collaborative design and decision-making work. In both studies, we built an interface that enabled a human to work on a real-world design problem by arranging a set of markers representing design components. In the first study, we built a physical interface in the form of a table. A robotic arm sat directly across from the human, constantly searching for improvements to the human's design, and occasionally rearranging the shared set of blocks to realize an improvement that met a set of criteria. In the second study, we designed a screen interface to test whether the AI agent can infer the human designers' intentions from the trail of solutions that the human explored.

In the study with our physical interface and the robot, we observed a series of negotiations that arose between human and robot, around the physical, creative, and social aspects of working together. We had

12 participants work with the robot to design an earth-observing satellite system. Participants were presented with 12 different types of sensors and five different orbits in which they could deploy any of the sensors. The sensors interacted in complex ways with the orbits and one another. For example, an active sensor that required more power might be more effective in an orbit with more exposure to the sun, and two sensors might share underlying subsystems that would make deploying them together more efficient. Solutions were constructed by arranging blocks representing sensors into different regions of the table, representing different orbits. For each solution, a life-cycle cost was predicted, as well as a score indicating how well the system satisfied 371 data-collection criteria from the World Meteorological Society. Participants were asked to work with the robot to construct a solution that maximized this score while minimizing the life-cycle cost.

Through a thematic analysis of think-aloud video data and poststudy semistructured interviews, we observed participants negotiating turns and use of space with the robot in the shared physical interface, as well as roles, goals, and strategies in approaching the shared creative task. Concurrently, we observed participants engaging with the robot in social ways, addressing it directly or ascribing social meaning to its actions, even though the robot was not designed to behave in a deliberately social way.

On the surface level, the human participants felt the need to negotiate with the robot over the shared physical workspace and design representation. This could be as simple as working out turns to access the workspace or where to place object blocks. Sharing a workspace affords shared reference objects, visibility, and so forth, but also creates constraints on simultaneous work, as each collaborator's actions affect the other's. For some, negotiating the physical space resulted in issues around creative control. Issues around turn-taking also manifested themselves in terms of pacing and compatibility with different cognitive styles. Some of our participants adopted a rapid pace, trying as many solutions as possible to see what insights fell out about the design space. Others were much more deliberate, reflecting a "comprehensive" style of top-down idea organization. For participants on either end of this spectrum, the robot's interventions were disturbing, as it either moved too slowly for them and subsequently suggested stale ideas, or moved too quickly, upsetting the participant's deliberate train of thought. We also noticed that some of our participants used the physical space to spatially represent ideas. For example, one participant adopted a pattern of moving all the components he liked to the right side of the table. This reflects studies of human–human design collaboration, where designers use objects, gesture, and space to externalize and organize ideas together (Cash & Maier, 2016; Tuikka & Kuutti, 2000). This tendency presents an opportunity for a robot to understand more about what a human is thinking. At the same time, it creates a challenge in terms of recognizing what spatial structures are significant and not disturbing them unintentionally.

We also recognized tendencies among our participants to gravitate toward different collaborative roles with respect to the shared design task. Some participants sought control, verbally telling the robot what to do or granting it permission to change the configuration on the table. One participant expressed a belief that a robot's use should be limited to "hard labor." Other participants viewed themselves as subordinates to the robot, perceiving its expertise at the task to be superior. These participants tried to follow the robot's lead, even if they felt that some changes it made were suboptimal. One participant told us that she felt like the robot was her colleague, listening and responding to her. This seemed to have a lot to do with timing, with the robot sometimes incidentally reaching for components that she was considering or dropping blocks when it made a change she didn't like. She described its role as more than assistance, rather "a colleague . . . discussing through this process." Finally, some participants perceived an adversarial relationship with the robot, which they variously saw as annoying, "trying to confuse me" or "playing with me."

Whether these perceived roles led to collaboration or conflict, we found the participant's ability to establish shared intentions with the robot to play a key factor. In decision making or other open-ended tasks, framing the problem through a set of intentions is prerequisite to seeking a solution. In our study, conflicts arose over both longer-term goals and intermediate strategies. Participants struggled to understand the robot's intentions. This created conflicts with participants who favored one objective or adopted different search strategies. Failure to do so could result in annoyance or a breakdown in trust. As one participant put it, "I wouldn't say it was stupid for wanting its own agenda, but it was annoying to me." This is complicated by the fact that, in unstructured tasks, intentions can evolve as the actor learns more about the problem. While it is not necessary for human and robot goals to exactly align at all times, understanding a robot's goals makes it easier for a human to collaborate with it, and the converse should apply as well. This motivates further work toward both explainable AI for robots and models that infer the intentions of human decision-makers.

Our last study therefore focused on the challenge of inferring human intentions by an AI algorithm (Law et al., 2020). In this study, we tackled another multiobjective decision-making task. Participants were asked to draw fair voting districts in a U.S. state. We chose this problem because it represents a decision-making challenge where the very definition of the "fairness" objective is open to interpretation. We enabled participants to choose among three fairness objectives and any combination thereof: (a) the population balance between districts, (b) the voter efficiency of districts, and (c) the compactness of district shapes.

Within this formulation, we explored the question of whether an AI agent could predict which combination of these fairness objectives a human was focused on, based on the trajectory of fairness outcomes as they explored possible solutions. To do so, we used a deep neural network (DNN) to model the mapping between outcomes and intentions. The DNN was a combination of a long-short-term-memory architecture with a fully convolutional network adapted from Karim, Majumdar, Darabi, and Harford (2019), chosen for its ability to represent complex relationships, including temporal connections between data points (Karim, Majumdar, & Darabi, 2019). The intention–inference problem thus became a supervised learning problem. In theory, if we had had enough ground truth data of humans exploring voting district solutions with knowledge of the human's design intentions in each case, our DNN could have been able to learn this mapping. However, DNNs require large amounts of data, and data generated by humans working on such a task is expensive to collect. Furthermore, in many cases, humans might not be explicitly aware of their exact intentions and preference structure.

In consideration of this challenge, we trained our model on data generated by another AI agent, simulating the human decision-making process via a search optimization algorithm. We set a fixed preference distribution of the three fairness goals and built a human-simulating agent to find optimal districting solutions given these preferences. This resulted in synthetic training data that mapped preferences to exploration trajectories, consisting of outcomes from 130 episodes of 100 districting steps each for a total of 73,710 training samples.

This approach, of course, relies on the extent to which simulated explorations can accurately model human behavior for a given complex decision-making task. To test this, we evaluated our model, trained on the synthetic data, on "real" data, collected from four humans whom we asked to delineate districts for each of the possible combinations of intentions. The human test data consisted of an average of 689.4 steps per task across the four participants.

The core metric we used to evaluate our trained model was its prediction accuracy on the intentions specified for each task. Our model was able to infer the complete combination of intentions across three objectives 31.3% of the time (vs. 14.3% random accuracy). The model inferred the complete intention set in its top two predictions 50.9% of the time, and 67.2% of the time in its top three (vs. 42.9% and 57.1% random accuracy). For the three individual intentions, *balancing population*, *improving voter efficiency*, and *maximizing compactness*, the network predictions' average precision was 0.739, and its average recall was 0.700 (F1-score 0.719). In other words, when the network predicted a human was holding any one of the three intentions, it was correct, on average, 73.9% of the time, and it detected when the human was holding an intention, on average, 70.0% of the time. A random prediction would select a particular intention 57.1% of the time.

This study is an early investigation of the ability of an artificial agent to infer intentions in complex decision-making tasks. However, in such decision-making tasks, possible intentions can evolve and may not even be clear to the human worker at the start. To support collaboration, this dynamic necessitates the development of more complex models and representations.

Future work will seek to improve on this model using more contextual information about the task itself. This approach should be evaluated in physical interactions with a shared interface and a robot.

Once the human creative intention is correctly inferred, there remain additional questions: What is the best way to make use of knowledge about a human collaborator's intentions in an unstructured task? Should the robot simply adopt the human's intentions? Should it use information about the human's goals to optimize the likelihood of achieving its own goals using a game-theoretic approach?

In the two studies described, we identified challenges in human–robot collaborative decision making and proposed a model to give a robot more information about its partner's intentions. Given the increasing importance of cognitive and decision-making work in the future workplace, continuing to study this problem will remain a central challenge for the HRI research community.

CONCLUSION

Robots entering the workplace could be thought of as merely another type of machine or tool, a capital investment made by an employer to improve the efficiency of processes and tasks. However, years of laboratory research in human–robot interaction has shown that in many cases, people treat robots as social actors (Fischer, 2011; e.g., Fussell et al., 2008; Jung & Lee, 2004; Lombard & Xu, 2021), which might place robots in a category more akin to a team member. The subsequent relationship between robotics and labor puts the workplace robot in a societal and economic context that might be different from other workplace tools. Economists are still divided on how robots might affect labor in the long term, but the evidence shows that they already do so, reducing both workforce participation and wages (Acemoglu & Restrepo, 2017).

Whether robots replace workers or enter human teams as additional labor, robots in the workplace will interact with humans in psychologically intricate ways. The research field of HRI tackles both the engineering and the social aspects of this interaction, but it has predominantly focused on interaction in the home, in public places, and with service customers. This chapter argues that there will be unique and worthwhile research questions that specifically address the psychological needs and outcomes of workers interacting with robots, be it in offices, in warehouses, in factories, or in service locations, such as care facilities or retail centers.

Our own research presented here examined three such challenges. In one study, we evaluated how monetary competition with robots affects people's motivation to do well on a mundane and repetitive task. We found a demotivating effect, as well as a decrease in people's liking of the robot and their self-competency assessment when a robot was doing better at the task. In a second study, we saw that when the task is anything but mundane, complex dynamics need to be negotiated. Questions such as whose turn it is to act, what the goal of the team is, who leads and who follows, and how certain gestures might be perceived, all played a role when a human and a robot tried to solve a complex decision-making problem. A third study asked whether we could develop AI algorithms that can correctly infer a human's intentions in such a cognitive work task, and whether simulated data from other AI algorithms can serve as adequate models for human behavior. We found that even when training on simulated decision-making, our system was able to infer real human participants' intentions better than chance, suggesting promising ways in which robots could support workers in contexts outside of manufacturing.

These studies address only three of the many workplace-related psychological questions the field of HRI is sure to tackle in the years to come. Other issues may relate to the question of *where* robots would be most beneficial, not just economically, but also from a psychological perspective. This requires both an analysis of the labor market and a public discussion on the values and goals we set as a society. Researchers and members of the public have to ask themselves what kinds of work we would want robots to do (Ju & Takayama, 2011; Takayama et al., 2008). Do we want robot teachers, psychologists, or police investigators? Will future robots work in collaboration or in competition with human workers? As the bulk of the economy is moving from manufacturing and agriculture to services, and more and more workers are doing creative and open-ended work, the field of HRI has to also adapt to solve some of the complex problems in artificial intelligence that the research and development domains carry.

Finally, future research will likely be grounded in questions surrounding workplace social dynamics, organizational structures, and incentives. This will include ethnographic research observing workplace processes and behaviors, and also the adaptation of existing HRI research that was studied in other contexts. Researchers could study psychological issues such as social hierarchies, personality, dialog, trust, cognitive biases, and many others, in the relatively unexplored context of HRI in the workplace.

To summarize, research in human–robot interaction has made great strides in the nearly two decades that it has existed as a field. Many of its findings and developments are already relevant to offices, warehouses, factories, and grocery stores. With continued advances in artificial intelligence and robotics, HRI studies could make an increasing impact, as robots are more often used in the spaces in which humans work.

REFERENCES

Aaltonen, I., Arvola, A., Heikkilä, P., & Lammi, H. (2017). Hello Pepper, may I tickle you? Children's and adults' responses to an entertainment robot at a shopping mall. In *HR'17: Proceedings of the Companion of the 2017 ACM/IEEE International Conference on Human–Robot Interaction* (pp. 53–54). Association for Computing Machinery. https://doi.org/10.1145/3029798.3038362

Acemoglu, D., & Restrepo, P. (2017). *Robots and jobs: Evidence from US labor markets* (Working Paper No. 23285). National Bureau of Economic Research. https://www.nber.org/papers/w23285

Admoni, H., & Scassellati, B. (2017). Social eye gaze in human–robot interaction: A review. *Journal of Human–Robot Interaction*, 6(1), 25–63. https://doi.org/10.5898/JHRI.6.1.Admoni

Ambrose, R. O., Aldridge, H., Askew, R. S., Burridge, R. R., Bluethmann, W., Diftler, M., Lovchik, C., Magruder, D., & Rehnmark, F. (2000). Robonaut: NASA's space humanoid. *IEEE Intelligent Systems & their Applications*, 15(4), 57–63. https://doi.org/10.1109/5254.867913

Aylett, R. S., Castellano, G., Raducanu, B., Paiva, A., & Hanheide, M. (2011). Long-term socially perceptive and interactive robot companions: Challenges and future perspectives. In *ICMA'13: Proceedings of the 13th International Conference on Multimodal Interfaces* (pp. 323–326). Association for Computing Machinery. https://doi.org/10.1145/2070481.2070543

Babbar-Sebens, M., & Minsker, B. S. (2012). Interactive genetic algorithm with mixed initiative interaction for multi-criteria ground water monitoring design. *Applied Soft Computing, 12*(1), 182–195. https://doi.org/10.1016/j.asoc.2011.08.054

Bacciu, D., Gallicchio, C., Micheli, A., Di Rocco, M., & Saffiotti, A. (2014). Learning context-aware mobile robot navigation in home environments. In *IISA 2014: 5th International Conference on Information, Intelligence, Systems and Applications* (pp. 57–62). IEEE. https://doi.org/10.1109/iisa.2014.6878733

Bajones, M. (2016). Enabling long-term human–robot interaction through adaptive behavior coordination. In *2016 11th ACM/IEEE International Conference on Human–Robot Interaction (HRI)* (pp. 597–598). IEEE. https://doi.org/10.1109/hri.2016.7451874

Banerjee, A., Quiroz, J. C., & Louis, S. J. (2008). A model of creative design using collaborative interactive genetic algorithms. In J. S. Gero & A. K. Goel (Eds.), *Design computing and cognition '08* (pp. 397–416). Springer. https://doi.org/10.1007/978-1-4020-8728-8_21

Barnes, M. J., & Evans, A. W., III. (2010). Soldier–robot teams in future battlefields: An overview. In F. Jentsch & M. Barnes (Eds.), *Human–robot interactions in future military operations* (pp. 9–29). CRC Press. https://doi.org/10.4324/9781315587622

Bauer, A., Wollherr, D., & Buss, M. (2008). Human–robot collaboration: A survey. *International Journal of Humanoid Robotics, 5*(01), 47–66. https://doi.org/10.1142/S0219843608001303

Baxter, P., Kennedy, J., Senft, E., Lemaignan, S., & Belpaeme, T. (2016). From characterising three years of HRI to methodology and reporting recommendations. In *2016 11th ACM/IEEE International Conference on Human–Robot Interaction (HRI)* (pp. 391–398). IEEE.

Bera, A., Randhavane, T., Prinja, R., & Manocha, D. (2017). Sociosense: Robot navigation amongst pedestrians with social and psychological constraints. In *2017 IEEE/RSJ International Conference on Intelligent Robots and Systems (IROS)* (pp. 7018–7025). IEEE. https://doi.org/10.1109/IROS.2017.8206628

Bethel, C. L., & Murphy, R. R. (2008). Survey of non-facial/non-verbal affective expressions for appearance-constrained robots. In V. Marik (Ed.), *IEEE transactions on systems, man and cybernetics: Part c. Applications and reviews, 38*(1) (pp. 83–92). IEEE. https://doi.org/10.1109/TSMCC.2007.905845

Bhattacharjee, T., Lee, G., Song, H., & Srinivasa, S. S. (2019). Towards robotic feeding: Role of haptics in fork-based food manipulation. *IEEE Robotics and Automation Letters, 4*(2), 1485–1492. https://doi.org/10.1109/LRA.2019.2894592

Billard, A., & Kragic, D. (2019). Trends and challenges in robot manipulation. *Science, 364*(6446), eaat8414. https://doi.org/10.1126/science.aat8414

Birnbaum, G. E., Mizrahi, M., Hoffman, G., Reis, H. T., Finkel, E. J., & Sass, O. (2016). What robots can teach us about intimacy: The reassuring effects of robot responsiveness to human disclosure. *Computers in Human Behavior, 63,* 416–423. https://doi.org/10.1016/j.chb.2016.05.064

Biswas, M., & Murray, J. C. (2015). Towards an imperfect robot for long-term companionship: Case studies using cognitive biases. In *2015 IEEE/RSJ International Conference on Intelligent Robots and Systems (IROS)* (pp. 5978–5983). IEEE. https://doi.org/10.1109/iros.2015.7354228

Bogue, R. (2019). Strong prospects for robots in retail. *Industrial Robot: The International Journal of Robotics Research and Application, 46,* 326–331. https://doi.org/10.1108/IR-01-2019-0023

Boroush, M. (2020). Research and development: U.S. trends and international comparisons. *Science and Engineering Indicators*. National Science Board.

Breazeal, C., Dautenhahn, K., & Kanda, T. (2016). Social robotics. In B. Siciliano & O. Khatib (Eds.), *Springer handbook of robotics* (pp. 1935–1972). Springer. https://doi.org/10.1007/978-3-319-32552-1_72

Broekens, J., Heerink, M., & Rosendal, H. (2009). Assistive social robots in elderly care: A review. *Gerontechnology, 8*(2), 94–103. https://journal.gerontechnology.org/currentIssueContent.aspx?aid=986

Cabibihan, J.-J., Javed, H., Ang, M., Jr., & Aljunied, S. M. (2013). Why robots? A survey on the roles and benefits of social robots in the therapy of children with autism. *International Journal of Social Robotics, 5*(4), 593–618. https://doi.org/10.1007/s12369-013-0202-2

Cakmak, M., & Takayama, L. (2013, March). Towards a comprehensive chore list for domestic robots. In *2013 8th ACM/IEEE International Conference on Human–Robot Interaction (HRI)* (pp. 23–94). IEEE. https://doi.org/10.1109/hri.2013.6483517

Campos, T., Pacheck, A., Hoffman, G., & Kress-Gazit, H. (2019). SMT-based control and feedback for social navigation. In *2019 International Conference on Robotics and Automation (ICRA)* (pp. 5005–5011). https://doi.org/10.1109/ICRA.2019.8794208

Capezio, F., Mastrogiovanni, F., Sgorbissa, A., & Zaccaria, R. (2007). The ANSER project: Airport nonstop surveillance expert robot. *2007 IEEE/RSJ International Conference on Intelligent Robots and Systems* (pp. 991–996). IEEE. https://doi.org/10.1109/IROS.2007.4399319

Cash, P., & Maier, A. (2016). Prototyping with your hands: The many roles of gesture in the communication of design concepts. *Journal of Engineering Design, 27*(1–3), 118–145. https://doi.org/10.1080/09544828.2015.1126702

Cha, E., Kim, Y., Fong, T., & Mataric, M. J. (2018). A survey of nonverbal signaling methods for non-humanoid robots. *Foundations and Trends in Robotics*, 6(4), 211–323. https://doi.org/10.1561/2300000057

Chen, Y. F., Everett, M., Liu, M., & How, J. P. (2017). Socially aware motion planning with deep reinforcement learning. *2017 IEEE/RSJ International Conference on Intelligent Robots and Systems (IROS)* (pp. 1343–1350). IEEE. https://doi.org/10.1109/IROS.2017.8202312

Cho, S.-B. (2002). Towards creative evolutionary systems with interactive genetic algorithm. *Applied Intelligence*, 16(2), 129–138. https://doi.org/10.1023/A:1013614519179

Clabaugh, C., & Matarić, M. (2018). Robots for the people, by the people: Personalizing human-machine interaction. *Science Robotics*, 3(21), eaat7451. https://doi.org/10.1126/scirobotics.aat7451

D'Andrea, R. (2012). A revolution in the warehouse: A retrospective on Kiva systems and the grand challenges ahead. *IEEE Transactions on Automation Science and Engineering*, 9(4), 638–639. https://doi.org/10.1109/TASE.2012.2214676

Dautenhahn, K., Woods, S., Kaouri, C., Walters, M. L., Koay, K. L., & Werry, I. (2005). What is a robot companion—Friend, assistant or butler? In *2005 IEEE/RSJ International Conference on Intelligent Robots and Systems* (pp. 1192–1197). IEEE. https://doi.org/10.1109/iros.2005.1545189

Davis, N., Hsiao, C.-P., Popova, Y., & Magerko, B. (2015). An enactive model of creativity for computational collaboration and co-creation. In N. Zagalo & P. Branco (Eds.), *Creativity in the digital age* (pp. 109–133). Springer. https://doi.org/10.1007/978-1-4471-6681-8_7

Deutsch, I., Erel, H., Paz, M., Hoffman, G., & Zuckerman, O. (2019). Home robotic devices for older adults: Opportunities and concerns. *Computers in Human Behavior*, 98, 122–133. https://doi.org/10.1016/j.chb.2019.04.002

Djuric, A. M., Urbanic, R., & Rickli, J. (2016). A framework for collaborative robot (CoBot) integration in advanced manufacturing systems. *SAE International Journal of Materials and Manufacturing*, 9(2), 457–464. https://doi.org/10.4271/2016-01-0337

Dorst, K. (2011). The core of "design thinking" and its application. *Design Studies*, 32(6), 521–532. https://doi.org/10.1016/j.destud.2011.07.006

Duchaine, V., & Gosselin, C. (2009). Safe, stable and intuitive control for physical human–robot interaction. *2009 IEEE International Conference on Robotics and Automation* (pp. 3383–3388). IEEE. https://doi.org/10.1109/ROBOT.2009.5152664

Egan, P., & Cagan, J. (2016). Human and computational approaches for design problem-solving. In P. Cash, T. Stanković, & M. Štorga (Eds.), *Experimental design research* (pp. 187–205). Springer. https://doi.org/10.1007/978-3-319-33781-4_11

Ekman, P. (1999). Basic emotions. *Handbook of cognition and emotion* (pp. 45–60). John Wiley & Sons Ltd. https://doi.org/10.1002/0470013494.ch3

Ekman, P., & Friesen, W. V. (1969). The repertoire of nonverbal behavior: Categories, origins, usage, and coding. *Semiotica, 1*(1), 49–98. https://doi.org/10.1515/semi.1969.1.1.49

El Ayadi, M., Kamel, M. S., & Karray, F. (2011). Survey on speech emotion recognition: Features, classification schemes, and databases. *Pattern Recognition, 44*(3), 572–587. https://doi.org/10.1016/j.patcog.2010.09.020

Eppner, C., Höfer, S., Jonschkowski, R., Martín-Martín, R., Sieverling, A., Wall, V., & Brock, O. (2016). Lessons from the amazon picking challenge: Four aspects of building robotic systems. In D. Hsu, N. Amato, S. Berman, & S. Jacobs (Eds.), *Proceedings of Robotics: Science and Systems XII.* Advance online publication. https://doi.org/10.15607/RSS.2016.XII.036

Estevez, D., Victores, J. G., Fernandez-Fernandez, R., & Balaguer, C. (2020). Enabling garment-agnostic laundry tasks for a robot household companion. *Robotics and Autonomous Systems, 123*, 103330. https://doi.org/10.1016/j.robot.2019.103330

Fasola, J., & Mataric, M. J. (2012). Using socially assistive human–robot interaction to motivate physical exercise for older adults. *Proceedings of the IEEE, 100*(8), 2512–2526. https://doi.org/10.1109/JPROC.2012.2200539

Fasola, J., & Mataric, M. J. (2013). A socially assistive robot exercise coach for the elderly. *Journal of Human–Robot Interaction, 2*(2), 3–32. https://doi.org/10.5898/JHRI.2.2.Fasola

Ferrer, G., Zulueta, A. G., Herrero Cotarelo, F., & Sanfeliu, A. (2017). Robot social-aware navigation framework to accompany people walking side-by-side. *Autonomous Robots, 41*(4), 775–793. https://doi.org/10.1007/s10514-016-9584-y

Fiorini, P., & Prassler, E. (2000). Cleaning and household robots: A technology survey. *Autonomous Robots, 9*(3), 227–235. https://doi.org/10.1023/A:1008954632763

Fischer, K. (2011). Interpersonal variation in understanding robots as social actors. In *HRI'11: Proceedings of the 6th ACM/IEEE International Conference on Human–Robot Interaction* (pp. 53–60). Association for Computing Machinery. https://doi.org/10.1145/1957656.1957672

Fong, T., Nourbakhsh, I., & Dautenhahn, K. (2003). A survey of socially interactive robots. *Robotics and Autonomous Systems, 42*(3–4), 143–166. https://doi.org/10.1016/S0921-8890(02)00372-X

Foster, M. E., Alami, R., Gestranius, O., Lemon, O., Niemelä, M., Odobez, J.-M., & Pandey, A. K. (2016). The MuMMER project: Engaging human–robot interaction in real-world public spaces. In A. Agah, J.-J. Cabibihan, A. M. Howard, M. A. Salichs, & H. He (Eds.), *Social Robotics: 8th International Conference, ICSR 2016* (pp. 753–763). Springer.

Fraune, M. R., Kawakami, S., Sabanovic, S., De Silva, P. R. S., & Okada, M. (2015). Three's company, or a crowd?: The effects of robot number and behavior on HRI in Japan and the USA. In L. E. Kavraki, D. Hsu, & J. Buchli (Eds.), *Proceedings of Robotics: Science and Systems XI*. Advance online publication. https://doi.org/10.15607/RSS.2015.XI.033

Fraune, M. R., Sherrin, S., Šabanović, S., & Smith, E. R. (2019). Is human–robot interaction more competitive between groups than between individuals? In *HRI'19: 14th ACM/IEEE International Conference on Human–Robot Interaction* (pp. 104–113). IEEE. https://doi.org/10.1109/HRI.2019.8673241

Froman, L. (2010). Positive psychology in the workplace. *Journal of Adult Development, 17*(2), 59–69. https://doi.org/10.1007/s10804-009-9080-0

Fussell, S. R., Kiesler, S., Setlock, L. D., & Yew, V. (2008). How people anthropomorphize robots. In *HRI '08: Proceedings of the 3rd ACM/IEEE International Conference on Human–Robot Interaction* (pp. 145–152). Association for Computing Machinery. https://doi.org/10.1145/1349822.1349842

Gallenberger, D., Bhattacharjee, T., Kim, Y., & Srinivasa, S. S. (2019). Transfer depends on acquisition: Analyzing manipulation strategies for robotic feeding. In *2019 14th ACM/IEEE International Conference on Human–Robot Interaction (HRI)* (pp. 267–276). IEEE. https://doi.org/10.1109/hri.2019.8673309

Gemser, G., & Leenders, M. A. (2001). How integrating industrial design in the product development process impacts on company performance. *Journal of Product Innovation Management, 18*(1), 28–38. https://doi.org/10.1111/1540-5885.1810028

Gibson, T. (2021). Service with a robot smile. *Mechanical Engineering, 143*(4), 52–57. https://doi.org/10.1115/1.2021-JUL4

Gill, D., & Prowse, V. (2012). A structural analysis of disappointment aversion in a real effort competition. *The American Economic Review, 102*(1), 469–503. https://doi.org/10.1257/aer.102.1.469

Gleeson, B., MacLean, K., Croft, E., & Alcazar, J. (2013). *Human–robot communication for collaborative assembly* [Paper presentation]. GRAND General Meeting. Toronto, Canada.

Gockley, R., Bruce, A., Forlizzi, J., Michalowski, M., Mundell, A., Rosenthal, S., Sellner, B., Simmons, R., Snipes, K., Schultz, A. C., & Wang, J. (2005). Designing robots for long-term social interaction. In *2005 IEEE/RSJ International Conference on Intelligent Robots and Systems* (pp. 1338–1343). IEEE. https://doi.org/10.1109/IROS.2005.1545303

Gombolay, M., Yang, X. J., Hayes, B., Seo, N., Liu, Z., Wadhwania, S., Yu, T., Shah, N., Golen, T., & Shah, J. (2018). Robotic assistance in the coordination of patient care. *The International Journal of Robotics Research, 37*(10), 1300–1316. https://doi.org/10.1177/0278364918778344

Gonzalez, M. E. Q., & Haselager, W. F. G. (2005). Creativity: Surprise and abductive reasoning. *Semiotica, 2005*(153), 325–342. https://doi.org/10.1515/semi.2005.2005.153-1-4.325

Goodrich, M. A., Crandall, J. W., Oudah, M., & Mathema, N. (2018, March). *Using narrative to enable longitudinal human–robot interactions* [Paper presentation]. HRI2018 Workshop on Longitudinal Human–Robot Teaming, Chicago, IL, United States. https://faculty.cs.byu.edu/~crandall/papers/LongitudinalHRIWorkshop2018.pdf

Goodrich, M. A., & Schultz, A. C. (2007). Human–robot interaction: A survey. *Foundations and Trends in Human–Computer Interaction, 1*(3), 203–275. https://doi.org/10.1561/1100000005

Gordon, G., & Breazeal, C. (2015). Bayesian active learning-based robot tutor for children's word-reading skills. In *Proceedings of the Twenty-Ninth AAAI Conference on Artificial Intelligence* (pp. 1343–1349). Association for the Advancement of Artificial Intelligence. https://www.aaai.org/ocs/index.php/AAAI/AAAI15/paper/view/9280

Gordon, G., Spaulding, S., Westlund, J. K., Lee, J. J., Plummer, L., Martinez, M., Das, M., & Breazeal, C. (2016). Affective personalization of a social robot tutor for children's second language skills. In *Proceedings of the Thirtieth AAAI Conference on Artificial Intelligence* (pp. 3951–3957). Association for the Advancement of Artificial Intelligence. https://dl.acm.org/doi/10.5555/3016387.3016461

Graether, E., & Mueller, F. (2012). Joggobot: A flying robot as jogging companion. In *CHI EA'12: Proceedings of the 2012 ACM Annual Conference extended abstracts on human factors in computing systems extended abstracts* (pp. 1063–1066). Association for Computing Machinery. https://doi.org/10.1145/2212776.2212386

Hancock, P. A., Billings, D. R., & Schaefer, K. E. (2011). Can you trust your robot? *Ergonomics in Design, 19*(3), 24–29. https://doi.org/10.1177/1064804611415045

Hancock, P. A., Billings, D. R., Schaefer, K. E., Chen, J. Y. C., de Visser, E. J., & Parasuraman, R. (2011). A meta-analysis of factors affecting trust in human–robot interaction. *Human Factors, 53*(5), 517–527. https://doi.org/10.1177/0018720811417254

Hayashi, K., Sakamoto, D., Kanda, T., Shiomi, M., Koizumi, S., Ishiguro, H., Ogasawara, T., & Hagita, N. (2007). Humanoid robots as a passive-social medium—A field experiment at a train station. In *HRI'07: Proceedings of the ACM/IEEE International Conference on Human–Robot Interaction (HRI)* (pp. 137–144). Association for Computing Machinery. https://doi.org/10.1145/1228716.1228735

Hayes, B., & Scassellati, B. (2013, March 3–6). *Challenges in shared-environment human–robot collaboration* [Paper presentation]. Collaborative Manipulation

Workshop at the 8th ACM/IEEE International Conference on Human–Robot Interaction, Tokyo, Japan. http://bradhayes.info/papers/hayes_challenges_HRI_collab_2013.pdf

Heerink, M., Kröse, B., Evers, V., & Wielinga, B. (2008). The influence of social presence on acceptance of a companion robot by older people. *Journal of Physical Agents, 2*(2), 33–40. https://doi.org/10.14198/JoPha.2008.2.2.05

Heerink, M., Krose, B., Evers, V., & Wielinga, B. (2009, September). Measuring acceptance of an assistive social robot: A suggested toolkit. In *Proceedings of RO-MAN 2009: The 18th IEEE International Symposium on Robot and Human Interactive Communication* (pp. 528–533). IEEE. https://doi.org/10.1109/roman.2009.5326320

Hoffman, G., & Vanunu, K. (2013). Effects of robotic companionship on music enjoyment and agent perception. In *Proceedings of 2013 8th ACM/IEEE International Conference on Human–Robot Interaction (HRI)* (pp. 317–324). IEEE. https://doi.org/10.1109/hri.2013.6483605

Huang, C.-M., Cakmak, M., & Mutlu, B. (2015). Adaptive coordination strategies for human–robot handovers. In L. E. Kavraki, D. Hsu, & J. Buchli (Eds.), *Proceedings of Robotics: Science and Systems XI*. Advanced online publication. https://doi.org/10.15607/RSS.2015.XI.031

Ju, W., & Takayama, L. (2011). Should robots or people do these jobs? A survey of robotics experts and non-experts about which jobs robots should do. In *2011 IEEE/RSJ International Conference on Intelligent Robots and Systems* (pp. 2452–2459). IEEE. https://doi.org/10.1109/iros.2011.6094759

Jung, Y., & Lee, K. M. (2004). Effects of physical embodiment on social presence of social robots. In *Proceedings of PRESENCE: 7th Annual International Workshop on Presence* (pp. 80–87). International Society for Presence Research. http://matthewlombard.com/ISPR/Proceedings/2004/Jung%20and%20Lee.pdf

Kachouie, R., Sedighadeli, S., Khosla, R., & Chu, M.-T. (2014). Socially assistive robots in elderly care: A mixed-method systematic literature review. *International Journal of Human-Computer Interaction, 30*(5), 369–393. https://doi.org/10.1080/10447318.2013.873278

Kahneman, D., & Tversky, A. (2013). Prospect theory: An analysis of decision under risk. In L. C. MacLean & W. T. Ziemba (Eds.), *Handbook of the fundamentals of financial decision making: Part I* (pp. 99–127). World Scientific. https://doi.org/10.1142/9789814417358_0006

Kamei, K., Shinozawa, K., Ikeda, T., Utsumi, A., Miyashita, T., & Hagita, N. (2010). Recommendation from robots in a real-world retail shop. In *Proceedings of ICMI-MLMI '10: International Conference on Multimodal Interfaces and the Workshop on Machine Learning for Multimodal Interaction* (pp. 1–8). Association for Computing Machinery. https://doi.org/10.1145/1891903.1891929

Kanda, T., Shiomi, M., Miyashita, Z., Ishiguro, H., & Hagita, N. (2009). An affective guide robot in a shopping mall. In *Proceedings of the 4th ACM/IEEE International Conference on Human–Robot Interaction* (pp. 173–180). Association for Computing Machinery. https://doi.org/10.1145/1514095.1514127

Karim, F., Majumdar, S., & Darabi, H. (2019). Insights into LSTM fully convolutional networks for time series classification. *IEEE, 7*, 67718–67725. https://doi.org/10.1109/ACCESS.2019.2916828

Karim, F., Majumdar, S., Darabi, H., & Harford, S. (2019). Multivariate LSTM-FCNs for time series classification. *Neural Networks, 116*, 237–245. https://doi.org/10.1016/j.neunet.2019.04.014

Karimi, P., Grace, K., Davis, N., & Maher, M. L. (2018). Creative sketching apprentice: Supporting conceptual shifts in sketch ideation. In J. Gero (Ed.), *Design Computing and Cognition '18. DCC 2018* (pp. 721–738). Springer. https://doi.org/10.1007/978-3-030-05363-5_39

Kemp, C. C., Edsinger, A., & Torres-Jara, E. (2007). Challenges for robot manipulation in human environments [grand challenges of robotics]. *IEEE Robotics & Automation Magazine, 14*(1), 20–29. https://doi.org/10.1109/MRA.2007.339604

Kennedy, J., Baxter, P., & Belpaeme, T. (2015). The robot who tried too hard: Social behaviour of a robot tutor can negatively affect child learning. In *HRI'15: Proceedings of the Tenth Annual ACM/IEEE International Conference on Human–Robot Interaction* (pp. 67–74). Association for Computing Machinery. https://doi.org/10.1145/2696454.2696457

Kennedy, J., Baxter, P., Senft, E., & Belpaeme, T. (2016). Heart vs. hard drive: Children learn more from a human tutor than a social robot. In *Proceedings of the 11th ACM/IEEE international conference on human–robot interaction* (pp. 451–452). Association for Computing Machinery. https://doi.org/10.1109/hri.2016.7451801

Khalastchi, E., & Kalech, M. (2018). On fault detection and diagnosis in robotic systems. *ACM Computing Surveys, 51*(1), 1–24. https://doi.org/10.1145/3146389

Kim, H.-S., & Cho, S.-B. (2000). Application of interactive genetic algorithm to fashion design. *Engineering Applications of Artificial Intelligence, 13*(6), 635–644. https://doi.org/10.1016/S0952-1976(00)00045-2

Ko, B. C. (2018). A brief review of facial emotion recognition based on visual information. *Sensors, 18*(2), 401. https://doi.org/10.3390/s18020401

Kolko, J. (2010). Abductive thinking and sensemaking: The drivers of design synthesis. *Design Issues, 26*(1), 15–28. https://doi.org/10.1162/desi.2010.26.1.15

Kong, Y., & Fu, Y. (2018). *Human action recognition and prediction: A survey*. arXiv. https://arxiv.org/abs/1806.11230

Kőszegi, B., & Rabin, M. (2006). A model of reference-dependent preferences. *The Quarterly Journal of Economics, 121*(4), 1133–1165. https://www.jstor.org/stable/25098823

Kőszegi, B., & Rabin, M. (2007). Reference-dependent risk attitudes. *The American Economic Review, 97*(4), 1047–1073. https://doi.org/10.1257/aer.97.4.1047

Kőszegi, B., & Rabin, M. (2009). Reference-dependent consumption plans. *The American Economic Review, 99*(3), 909–936. https://doi.org/10.1257/aer.99.3.909

Kruse, T., Pandey, A. K., Alami, R., & Kirsch, A. (2013). Human-aware robot navigation: A survey. *Robotics and Autonomous Systems, 61*(12), 1726–1743. https://doi.org/10.1016/j.robot.2013.05.007

Kshirsagar, A., Dreyfuss, B., Ishai, G., Heffetz, O., & Hoffman, G. (2019). Monetary-incentive competition between humans and robots: Experimental results. *HRI'19:2019 14th ACM/IEEE International Conference on Human-Robot Interaction* (pp. 95–103). IEEE.

Kshirsagar, A., Lim, M., Christian, S., & Hoffman, G. (2020). Robot gaze behaviors in human-to-robot handovers. *IEEE Robotics and Automation Letters, 5*(4), 6552–6558. https://doi.org/10.1109/LRA.2020.3015692

Kuno, Y., Sadazuka, K., Kawashima, M., Yamazaki, K., Yamazaki, A., & Kuzuoka, H. (2007). Museum guide robot based on sociological interaction analysis. In *CHI '07: Proceedings of the SIGCHI Conference on Human Factors in Computing Systems* (pp. 1191–1194). Association for Computing Machinery. https://doi.org/10.1145/1240624.1240804

Law, M. V., Jeong, J., Kwatra, A., Jung, M. F., & Hoffman, G. (2019). Negotiating the creative space in human–robot collaborative design. In *Proceedings of the 2019 on Designing Interactive Systems Conference* (pp. 645–657). Association for Computing Machinery. https://doi.org/10.1145/3322276.3322343

Law, M. V., Kwatra, A., Dhawan, N., Einhorn, M., Rajesh, A., & Hoffman, G. (2020). Design intention inference for virtual co-design agents. In *Proceedings of the 20th ACM International Conference on Intelligent Virtual Agents* (pp. 1–8). Association for Computing Machinery. https://doi.org/10.1145/3383652.3423861

Lee, J. J., Knox, W. B., Wormwood, J. B., Breazeal, C., & Desteno, D. (2013). Computationally modeling interpersonal trust. *Frontiers in Psychology, 4*, 893. https://doi.org/10.3389/fpsyg.2013.00893

Lee, K. M., Jung, Y., Kim, J., & Kim, S. R. (2006). Are physically embodied social agents better than disembodied social agents?: The effects of physical embodiment, tactile interaction, and people's loneliness in human–robot interaction. *International Journal of Human-Computer Studies, 64*(10), 962–973. https://doi.org/10.1016/j.ijhcs.2006.05.002

Leite, I., Martinho, C., & Paiva, A. (2013). Social robots for long-term interaction: A survey. *International Journal of Social Robotics, 5*(2), 291–308. https://doi.org/10.1007/s12369-013-0178-y

Liang, Y., & Lee, S. A. (2017). Fear of autonomous robots and artificial intelligence: Evidence from national representative data with probability sampling.

International Journal of Social Robotics, 9(3), 379–384. https://doi.org/10.1007/s12369-017-0401-3

Lichtenthäler, C., Peters, A., Griffiths, S., & Kirsch, A. (2013). Be a robot! Robot navigation patterns in a path crossing scenario. In *Proceedings of the 2013 8th ACM/IEEE International Conference on Human–Robot Interaction (HRI)* (pp. 181–182). Association for Computing Machinery. https://doi.org/10.1109/HRI.2013.6483561

Ligthart, M. E. U., Neerincx, M. A., & Hindriks, K. V. (2020). Design patterns for an interactive storytelling robot to support children's engagement and agency. In *HRI '20: Proceedings of the 2020 ACM/IEEE International Conference on Human–Robot Interaction* (pp. 409–418). Association for Computing Machinery. https://doi.org/10.1145/3319502.3374826

Lombard, M., & Xu, K. (2021). Social responses to media technologies in the 21st century: The media are social actors paradigm. *Human–Machine Communication, 2*(1), 29–55. https://doi.org/10.30658/hmc.2.2

Mandal, F. B. (2014). Nonverbal communication in humans. *Journal of Human Behavior in the Social Environment, 24*(4), 417–421. https://doi.org/10.1080/10911359.2013.831288

Mann, S. (2007). Expectations of emotional display in the workplace. *Leadership and Organization Development Journal, 28*(6), 552–570. https://doi.org/10.1108/01437730710780985

Marti, P., & Giusti, L. (2010). A robot companion for inclusive games: A user-centred design perspective. In *Proceedings of the 2010 IEEE International Conference on Robotics and Automation* (pp. 4348–4353). IEEE. https://doi.org/10.1109/robot.2010.5509385

Martínez-Villaseñor, L., & Ponce, H. (2019). A concise review on sensor signal acquisition and transformation applied to human activity recognition and human–robot interaction. *International Journal of Distributed Sensor Networks, 15*(6), 1–12. https://doi.org/10.1177/1550147719853987

Mavridis, N., Datta, C., Emami, S., Tanoto, A., BenAbdelkader, C., & Rabie, T. (2009). FaceBots: Robots utilizing and publishing social information in Facebook. In *Proceedings of the 4th ACM/IEEE International Conference on Human–Robot Interaction* (pp. 273–274). Association for Computing Machinery. https://doi.org/10.1145/1514095.1514172

Mead, R., & Matarić, M. J. (2017). Autonomous human–robot proxemics: Socially aware navigation based on interaction potential. *Autonomous Robots, 41*(5), 1189–1201. https://doi.org/10.1007/s10514-016-9572-2

Mehrabian, A. (1980). *Basic dimensions for a general psychological theory: Implications for personality, social, environmental, and developmental studies*. Oelgeschlager, Gunn & Hain.

Moniz, A. B., & Krings, B.-J. (2016). Robots working with humans or humans working with robots? Searching for social dimensions in new human–robot interaction in industry. *Societies, 6*(3), 23. https://doi.org/10.3390/soc6030023

Murakami, R., Morales Saiki, L. Y., Satake, S., Kanda, T., & Ishiguro, H. (2014). Destination unknown: Walking side-by-side without knowing the goal. In *Proceedings of the 2014 ACM/IEEE International Conference on Human–Robot Interaction* (pp. 471–478). Association for Computing Machinery. https://doi.org/10.1145/2559636.2559665

Mussakhojayeva, S., & Sandygulova, A. (2017). Cross-cultural differences for adaptive strategies of robots in public spaces. In *2017 26th IEEE International Symposium on Robot and Human Interactive Communication (RO-MAN)* (pp. 573–578). https://doi.org/10.1109/ROMAN.2017.8172360

Mutlu, B., & Forlizzi, J. (2008). Robots in organizations. In *HRI'08: Proceedings of the 3rd International Conference on Human Robot Interaction* (pp. 287–294). Association for Computing Machinery. https://doi.org/10.1145/1349822.1349860

Mutlu, B., Osman, S., Forlizzi, J., Hodgins, J., & Kiesler, S. (2006). Perceptions of ASIMO: An exploration on co-operation and competition with humans and humanoid robots. In *Proceedings of the 1st ACM SIGCHI/SIGART Conference on Human–Robot Interaction* (pp. 351–352). Association for Computing Machinery. https://doi.org/10.1145/1121241.1121311

Nehaniv, C. L., Dautenhahn, K., Kubacki, J., Haegele, M., Parlitz, C., & Alami, R. (2005). A methodological approach relating the classification of gesture to identification of human intent in the context of human–robot interaction. In *ROMAN 2005: IEEE International Workshop on Robot and Human Interactive Communication* (pp. 371–377). IEEE. https://doi.org/10.1109/roman.2005.1513807

Niemelä, M., Heikkilä, P., & Lammi, H. (2017). A social service robot in a shopping mall: Expectations of the management, retailers and consumers. In *HR'17: Proceedings of the Companion of the 2017 ACM/IEEE International Conference on Human–Robot Interaction* (pp. 227–228). Association for Computing Machinery. https://doi.org/10.1145/3029798.3038301

Noroozi, F., Corneanu, C., Kaminska, D., Sapinski, T., Escalera, S., & Anbarjafari, G. (2019). Survey on emotional body gesture recognition. *IEEE Transactions on Affective Computing, 12*(2), 505–523. https://doi.org/10.1109/taffc.2018.2874986

Odekerken-Schröder, G., Mele, C., Russo-Spena, T., Mahr, D., & Ruggiero, A. (2020). Mitigating loneliness with companion robots in the COVID-19 pandemic and beyond: An integrative framework and research agenda. *Journal of Service Management, 31*(6), 1149–1162. https://doi.org/10.1108/josm-05-2020-0148

Ozawa, R., & Tahara, K. (2017). Grasp and dexterous manipulation of multi-fingered robotic hands: A review from a control view point. *Advanced Robotics*, *31*(19–20), 1030–1050. https://doi.org/10.1080/01691864.2017.1365011

Pantofaru, C., Takayama, L., Foote, T., & Soto, B. (2012). Exploring the role of robots in home organization. In *HRI '12: Proceedings of the Seventh Annual ACM/IEEE International Conference on Human–Robot Interaction* (pp. 327–334). Association for Computing Machinery. https://doi.org/10.1145/2157689.2157805

Paolanti, M., Romeo, L., Martini, M., Mancini, A., Frontoni, E., & Zingaretti, P. (2019). Robotic retail surveying by deep learning visual and textual data. *Robotics and Autonomous Systems*, *118*, 179–188. https://doi.org/10.1016/j.robot.2019.01.021

Parker, L. E., & Draper, J. V. (1999). Robotics applications in maintenance and repair. In S. Y. Nof (Ed.), *Handbook of industrial robotics* (2nd ed., pp. 1023–1036). John Wiley & Sons.

Pham, T. X. N., Hayashi, K., Becker-Asano, C., Lacher, S., & Mizuuchi, I. (2017). Evaluating the usability and users' acceptance of a kitchen assistant robot in household environment. In *2017 26th IEEE International Symposium on Robot and Human Interactive Communication (RO-MAN)* (pp. 987–992). IEEE. https://doi.org/10.1109/roman.2017.8172423

Plutchik, R. (2001). The nature of emotions: Human emotions have deep evolutionary roots, a fact that may explain their complexity and provide tools for clinical practice. *American Scientist*, *89*(4), 344–350. https://www.jstor.org/stable/27857503

Posner, J., Russell, J. A., & Peterson, B. S. (2005). The circumplex model of affect: An integrative approach to affective neuroscience, cognitive development, and psychopathology. *Development and Psychopathology*, *17*(3), 715–734. https://doi.org/10.1017/S0954579405050340

Prassler, E., Ritter, A., Schaeffer, C., & Fiorini, P. (2000). A short history of cleaning robots. *Autonomous Robots*, *9*(3), 211–226. https://doi.org/10.1023/A:1008974515925

Rios-Martinez, J., Spalanzani, A., & Laugier, C. (2015). From proxemics theory to socially-aware navigation: A survey. *International Journal of Social Robotics*, *7*(2), 137–153. https://doi.org/10.1007/s12369-014-0251-1

Robinette, P., Wagner, A. R., & Howard, A. M. (2014). Assessment of robot guidance modalities conveying instructions to humans in emergency situations. In *The 23rd IEEE International Symposium on Robot and Human Interactive Communication* (pp. 1043–1049). https://doi.org/10.1109/roman.2014.6926390

Robinson, H., Macdonald, B., Kerse, N., & Broadbent, E. (2013). The psychosocial effects of a companion robot: A randomized controlled trial. *Journal of the*

American Medical Directors Association, 14(9), 661–667. https://doi.org/10.1016/j.jamda.2013.02.007

Rosenfeld, A., Maksimov, O., Kraus, S., & Agmon, N. (2016, May). *Human-multi-robot team collaboration for efficient warehouse operation* [Paper presentation]. Autonomous Robots and Multirobot Systems (ARMS) Workshop at AAMAS International Conference, Singapore.

Roy, R., & Potter, S. (1993). The commercial impacts of investment in design. *Design Studies, 14*(2), 171–193. https://doi.org/10.1016/0142-694X(93)80046-F

Sakamoto, D., Hayashi, K., Kanda, T., Shiomi, M., Koizumi, S., Ishiguro, H., Ogasawara, T., & Hagita, N. (2009). Humanoid robots as a broadcasting communication medium in open public spaces. *International Journal of Social Robotics, 1*(2), 157–169. https://doi.org/10.1007/s12369-009-0015-5

Salem, M., Lakatos, G., Amirabdollahian, F., & Dautenhahn, K. (2015). Would you trust a (faulty) robot? Effects of error, task type and personality on human-robot cooperation and trust. In *HRI'15: Proceedings of the Tenth Annual ACM/IEEE International Conference on Human–Robot Interaction* (pp. 141–148). Association for Computing Machinery. https://doi.org/10.1145/2696454.2696497

Sanneman, L., Fourie, C., & Shah, J. A. (2020). *The state of industrial robotics: Emerging technologies, challenges, and key research directions.* arXiv. https://arxiv.org/abs/2010.14537

Saunderson, S., & Nejat, G. (2019). How robots influence humans: A survey of nonverbal communication in social human–robot interaction. *International Journal of Social Robotics, 11*(4), 575–608. https://doi.org/10.1007/s12369-019-00523-0

Savery, R., & Weinberg, G. (2020). *A survey of robotics and emotion: Classifications and models of emotional interaction.* arXiv. https://arxiv.org/abs/2007.14838

Schneider, S., & Kummert, F. (2016, November). Exercising with a humanoid companion is more effective than exercising alone. In *2016 IEEE-RAS 16th International Conference on Humanoid Robots (Humanoids)* (pp. 495–501). IEEE. https://doi.org/10.1109/humanoids.2016.7803321

Schroeter, Ch., Mueller, S., Volkhardt, M., Einhorn, E., Huijnen, C., van den Heuvel, H., van Berlo, A., Bley, A., & Gross, H.-M. (2013). Realization and user evaluation of a companion robot for people with mild cognitive impairments. *2013 IEEE International Conference on Robotics and Automation* (pp. 1153–1159). IEEE. https://doi.org/10.1109/icra.2013.6630717

Shi, C., Satake, S., Kanda, T., & Ishiguro, H. (2016). How would store managers employ social robots? In *Proceedings of the 2016 11th ACM/IEEE International*

Conference on Human–Robot Interaction (HRI) (pp. 519–520). Association for Computing Machinery. https://doi.org/10.1109/HRI.2016.7451835

Shiomi, M., Sakamoto, D., Kanda, T., Ishi, C. T., Ishiguro, H., & Hagita, N. (2008). A semi-autonomous communication robot—A field trial at a train station. In *Proceedings of the 2008 3rd ACM/IEEE International Conference on Human–Robot Interaction (HRI)* (pp. 303–310). Association for Computing Machinery. https://doi.org/10.1145/1349822.1349862

Short, E., Hart, J., Vu, M., & Scassellati, B. (2010). No fair!! An interaction with a cheating robot. In *Proceedings of the 2010 5th ACM/IEEE International Conference on Human–Robot Interaction (HRI)* (pp. 219–226). Association for Computing Machinery. https://doi.org/10.1109/HRI.2010.5453193

Sisbot, E. A., Marin-Urias, L. F., Alami, R., & Simeon, T. (2007). A human aware mobile robot motion planner. *IEEE Transactions on Robotics, 23*(5), 874–883. https://doi.org/10.1109/TRO.2007.904911

Smedegaard, C. V. (2019). Reframing the role of novelty within social HRI: From noise to information. In *Proceedings of the 2019 14th ACM/IEEE International Conference on Human–Robot Interaction (HRI)* (pp. 411–420). Association for Computing Machinery. https://doi.org/10.1109/HRI.2019.8673219

Smith, G., Whitehead, J., & Mateas, M. (2010). Tanagra: A mixed-initiative level design tool. In *FDG '10: Proceedings of the Fifth International Conference on the Foundations of Digital Games* (pp. 209–216). Association for Computing Machinery. https://doi.org/10.1145/1822348.1822376

Szollosy, M. (2017). Freud, Frankenstein and our fear of robots: Projection in our cultural perception of technology. *AI & Society, 32*(3), 433–439. https://doi.org/10.1007/s00146-016-0654-7

Takayama, L., Ju, W., & Nass, C. (2008). Beyond dirty, dangerous and dull: What everyday people think robots should do. In *HRI '08: Proceedings of the 3rd International Conference on Human–Robot Interaction* (pp. 25–32). Association for Computing Machinery. https://doi.org/10.1145/1349822.1349827

Thomaz, A. (2021). *Meet Moxi: Our socially intelligent robot supporting health care teams.* Diligent Robotics. https://www.diligentrobots.com/blog/2020-2

Thomaz, A., Hoffman, G., & Cakmak, M. (2016). Computational human–robot interaction. *Foundations and Trends in Robotics, 4*(2–3), 105–223. https://doi.org/10.1561/2300000049

Thrun, S., Bennewitz, M., Burgard, W., Cremers, A. B., Dellaert, F., Fox, D., Hahnel, D., Rosenberg, C., Roy, N., Schulte, J., & Schulz, D. (1999). MINERVA: A second-generation museum tour-guide robot. In *Proceedings of the Proceedings 1999 IEEE International Conference on Robotics and Automation* (Vol. 3, pp. 1999–2005). IEEE. https://doi.org/10.1109/ROBOT.1999.770401

Tonkin, M., Vitale, J., Herse, S., Williams, M.-A., Judge, W., & Wang, X. (2018). Design methodology for the UX of HRI: A field study of a commercial social robot at an airport. In *Proceedings of the 2018 ACM/IEEE International Conference on Human–Robot Interaction* (pp. 407–415). Association for Computing Machinery. https://doi.org/10.1145/3171221.3171270

Triebel, R., Arras, K., Alami, R., Beyer, L., Breuers, S., Chatila, R., Chetouani, M., Cremers, D., Evers, V., Fiore, M., Hung, H., Islas Ramírez, O. A., Joosse, M., Khambhaita, H., Kucner, T., Leibe, B., Lilienthal, A. J., Linder, T., Lohse, M., . . . Zhang, L. (2016). SPENCER: A socially aware service robot for passenger guidance and help in busy airports. In D. Wittergreen & T. Barfoot (Eds.), *Field and service robotics* (pp. 607–622). Springer. https://doi.org/10.1007/978-3-319-27702-8_40

Tsarouchi, P., Makris, S., & Chryssolouris, G. (2016). Human–robot interaction review and challenges on task planning and programming. *International Journal of Computer Integrated Manufacturing, 29*(8), 916–931. https://doi.org/10.1080/0951192X.2015.1130251

Tuikka, T., & Kuutti, K. (2000). Making new design ideas more concrete. *Knowledge-Based Systems, 13*(6), 395–402. https://doi.org/10.1016/S0950-7051(00)00080-0

Turja, T., Rantanen, T., & Oksanen, A. (2019). Robot use self-efficacy in health care work (RUSH): Development and validation of a new measure. *AI & Society, 34*(1), 137–143. https://doi.org/10.1007/s00146-017-0751-2

U.S. Bureau of Economic Analysis. (2019). *Employment by industry.* https://www.bea.gov/data/employment/employment-by-industry

U.S. Bureau of Economic Analysis. (2021). *GDP by industry.* https://www.bea.gov/data/gdp/gdp-industr

Volkhardt, M., Mueller, S., Schroeter, C., & Gross, H.-M. (2011). Playing hide and seek with a mobile companion robot. In *Proceedings of the 2011 11th IEEE-RAS International Conference on Humanoid Robots* (pp. 40–46). IEEE. https://doi.org/10.1109/humanoids.2011.6100809

Wada, K., & Shibata, T. (2007). Living with seal robots—Its sociopsychological and physiological influences on the elderly at a care house. *IEEE Transactions on Robotics, 23*(5), 972–980. https://doi.org/10.1109/TRO.2007.906261

Wada, K., Shibata, T., Musha, T., & Kimura, S. (2005). Effects of robot therapy for demented patients evaluated by EEG. *2005 IEEE/RSJ International Conference on Intelligent Robots and Systems* (pp. 1552–1557). IEEE. https://doi.org/10.1109/IROS.2005.1545304

Wasen, K. (2010). Replacement of highly educated surgical assistants by robot technology in working life: Paradigm shift in the service sector. *International Journal of Social Robotics, 2*(4), 431–438. https://doi.org/10.1007/s12369-010-0062-y

Wilcox, R., Nikolaidis, S., & Shah, J. (2013). Optimization of temporal dynamics for adaptive human–robot interaction in assembly manufacturing. In N. Roy, P. Newman, & S. Srinivasa (Eds.), *Proceedings of Robotics: Science and Systems VII*. https://doi.org/10.15607/RSS.2012.VIII.056

Wiltshire, T. J., Lobato, E. J. C., Wedell, A. V., Huang, W., Axelrod, B., & Fiore, S. M. (2013). Effects of robot gaze and proxemic behavior on perceived social presence during a hallway navigation scenario. *Proceedings of the Human Factors and Ergonomics Society Annual Meeting*, 57(1), 1273–1277. https://doi.org/10.1177/1541931213571282

Yan, H., Ang, M. H., Jr., & Poo, A. N. (2014). A survey on perception methods for human–robot interaction in social robots. *International Journal of Social Robotics*, 6(1), 85–119. https://doi.org/10.1007/s12369-013-0199-6

Zanchettin, A. M., Ceriani, N. M., Rocco, P., Ding, H., & Matthias, B. (2016). Safety in human–robot collaborative manufacturing environments: Metrics and control. *IEEE Transactions on Automation Science and Engineering*, 13(2), 882–893. https://doi.org/10.1109/TASE.2015.2412256

10

The Psychology of Big Data: Developing a "Theory of Machine" to Examine Perceptions of Algorithms

Jennifer M. Logg

Historically, people have informed their decisions with advice from other people. However, the rise of "Big Data" has increased both the availability and utility of a new source of advice: *algorithms*. These scripts for mathematical calculations cull through massive amounts of data and produce insights that can improve decision making. Many organizations are already trying to capture this potential, using algorithms to hire promising applicants (e.g., Amazon), predict performance of current employees (e.g., Navy Seals, National Football League teams, and Premier League soccer teams), and identify individuals who are likely to leave in order to improve retention (e.g., Johnson & Johnson).

DATA ANALYTICS NEEDS PSYCHOLOGY

While organizations are swimming in data and investing in algorithms, many are trying to understand how to maximize the benefits of algorithmic advice. And while companies focus on producing more analytical insights, it is not clear how well those insights are utilized. What happens when algorithmic advice lands in the hands of managers and other decision makers? When do they listen to it, and when do they disregard it?

One article aptly labeled the gap between producing and utilizing insights from algorithms as the "last mile problem" (Berinato, 2019). The "last mile" concept is commonly used in supply chain management to describe how goods are transported from a centralized hub to the final end user (Rodrigue et al., 2009). In data analytics, the last mile problem describes the issue of producing analytical insights but (a) failing to communicate them at all, leading to wasted information; or (b) failing to communicate them clearly, leading to misapplied information. This problem is similar to a soccer player with great footwork who fails to convert plays into goals or assists. Failing to communicate analytical results clearly hinders decision makers from acting on them. In an article describing predictive analytics, Schrage (2016) eloquently stated, "Effectively communicating and sharing analytic insights is as important as finding them" (p. 2).

Data analytics needs psychology. Organizations cannot realize the full potential of algorithms until they address the last mile problem and consider how people respond to algorithmic advice. Algorithms have the potential to greatly improve human judgment and decision making, as they generally outperform the accuracy of experts when the two are directly compared (Dawes et al., 1989; Meehl, 1954). But people can only leverage the accuracy of algorithms if they are willing to listen. Should they ignore algorithmic advice, the resources invested into data analytics, both within academia and industry, will go to waste. While the field of data analytics (the systematic computation of data, most commonly using algorithms) continues to evolve at a rapid rate, most overlook the important connection between producing and utilizing insights.

Thus, this chapter calls for research on the psychology of Big Data[1] in order to bring a psychological perspective to the field of data analytics. I propose this perspective in order to reexamine and extend what researchers know about human behavior and decision making by examining decisions in the age of Big Data. Additionally, this chapter provides a theoretical framework to help guide research. I call the framework *theory of machine*, which examines how people expect algorithmic and human judgment to differ. People's expectations for each of these likely influence how they respond to information generated by both. Understanding how people respond to algorithmic output is crucial for helping organizations leverage the power of Big Data to improve their decisions.

OVERVIEW OF CURRENT WORK

This chapter discusses (a) the last mile problem, data analytics' biggest hurdle to improving decisions; (b) how the field of psychology is well-positioned to solve this issue; and (c) specific avenues that are ripe for basic psychological research. First, I outline how siloes in organizations contribute to the last mile problem and how computer science has previously discussed combining algorithmic and human judgment. Then, I contextualize the importance of the last mile problem by reviewing literature documenting the superior predictive accuracy of algorithms compared with expert human judgment.

The focus of this chapter is the theoretical framework I developed in a paper entitled "Algorithm Appreciation" (Logg et al., 2019). Here, my colleagues and I empirically examined how people respond to algorithmic advice relative to advice from other people. We found that people relied more on identical advice when they thought it came from an algorithm than when they thought it came from a person.[2] In that paper,

---

[1] I created a class on this topic. More information can be found online (https://www.jennlogg.com/psychology-of-big-data.html).

[2] In this type of research, what matters more than how researchers conceptualize the term "algorithm" is how lay people define it. When participants were asked, 42% said it was math, an equation, or calculation, 26% said a step-by-step procedure, 14% said logic or a formula and the remaining fell into a category of descriptions that tended to mention computers.

I introduced the framework theory of machine to document people's lay theories about algorithmic judgment.

Theory of machine examines how people expect algorithmic and human judgment, at their finest, to differ from each other. It is a twist on the classic *theory of mind*. Philosophical work on theory of mind examines how people infer other people's intentions and beliefs (Dennett, 1987; Saxe, 2013). Theory of machine similarly examines people's expectations of another agent's internal processes. But in contrast to theory of mind, theory of machine considers how people expect algorithmic judgment and human judgment to differ from each other in their input (the information used), process (how the same information is utilized), and output (the predictions, advice, and feedback that are produced). As people receive more and more information from algorithms, both in their jobs and personal lives, science needs to understand how people develop their theory of machine.

I further develop theory of machine in this chapter by proposing specific, testable predictions about (a) people's expectations of algorithmic and human judgment and (b) how that in turn influences when people are most likely to utilize algorithmic output. I consider how the type of decision (whether the decision-maker is considering a prediction, a performance assessment, or feedback) may influence people's theory of machine. Throughout this chapter, I introduce current research from colleagues that the theory of machine framework can speak to.

Psychology additionally has the potential to improve the process of generating algorithms, which could further increase use of algorithmic advice. Thus, I break down the process of building an algorithm into three key decision stages in the section Using Psychology to Create Better Algorithms. Within each stage, I identify questions that can help analytics teams decrease bias in order to improve the quality of algorithmic output. Finally, I conclude by differentiating research on the psychology of Big Data from work on innovation (in economics and sociology), anthropomorphism (in psychology), and robots (in human–machine interaction) and describing how understanding the development of theory of mind research (in philosophy and psychology) can help guide future research.

This chapter uses three key terms that are important to clarify:

- *last mile problem*: When applied to data analytics, the communication gap between connecting the production of analytical insights and the application of those insights. Psychological research is needed to help solve the last mile problem in data analytics.
- *psychology of Big Data*: An area of research that takes a psychological perspective to the field of data analytics in order to reexamine what we know about human behavior and decision making in the age of Big Data.
- *theory of machine*: A theoretical research framework to examine how people expect algorithmic and human judgment (at their finest) to differ from each other in their input, process, and output.

DATA ANALYTICS' LAST MILE PROBLEM

In organizations, data analytics teams and decision makers such as managers are frequently siloed. This structural divide easily reinforces, if not generates, the last mile problem, the failure to convert the generation to utilization of insights. Consider an example in the field of policy. National security analysts who build predictive models with the goal of informing policy may benefit from knowing how to present information in a way that ensures others use it appropriately (see Chapter 3 in this book for an introduction to Big Data in public policy). Meanwhile, policy makers benefit from knowing which information they receive is most important and actionable. People on data analytics teams often ask: "How can I present my results so that people will (a) understand them and (b) listen to them?" While others may not even consider how likely others are to act on the information they present. Decision makers often ask: (a) "How can our organization obtain more buy-in for investment in data analytics?" and (b) "What questions should I ask my data analytics team?" The answer to these questions is context-dependent and may vary from decision to decision. Within the private sector, some companies have started trying to bridge the structural gap between the role of analyst and decision maker with a new role: *data liaison* (https://www.Glassdoor.com). The goal of this role is to translate output generated by data analytics teams to audiences who do not have the same technical skills. Understanding how decision makers view algorithmic output can also help address this communication gap. This is where the psychology of Big Data comes in.

Research in computer science (the more established field under which data analytics falls) has long called for a combination of human and machine judgment (referred to as the "loop of interaction" [Card et al., 1983]). However, this work primarily focuses on optimally combining the two. Similar to classical economics, it assumes that people fully utilize the information they receive. In recent thought pieces, leaders in computer science have called for the field to extend its focus beyond optimization. They have called for attention beyond (a) the appropriate types of algorithms for specific tasks and (b) the stage at which algorithmic output is inserted into a decision making process in order to (c) provide greater precision about what artificial intelligence (AI) systems can and cannot do (Rahwan et al., 2019). They have additionally considered effective communication about algorithms and their advice, including the role of a computer as a teammate rather than assistant (Guszcza & Schwartz, 2020) and how to best audit algorithmic output (Guszcza et al., 2018).

Some economists have considered how to improve the production of analytical insights, highlighting how human judgment is useful to improve the input data fed to algorithms (Luca et al., 2016). Choosing data sets that are "wide" (i.e., have multiple features for each entry/person in the data set) and diverse (i.e., include different kinds of people within the sample), for example, could yield results that are more relevant. The psychology of Big Data is needed in each of these research areas to measure how analytical insights are introduced and how people respond to them. Indeed, such challenges posed to the field have culminated in explicit calls for interdisciplinary collaboration, including between computer scientists and psychologists (Guszcza et al., 2018; Rahwan et al., 2019).

THE POWER OF ALGORITHMS

Algorithmic advice holds tremendous potential for our decision making. The last mile problem is important to solve because algorithms generally produce more accurate judgments than people who are experts. In this section, I review the classic research on clinical versus actuarial judgment (clinical diagnoses from doctors vs. algorithms), the anecdotal evidence on doctor's reactions to the news that algorithms outperform doctors, and the boundaries to algorithmic capabilities.

Algorithmic Accuracy

Literature dating back to the 1950s compares the accuracy of clinical (human) versus actuarial (statistical) predictions and shows that algorithms often make more accurate forecasts than experts (Dawes, 1979; Dawes et al., 1989; Meehl, 1954). Simple aggregation can outperform experts in domains with life-altering outcomes: Algorithms are more accurate than doctors and pathologists in predicting the survival of cancer patients (Einhorn, 1972), more accurate than parole boards in predicting recidivism of parolees (Carroll et al., 1982), and more accurate than loan officers in predicting whether a business will go bankrupt (Libby, 1976). In fact, when forecasting a variety of future events, aggregation of individual forecasts can outperform those made by individuals (Hastie & Kameda, 2005; Larrick & Soll, 2006), even for geopolitical events including terrorist attacks and the outbreak of war (Ungar et al., 2012). A long literature on the wisdom of crowds explains why aggregation consistently works so well. Even the simplest algorithm (averaging across people) improves accuracy because it cancels out errors from each of the individuals (Galton, 1907; Hastie & Kameda, 2005; Soll & Larrick, 2009; Surowiecki, 2004).

When simple aggregation methods are compared with more complex algorithms, the latter outperform simple averaging in terms of accuracy (Baron et al., 2014). Some research compared the accuracy of a complex algorithm against a person (usually an expert in that field) by presenting both the algorithm and person with the same information, or cues, to make their prediction. Here, the algorithms outperformed experts in assessments of different pathologies (Beck et al., 2011; Goldman et al., 1977; Hedén et al., 1997) and even operational risk within business (Tazelaar & Snijders, 2013). Research has even found that algorithms are more accurate at answering trivia questions (Tesauro et al., 2013).

In these instances, algorithms surpassed human judgment because the algorithm weighed identical cues more appropriately than the person did (Dawes, 1979). In other words, algorithms are better than people at identifying which cues are most useful, or predictive. For example, when predicting heart attacks, an algorithm relied on different cues than a widely used medical program and expert cardiologists. The cues used by the algorithm better predicted heart attacks than the cues used by both the program and experts (Hedén et al., 1997). Algorithms can identify predictive cues that were not previously identified by people. When pathologists and an algorithm were given the same biopsy images, the algorithm predicted the severity of breast cancer better than pathologists partly because it identified cues other than those the pathologists were trained to use (Beck et al., 2011).

Distrust of algorithmic decision making can result in severe consequences. To pick one tragic example, in 2004, Flash Airlines Flight 604 crashed into the Red Sea, resulting in the largest death toll in Egypt's aviation history (Egyptian Ministry of Civil Aviation, 2004). The captain, who was experiencing a condition known as spatial disorientation, trusted his own flawed judgment over the aircraft's instruments. In this case, reliance on human judgment over calculations by a machine resulted in disaster and cost to human life.

How People Respond to Algorithmic Advice Is an Open Question

This long literature documenting the superior accuracy of algorithms to human judgment points to a simple prescription: Listen to algorithmic advice to improve decision making. What is conspicuously missing from this empirical evidence on accuracy, however, is how people actually respond to algorithmic advice relative to another person's advice. Regardless of their accuracy, algorithms are a tool that can only benefit decision makers who are willing to use them.

In the 1950s, when Meehl shared his empirical results on the accuracy of algorithms with clinicians, they were reluctant to believe that a simple mathematical calculation could outperform their own professional judgment. These conversations struck Meehl; he included them as the anecdotes in his classic book on the accuracy of algorithms relative to experts (Meehl, 1954). Other papers empirically testing accuracy included similar stories that would have us believe people distrust algorithms completely (Dawes, 1979; Dawes et al., 1989; Kleinmuntz, 1990; Meehl, 1957; see also Kleinmuntz & Schkade, 1993). Such skepticism persisted and morphed into the received wisdom that people do not trust advice from an algorithm. A number of articles still advise leaders about how they can overcome resistance to algorithms (Harrell, 2016). Only recently have researchers begun empirically testing people's perceptions of algorithms (see the Theory of Machine section).

Presenting Algorithms as a Threat Rather Than a Tool

Any expert hearing that an algorithm produces more accurate predictions likely sees a threat to their job security. Indeed, researchers are currently predicting which jobs computers might usurp from humans (Frey & Osborne, 2013). Framing algorithms as competitors to human judgment, rather than complements, may exacerbate any potential distrust. For instance, controversy surrounds high frequency trading (HFT), financial trading based on algorithms that can substitute for human traders, in both news headlines and popular books (Koba, 2014; Lewis, 2014), and popular books present algorithms as a threat to human agency (Steiner, 2012). Future research could test whether framing algorithms as tools might help people overcome distrust where it exists.

Boundaries to Algorithmic Capabilities

While human judgment is far from perfect (for a review on human biases, see Kahneman, 2011), certainly no algorithm is perfect either. There are obviously limits to what algorithms can achieve. Although we know little about people's perceptions of algorithmic potential, work on expectations of computers and robots reflects that few people expect a computer to understand things like culture (Copeland, 2013), to sing, to dance, or to tell jokes.

Yet, the boundaries to what algorithms, and machines driven by algorithms, can do are rapidly expanding as organizations increase investments to improve them. What seemed impossible several years ago has quickly become reality. For instance, the U.S. Government invested in technology to address the biggest threat to troops in Afghanistan in 2009, roadside bombs or IEDs (improvised explosive devices; CNN, 2009). By 2012, it had utilized an estimated 3,500 robots (Hodge, 2012) to help detect IEDs and inspect vehicles at checkpoints (Axe, 2011). In medicine, algorithms in intensive care units (ICUs) monitor and then contact remote doctors who are on-call overnight when the algorithm detects that patients in hospital rooms need care (Mullen-Fortino et al., 2012). To be sure, these examples will soon sound outdated. But the basic problem of connecting the production and application of insights will likely remain until we know more about the psychology driving people's responses.

THE PSYCHOLOGY OF BIG DATA

Psychology is poised especially well to examine the last mile problem in data analytics, but only recently have researchers started publishing psychological research on people's responses to algorithms. The next sections describe a small but growing number of papers that measure how people respond to output from algorithms versus people and how people think about algorithms more generally. For instance, my work on algorithm appreciation examined people's willingness to rely on advice produced by algorithms by testing how people respond to advice when they think it comes from a person versus when they think it comes from an algorithm (Logg et al., 2019). We found this result across different subjects who made different types of predictions. For example, we asked some people to forecast the occurrence of business and geopolitical events (e.g., the probability of North America or the EU imposing sanctions on a country in response to cyberattacks); we asked others to predict the rank of songs on the Billboard Hot 100; and we had one group of participants play online matchmaker (they read a person's dating profile, saw a photograph of her potential date, and predicted how much she would enjoy a date with him).

Although nonexperts were happy to rely on algorithmic advice more than advice from a person, experts unequivocally disregarded both algorithmic and advice from other humans when making business and geopolitical predictions. Instead, the experts relied more on their own judgment. Ironically, when experts refused to listen to the algorithmic advice, this hurt the accuracy of their predictions relative to nonexperts. This result is concerning because our experts were national security professionals who made a living by making predictions about consequential events.

In this paper (Logg et al., 2019), we presented identical numeric advice to participants and merely changed the label of the source. Thus, we measured how much people changed their own numeric estimate after receiving the advice when presented with an algorithm lacking any description of its input variables or calculations required for processing that information (called a *black box* algorithm). Black box algorithms lack transparency and are very common in everyday life. For example, viewers of Netflix and listeners of Spotify are not privy to the details that drive the recommendations they receive.

In some ways, this is similar to how people are unable to access others' inner thoughts and instead need to make inferences about them. In both instances, what matters most are people's lay theories, or assumptions, about each source. Understanding these assumptions was the main goal of our research. Importantly, measuring how people responded to information from a black box allowed us to measure people's unadulterated expectations—the assumptions people brought to the table without the influence of additional information.

As people receive more information from algorithms, both in their workplaces and personal lives, we need to examine their perceptions of how algorithms operate. The work on algorithm appreciation tested how people respond to algorithmic advice. This work purposely focused on how people respond to algorithmic output. In organizations, responses to algorithmic output likely depend on what people know or assume about the information the algorithm uses (input or cues), and how it sorts through that information (its processing). Thus, I developed the theory of machine framework in order to think more broadly about how people view algorithms. This allowed me to consider how people may expect not only output but also input and processing to differ between algorithmic and human judgment.

THEORY OF MACHINE

Theory of machine describes people's lay theories about how algorithmic judgment works and, importantly, how it differs from human judgment. Inspired by work in philosophy and psychology on theory of mind, theory of machine examines people's expectations of the internal processes of another agent. Philosophical work on theory of mind considers how people infer the intentions and beliefs of others (Dennett, 1987). Theory of mind is necessary to understand your own and others' behavior and how their behavior affects others (Begeer et al., 2011). For example, when you see someone stare at a plate of mozzarella sticks, you may infer that they are hungry (or that they like mozzarella sticks). Similarly, if you see someone smile, you may infer that they are feeling happy (perhaps they ate those mozzarella sticks). Without theory of mind, people cannot pick up on social cues, for example, that they have hurt someone's feelings or that a friend needs comforting (even with theory of mind, people may find it difficult to accurately perceive these situations).

Theory of Machine Framework

Theory of machine is an organizing framework that can guide researchers as they examine lay perceptions of how algorithmic and human judgment (at their finest) differ in their *input* (the information used), *process* (how the same information is assessed), and *output* (predictions, advice, and feedback that are produced). People's expectations of these differences between algorithmic and human judgment will likely drive how they respond to output produced by algorithms, relative to advice from other people. Figure 10.1 organizes my predictions about the inferences people make regarding the quality and quantity of information that algorithmic and human judgment can handle in terms of input, process, and output.

Input

People likely expect that algorithmic and human capabilities require different types of information (or cues) as their input. It is possible that people expect their input to differ in terms of both quantity and quality

	Input	Process ???	Output
Quality	1. Algorithms utilize data that is less nuanced (more abstract/categorical)	Algorithms process less "holistically," without taking broader patterns into account	1. Algorithms cannot provide an explanation, which makes them less persuasive
	2. Algorithms cannot utilize data that is subjective/intangible		2. Algorithms produce less relevant data to an individual (vs. to the average person)
Quantity	Algorithms utilize larger amounts of data as input	Algorithms process fewer categories of cues	Algorithms produce less output at a time (people can provide information and explanation)

Figure 10.1

The "Theory of Machine" framework.

(e.g., "As a job applicant, I know I am special and do not expect an algorithm to account for nuanced aspects of my character that make me unique in the same way that a person can."). Researchers could test whether people expect algorithms to capture greater quantities of data, data that is less nuanced or more abstract, and data that is objective.

Process

People likely observe that algorithms are often built for specific tasks. This may lead them to expect an algorithm's processing capabilities to be more limited in scope compared with human judgment. Thus, people may expect that an algorithm is better suited to process fewer categories or types of cues. In turn, they may expect algorithms to process cues in a less holistic way than people (e.g., "As a patient, I expect that an algorithm cannot take the entirety of my medical history into account when making a diagnosis in the same way that a doctor can.").

Pattern detection is necessary to create a heuristic (rule of thumb) to apply to a decision. People may also expect algorithms to detect patterns in data more slowly than a person. If people had this perception, it would be ironic considering (a) the extensive literature in the cognitive sciences on the limited processing capacity of the human brain (e.g., Miller, 1956); and (b) the work on judgment and decision making that documents how people overapply heuristics, which leads us towards biased judgments and away from accurate ones (for a review, see Bazerman & Moore, 2012).

An especially insightful paper has examined features of an algorithm's processing (Poursabzi-Sangdeh et al., 2018). This work manipulated the transparency of the algorithm's mechanics (black box or transparent). Surprisingly, regardless of the algorithm's transparency, whether people saw the mathematics that drove it or not, people preferred algorithmic advice to advice from people.

Output

Although algorithms may generally produce more accurate output than human judgment, people may expect that algorithms cannot (a) produce as many answers at a time or (b) accompany output with an explanation (e.g., "As a student, I expect the algorithm can assess my writing assignment based on a limited number of objective criteria and cannot provide reasons for its scoring."). If people expect that algorithms cannot utilize more subjective data, they should expect that it will produce output that is less relevant to them. More specifically, they may expect that algorithms will produce less personalized output (e.g., "As a student, I may think that the algorithm can predict other students' performance pretty well, but because it cannot take into account my intangible characteristics, such as grit and determination, it cannot predict my semester GPA well."). Overall, theory of machine can organize a myriad of research predictions to encourage systematic research that can keep up with the rapid pace of technological advancement injecting algorithms into the many aspects of our lives. Future work in the area could benefit by considering how the decision context influences people's theory of machine.

Theory of Machine and Decision Context

People are never faced with algorithmic output in a vacuum and without context. For instance, algorithms provide people with more and more information in the form of predictions, assessments, and feedback every day (e.g., predictions of how well someone will perform, assessments of how well they previously performed, and feedback on how they performed with information on how to improve). The purpose of the algorithm likely influences inferences people make about them. Taking this context into account when measuring inferences about input, process, and output in theory of machine should help researchers further specify their research predictions. This section (a) details how taking context into account can help researchers understand relevant work more systematically; and (b) outlines new research insights from theory of machine when taking the three decision contexts of prediction, assessment, and feedback into account.

The following examples are published work and ongoing research that examine responses to algorithmic and human output and have potential to reconcile results by considering the context of theory of machine. As previously mentioned, the paper *Algorithm Appreciation* (Logg et al., 2019) shows that when making predictions about the world and other people, people listen to algorithmic advice more than advice from people. In contrast, in the project *Algorithmic Hiring* (Logg & Tinsley, 2021), people applying for a role on a team prefer to have their own or their potential teammates' performance assessed by a person rather than an algorithm. There is a key moderator here: When the hiring manager was seen as a member of the applicant's outgroup, the applicant preferred the algorithm. In a related but third common domain, people often have their performance assessed with the goal of receiving feedback (unlike hiring, assessment does not necessarily include any actionable feedback). In the project *Robo-Coaching* (Logg et al., 2021), although people say they prefer feedback on their writing from a person, when provided with the feedback, they listen just as much to feedback from an algorithm as from a person.

Related work by Raveendhran and Fast (2021) also found that people preferred to have their performance on an aptitude test tracked by a computer application than a person. Similarly, when MBAs consider a scenario with multiple job offers, they prefer the company that employs behavior tracking monitored by an analyst over an algorithm. Additionally, Schlund and Zitek (2021) showed that when their performance was monitored and evaluated on a creativity task, people perceived a greater invasion of privacy from an algorithm than person. Further, they found that the monitoring source affects performance itself; people monitored by algorithms generated fewer ideas than those monitored by a person. Future work might consider how people respond to algorithmic feedback based on when the feedback is delivered. People may respond differently to algorithmic versus human feedback when the feedback is provided on the performance process (during the task) compared to when it is provided on the final product (following the task).

In each context, the information provided has a different focus, either judgments about the world (i.e., predictions about geopolitical events) or others (i.e., predicting whether two people will get along romantically), the self (i.e., assessment of your job application; feedback on your writing), or others (i.e., predicting whether two people will get along romantically). Organizing the decision space based on the type and focus of the judgment can help understand potential moderators. For example, people rely more on algorithms when making predictions about the world. But when judgments shift to focus on their own performance or a potential teammates' performance, they prefer an assessment made by a person. In Table 10.1, I outline how the preference for algorithmic or human judgment depends on the focus of the judgment (self, other, the world) and type of judgment (prediction, assessment, or feedback). I utilized this framework further by building new predictions around the existing results regarding theory of machine.

Considering the context of judgments could additionally reconcile seemingly contradictory results and contribute to a more cohesive understanding of theory of machine. For instance, in the presence of algorithmic error, people are less likely to trust algorithmic judgment than their own, which hurts their accuracy (Dietvorst et al., 2015; Dzindolet et al., 2002). At first glance, these results might appear to contradict those from Algorithm Appreciation, but they are compatible. In Algorithm

Table 10.1
Outline of Results and Potential Predictions for People's Preference for Algorithmic Versus Human Judgment Based on the Judgment Type and Judgment Focus

World	Self	Other
Prediction		
(e.g., The date when your company will reach a benchmark of 300 clients total)	(e.g., How well you will present an idea to a client)	(e.g., How well a colleague will present an idea to a client)
Algorithm Appreciation shows people prefer an algorithm to predict geopolitical events (Logg et al., 2019).	I predict that: People may prefer a person to predict their performance if they see themselves as unique "special snowflakes" that an algorithm cannot fully understand.	Algorithm Appreciation shows people prefer an algorithm to predict whether two people will get along romantically (Logg et al., 2019).
Assessment		
	(e.g., How well you presented an idea to a client)	(e.g., How well a colleague presented an idea to a client)
	Algorithmic Hiring shows people prefer a person to assess their job application (Logg & Tinsley, 2021).	I predict that: People may prefer an algorithm because they have less stake in the outcome and do not see others as being as unique as they are.
Feedback		
	(e.g., How well you presented an idea to a client + ways to improve)	(e.g., How well a colleague presented an idea to a client + ways to improve)
	Robo-Coaching shows people say they prefer a person but listen equally to feedback from an algorithm and aor a person (Logg et al., 2021).	I predict that: People may prefer an algorithm because they have less stake in the outcome.

Appreciation, participants do not receive information on the algorithm's performance. Here, they trust the advice more from an algorithm than from a person. In Dietvorst's work, prior to seeing algorithms make any errors, those participants were also willing to listen to the algorithm (Figure 10.2). One potential venue for future work is to test how quickly and effectively people expect humans and algorithms to learn from mistakes and "self-correct." Another is how people respond to algorithmic imperfection depending on the presence of a benchmark to human accuracy. I discuss these ideas in the following section, Context of Error: Expectations of Learning and Alternatives.

Context can also help reconcile other work. For example, people were less likely to trust algorithms for moral decisions about self-driving cars and their own medical outcomes (Bigman & Gray, 2018; Longoni et al., 2019; Promberger & Baron, 2006). In some subjective domains governed by personal taste, people preferred to receive recommendations for jokes, books, and movies from a close friend over an algorithm, even though the algorithm did a better job at predicting their preferences (Sinha & Swearingen, 2001; Yeomans et al., 2019). However, it is useful to note here that the story is not so straightforward. Even in very subjective domains (predicting interpersonal attraction and the popularity of songs), we still

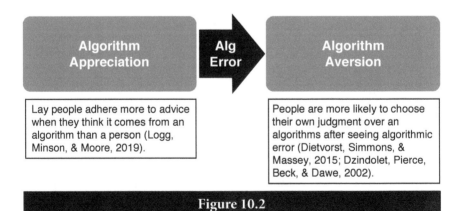

Figure 10.2

Algorithm appreciation versus algorithm aversion. Context matters: Prior to performance feedback, people show *Algorithm Appreciation* and after learning that an algorithm has made an error, people show *Algorithm Aversion*.

found algorithm appreciation (Logg et al., 2019). So although subjectivity of the domain may seem like an obvious moderator, perhaps what is more important is the focus of the judgment itself: the self (e.g., one's medical decisions, one's own taste for movies) versus the world (e.g., the popularity of a song in general, not my specific taste in music).

Some recent work examined algorithms in an organizational context where fairness is a strong concern. When people considered situations about promotions, layoffs, or hiring, they perceived the procedure as unfair when it included an algorithm (Newman et al., 2020). And when people imagined situations regarding job referral processes and hiring, they expected less discrimination from algorithms when considering the source of the judgment (Jago & Laurin, 2021). One key contextual moderator here might include whether people are assessing the procedures as a whole or the specific source making the decision.

CONTEXT OF ERROR: EXPECTATIONS OF LEARNING AND ALTERNATIVES

The context of error in judgment could inform another set of future research questions. One potential area to explore is how quickly people think algorithms can learn from mistakes relative to people. Once exposed to an error in judgment, do people expect that humans are better equipped to "course correct" and quickly update to more accurate answers than algorithms? Do people perceive algorithms as capable of updating at all or as purely static?

During discussions in my class The Psychology of Big Data, students read that 85% of the time, an algorithm accurately predicted which students are most likely to drop out of college (Treaster, 2017). They shared that they expected the predictive accuracy to be closer to 100%, and thus interpreted 85% as fairly low accuracy, especially for such an important domain. Focusing on 100% as the benchmark turned their focus away from another reasonable benchmark; the accuracy of a person making the same prediction. In fact, when algorithmic accuracy is reported in the media, the comparable accuracy benchmark from human judgment is often unreported. Would students expect an admissions officer to reach

60% accuracy? Once they make this comparison, 85% looks better. Here, research could examine how people view algorithmic advice depending on whether algorithmic accuracy is presented with or without a benchmark of human accuracy.

This section has focused on the psychological processes that shape people's willingness to utilize analytical insights. It leveraged the theory of machine framework both to organize new predictions and to start reconciling current work. The following section considers how psychology can help the process of creating algorithms.

Using Psychology to Create Better Algorithms

Although the accuracy of basic algorithms is well-documented, an algorithm is only as good as the input provided to it, as echoed in the common phrase found in computer science, "garbage in, garbage out." Injecting psychology into the process of building algorithms has potential to further improve data analytics, as improving the quality of algorithmic output can potentially further increase people's willingness to accept it. Luca et al. (2016) discussed how organizations can improve algorithms by improving the quality of their input. This section extends that work by developing strategies to improve algorithms through three stages of its construction: (a) preparing, (b) building, and (c) interpreting output from the algorithm. Thoughtful decisions at each of these stages should lead to higher quality information produced by an algorithm. The way an algorithm is constructed may influence how people respond to it, which can also address the last mile problem, something I discuss in the conclusion.

In my article "Using Algorithms to Understand the Biases in Your Organization" (Logg, 2019), I argued that organizations should leverage algorithms as magnifying glasses to search for and detect biases in past decisions. Often, past decisions (made by people) are compiled into a data set that is then fed into an algorithm as input. Once biases are detected in this historical data, organizations can learn where the biases originated and then work toward correcting them. The entire process of building the algorithm and auditing the input data opens the floodgates to many testable questions. Thus, considering the decisions at each stage should allow data analytics teams to (a) improve the quality of the algorithm's output and (b) better detect bias in past decisions made by people as reflected in the historical data that is used as input to algorithms. Taken from the article, the stages outlined in Table 10.2 include a variety of questions to ask and decisions to make that likely depend on the judgment context and goals of the organization.

This outline makes salient an important difference between human and algorithmic judgment: the necessity of numeric inputs for algorithms. Numeric inputs require specificity and can allow for greater transparency within a decision process. Thus, the act of constructing an algorithm has the potential to (a) clarify and better understand the goal of the prediction for which the algorithm is being built, (b) clarify the relative importance of the cues or factors being used to inform the decision, and (c) encourage clarification about what people value in their decisions more broadly. Whether building algorithms (regardless of implementing them) can improve the clarity and thus quality of decision-making processes is an empirical question I plan to test with a colleague shortly (D. Cain, personal communication, September 9, 2019).

Differentiation From Innovation, Anthropomorphism, and Robots

It is useful to differentiate the psychology of Big Data from related but separate work in the area of technology. First, work on innovation tends to focus on how people generate the ideas for new tools and devices, but rarely how people respond to new devices that supplant current tools (e.g., how do people adopt new technologies such as self-driving cars, which can replace the cars people currently drive themselves; for a review, see Gopalakrishnan & Damanpour, 1997). Work on the psychology of Big Data is especially exciting due to the fact that for the first time in human history, technology is replacing human judgment, rather than replacing current devices.

Likewise, at first blush the psychology of Big Data appears similar to work on anthropomorphism. Psychologists have examined the contexts in which people anthropomorphize machines, the consequences of doing

Table 10.2
Questions Analytics Teams Can Ask at Each Stage of Building an Algorithm

Team	Goal/Purpose	Input
Preparing to build the algorithm		
How diverse is the team building the algorithm? This is important because diversity helps unveil blind spots in our strategies (Nemeth & Kwan, 1987).	What is the goal of building this algorithm? Can it be quantified?	How objective are the input variables we are considering? If they are subjective, can we quantify them better?

Input		Testing
Building the algorithm		
How relevant is the input data to the prediction? Does each instance in the data have enough information? Is our data set rich (wide) enough? What types of decision makers produced this historical data? Is diversity represented in the input?		Was the algorithm tested by auditing the output data for bias?

Quality evaluation	Audit	Data-driven decision making
Interpreting output from the algorithm		
Does the algorithm predict "out-of-sample" with new input data?	Was the final output audited for bias and were proxy variables detected? Can unaccounted-for variables explain the output? What assumptions were made in the building of the algorithm?	What actions does the output suggest an organization should take? Does the output suggest that the team should try to revise the algorithm and start the process over?

Note. Data from Logg (2019).

so, and how they respond to new technologies (Epley et al., 2007; Gray et al., 2007; Waytz et al., 2010, 2014). That work examines how people impart human judgment to algorithms. In contrast, theory of machine examines how people expect the two to differ. Likewise, understanding how people interact with robots (Broadbent, 2017) is related to but qualitatively different from theory of machine, which examines how people respond to a specific, numeric output.

Broadly, the psychology of Big Data and theory of machine lies within the knowledge domain, often in situations where people make numeric judgments. In this domain, accuracy is easily measured. Thus, researchers can identify a normative standard for how much people should update to algorithmic advice. Such normative benchmarks can show that listening to advice allows people can improve the quality of their judgments in those specific instances. In contrast, it is difficult to determine how people should normatively respond to machines and robots during complex interactions, including contexts where there are virtually infinite ways of responding.

CONCLUSION: ALGORITHMS ARE TOOLS

Algorithms do not exist in a vacuum. Algorithms require an element of human judgment, whether to determine the input data or to interpret the output (in the case of advanced machine learning and beyond). These stages in the life of an algorithm allow multiple opportunities for a data analytics team to reflect on the goal of building the algorithm in the first place and audit the quality of the output as they iterate on the product. Data analytics needs psychology simply because a person must decide if the algorithmic insights produced are worth acting on. While the field focuses on refining more and more complex algorithms, this effort is in vain if people refuse to listen and act upon the recommendations made by these algorithms.

Organizations need to understand when people are willing to listen to algorithmic output to realize the full potential of algorithmic advice. Thus, this chapter calls for research on the psychology of Big Data. Taking a psychological perspective to the field of data analytics opens up a multitude of research avenues. I developed the theory of machine framework, which can organize this research space for a more systematic examination of how people respond to algorithms. Specifically, theory of machine examines how people expect algorithmic and human judgment to differ from each other in their input, process, and output. This chapter further develops this framework and offers specific predictions based on it in order to encourage a systematic examination of the topic.

Research on theory of machine is necessary to better understand and solve the last mile problem plaguing data analytics. We need to understand people's perceptions of how algorithms and human judgment differ in order to close the gap between producing and utilizing analytics insights. Examining lay theories of algorithmic output is especially important considering the accuracy of algorithms relative to human judgment. A long history of research suggests that if people are willing to listen to algorithms, they would, in most cases, make more accurate predictions and judgments.

Data analytics can doubly benefit from psychology by incorporating a psychological perspective in the process of constructing the algorithm, especially during three key stages: preparing, building, and interpreting output from the algorithm. Future research could consider how perceptions of algorithmic output is influenced by what people learn about the construction of an algorithm. It could measure how people respond to algorithms depending on who created it and how it was created (as outlined in the Using Psychology to Create Better Algorithms section). If data analytics teams improve the quality of their algorithms through more careful consideration in each stage, will their audience respond in turn by listening more to the algorithm? Do people listen to algorithmic output more depending on which input variables were weighted most heavily? For example, if an employee knows certain information about how an algorithm was created to decide promotions, are they more likely to view it as fair?

Work on theory of machine would do well to consider how the established research on theory of mind has evolved and advanced. For example, as people develop their theory of mind throughout childhood (Harris et al., 1989; Perner, 1991; Repacholi & Slaughter, 2003), so too are people likely currently developing a theory of machine as algorithmic information becomes more prevalent and embedded in daily life. Understanding the history of research related to theory of mind can help new research on theory of machine develop more efficiently.

Under the appropriate conditions, people are willing to listen to algorithmic predictions, assessments, and advice. Research should examine how organizations can shape the decision environment to help their employees and clients attain greater accuracy by listening to algorithmic advice. Without psychology, data analytics may continue investing time and resources into producing worthwhile insights that never find an audience.

REFERENCES

Axe, D. (2011, February 7). One in 50 troops in Afghanistan is a robot. *Wired*. https://www.wired.com/2011/02/1-in-50-troops-robots/

Baron, J., Mellers, B. A., Tetlock, P. E., Stone, E., & Ungar, L. H. (2014). Two reasons to make aggregated probability forecasts more extreme. *Decision Analysis*, *11*(2), 133–145. https://doi.org/10.1287/deca.2014.0293

Bazerman, M. H., & Moore, D. A. (2012). *Judgment in managerial decision making* (8th ed.). John Wiley & Sons.

Beck, A. H., Sangoi, A. R., Leung, S., Marinelli, R. J., Nielsen, T. O., Van De Vijver, M. J., West, R. B., van de Rijn, M., & Koller, D. (2011). Systematic analysis of breast cancer morphology uncovers stromal features associated with survival. *Science Translational Medicine*, *3*(108), 108ra113. https://doi.org/10.1126/scitranslmed.3002564

Begeer, S., Gevers, C., Clifford, P., Verhoeve, M., Kat, K., Hoddenbach, E., & Boer, F. (2011). Theory of Mind training in children with autism: A randomized controlled trial. *Journal of Autism and Developmental Disorders*, *41*(8), 997–1006. https://doi.org/10.1007/s10803-010-1121-9

Berinato, S. (2019, January–February). Data science and the art of persuasion. *Harvard Business Review*, *97*(1), 126–137. https://hbr.org/2019/01/data-science-and-the-art-of-persuasion

Bigman, Y. E., & Gray, K. (2018). People are averse to machines making moral decisions. *Cognition*, *181*, 21–34. https://doi.org/10.1016/j.cognition.2018.08.003

Broadbent, E. (2017). Interactions with robots: The truths we reveal about ourselves. *Annual Review of Psychology*, *68*(1), 627–652. https://doi.org/10.1146/annurev-psych-010416-043958

Card, S. K., Moran, T. P., & Newell, A. (1983). *The psychology of human–computer interaction*. Laurence Erlbaum Associates.

Carroll, J. S., Wiener, R. L., Coates, D., Galegher, J., & Alibrio, J. J. (1982). Evaluation, diagnosis, and prediction in parole decision making. *Law & Society Review, 17*(1), 199–228. https://doi.org/10.2307/3053536

Copeland, M. (2013, October 22). Where humans will always beat the robots. *The Atlantic,* https://www.theatlantic.com/business/archive/2013/10/where-humans-will-always-beat-the-robots/280762/

Dawes, R. M. (1979). The robust beauty of improper linear models in decision making. *American Psychologist, 34*(7), 571–582. https://doi.org/10.1037/0003-066X.34.7.571

Dawes, R. M., Faust, D., & Meehl, P. E. (1989). Clinical versus actuarial judgment. *Science, 243*(4899), 1668–1674. https://doi.org/10.1126/science.2648573

Dennett, D. (1987). *The intentional stance.* The MIT Press.

Dietvorst, B. J., Simmons, J. P., & Massey, C. (2015). Algorithm aversion: People erroneously avoid algorithms after seeing them err. *Journal of Experimental Psychology: General, 144*(1), 114–126. https://doi.org/10.1037/xge0000033

Dzindolet, M. T., Pierce, L. G., Beck, H. P., & Dawe, L. A. (2002). The perceived utility of human and automated aids in a visual detection task. *Human Factors, 44*(1), 79–94. https://doi.org/10.1518/0018720024494856

Egyptian Ministry of Civil Aviation. (2004). *Final report of the accident investigation: Flash Airlines flight 604.* https://www.bea.aero/docspa/2004/su-f040103a/pdf/su-f040103a.pdf

Einhorn, H. J. (1972). Expert measurement and mechanical combination. *Organizational Behavior and Human Performance, 7*(1), 86–106. https://doi.org/10.1016/0030-5073(72)90009-8

Epley, N., Waytz, A., & Cacioppo, J. T. (2007). On seeing human: A three-factor theory of anthropomorphism. *Psychological Review, 114*(4), 864–886. https://doi.org/10.1037/0033-295X.114.4.864

Frey, C. B., & Osborne, M. A. (2013). *The future of employment: How susceptible are jobs to computerization* [Working paper]. Oxford Martin Programme on Technology and Employment. https://www.oxfordmartin.ox.ac.uk/downloads/academic/future-of-employment.pdf

Galton, F. (1907). Vox populi. *Nature, 75*(1949), 450–451. https://doi.org/10.1038/075450a0

Goldman, L., Caldera, D. L., Nussbaum, S. R., Southwick, F. S., Krogstad, D., Murray, B., Burke, D. S., O'Malley, T. A., Goroll, A. H., Caplan, C. H., Nolan, J., Carabello, B., & Slater, E. E. (1977). Multifactorial index of cardiac risk in noncardiac surgical procedures. *The New England Journal of Medicine, 297*(16), 845–850. https://doi.org/10.1056/NEJM197710202971601

Gopalakrishnan, S., & Damanpour, F. (1997). A review of innovation research in economics, sociology and technology management. *Omega, 25*(1), 15–28. https://doi.org/10.1016/S0305-0483(96)00043-6

Gray, H. M., Gray, K., & Wegner, D. M. (2007). Dimensions of mind perception. *Science, 315*(5812), 619. https://doi.org/10.1126/science.1134475

Guszcza, J., Rahwan, I., Bible, W., Cebrian, M., & Katyal, V. (2018, November 28). Why we need to audit algorithms. *Harvard Business Review.* https://hbr.org/2018/11/why-we-need-to-audit-algorithms

Guszcza, J., & Schwartz, J. (2020). Superminds, not substitutes: Designing human-machine collaboration for a better future of work. *Deloitte Review, 27,* 25–46. https://www2.deloitte.com/content/dam/insights/us/articles/6672_superminds-not-substitutes/DI_DR27-Superminds.pdf

Harrell, E. (2016, September 7). Managers shouldn't fear algorithm-based decision making. *Harvard Business Review.* https://hbr.org/2016/09/managers-shouldnt-fear-algorithm-based-decision-making

Harris, P. L., Johnson, C. N., Hutton, D., Andrews, G., & Cooke, T. (1989). Young children's theory of mind and emotion. *Cognition and Emotion, 3*(4), 379–400. https://doi.org/10.1080/02699938908412713

Hastie, R., & Kameda, T. (2005). The robust beauty of majority rules in group decisions. *Psychological Review, 112*(2), 494–508. https://doi.org/10.1037/0033-295X.112.2.494

Hedén, B., Ohlin, H., Rittner, R., & Edenbrandt, L. (1997). Acute myocardial infarction detected in the 12-lead ECG by artificial neural networks. *Circulation, 96*(6), 1798–1802. https://doi.org/10.1161/01.CIR.96.6.1798

Hodge, N. (2012, June 13). In the Afghan War, a little robot can be a soldier's best friend. *The Wall Street Journal.* https://www.wsj.com/articles/SB10001424052702303901504577460372836320982

Jago, A. S., & Laurin, K. (2021). Assumptions about algorithms' capacity for discrimination. *Personality and Social Psychology Bulletin.* Advance online publication. https://doi.org/10.1177/01461672211016187

Kahneman, D. (2011). *Thinking, fast and slow.* Macmillan.

Kleinmuntz, B. (1990). Why we still use our heads instead of formulas: Toward an integrative approach. *Psychological Bulletin, 107*(3), 296–310. https://doi.org/10.1037/0033-2909.107.3.296

Kleinmuntz, D. N., & Schkade, D. A. (1993). Information displays and decision processes. *Psychological Science, 4*(4), 221–227. https://doi.org/10.1111/j.1467-9280.1993.tb00265.x

Koba, M. (2014, April 22). *High frequency trading: CNBC explains.* CNBC. https://www.cnbc.com/id/100405633

Larrick, R. P., & Soll, J. B. (2006). Intuitions about combining opinions: Misappreciation of the averaging principle. *Management Science, 52*(1), 111–127. https://doi.org/10.1287/mnsc.1050.0459

Lewis, M. (2014). *Flash boys: A Wall Street revolt*. W. W. Norton.

Libby, R. (1976). Man versus model of man: Some conflicting evidence. *Organizational Behavior and Human Performance, 16*(1), 1–12. https://doi.org/10.1016/0030-5073(76)90002-7

Logg, J. M. (2019, August 9). Using algorithms to understand the biases in your organization. *Harvard Business Review*. https://hbr.org/2019/08/using-algorithms-to-understand-the-biases-in-your-organization

Logg, J. M., Gino, F., & Minson, J. A. (2021). *Robo-coaching: When do people prefer performance assessments from algorithms versus people?* [Manuscript in preparation]. McDonough School of Business, Georgetown University.

Logg, J. M., Minson, J. A., & Moore, D. A. (2019). Algorithm appreciation: People prefer algorithmic to human judgment. *Organizational Behavior and Human Decision Processes, 151*, 90–103. https://doi.org/10.1016/j.obhdp.2018.12.005

Logg, J. M., & Tinsley, C. (2021). *Algorithmic hiring* [Manuscript in preparation]. McDonough School of Business, Georgetown University.

Longoni, C., Bonezzi, A., & Morewedge, C. K. (2019). Resistance to medical artificial intelligence. *The Journal of Consumer Research, 46*(4), 629–650. https://doi.org/10.1093/jcr/ucz013

Luca, M., Kleinberg, J., & Mullainathan, S. (2016, January–February). Algorithms need managers, too. *Harvard Business Review, 94*(1), 20. https://hbr.org/2016/01/algorithms-need-managers-too

Meehl, P. E. (1954). *Clinical versus statistical prediction: A theoretical analysis and a review of the evidence*. University of Minnesota Press. https://doi.org/10.1037/11281-000

Meehl, P. E. (1957). When shall we use our heads instead of the formula? *Journal of Counseling Psychology, 4*(4), 268–273. https://doi.org/10.1037/h0047554

Miller, G. A. (1956). The magical number seven, plus or minus two: Some limits on our capacity for processing information. *Psychological Review, 63*, 81–97. https://doi.org/10.1037/h0043158

Mullen-Fortino, M., DiMartino, J., Entrikin, L., Mulliner, S., Hanson, C. W., & Kahn, J. M. (2012). Bedside nurses' perceptions of intensive care unit telemedicine. *American Journal of Critical Care, 21*(1), 24–32. https://doi.org/10.4037/ajcc2012801

Nemeth, C. J., & Kwan, J. L. (1987). Minority influence, divergent thinking and detection of correct solutions. *Journal of Applied Social Psychology, 17*(9), 788–799. https://doi.org/10.1111/j.1559-1816.1987.tb00339.x

Newman, D. T., Fast, N. J., & Harmon, D. J. (2020). When eliminating bias isn't fair: Algorithmic reductionism and procedural justice in human resource decisions. *Organizational Behavior and Human Decision Processes, 160,* 149–167. https://doi.org/10.1016/j.obhdp.2020.03.008

Perner, J. (1991). *Understanding the representational mind.* The MIT Press.

Poursabzi-Sangdeh, F., Goldstein, D. G., Hofman, J. M., Vaughan, J. W., & Wallach, H. (2018). *Manipulating and measuring model interpretability.* arXiv. https://arxiv.org/abs/1802.07810

Promberger, M., & Baron, J. (2006). Do patients trust computers? *Journal of Behavioral Decision Making, 19*(5), 455–468. https://doi.org/10.1002/bdm.542

Rahwan, I., Cebrian, M., Obradovich, N., Bongard, J., Bonnefon, J. F., Breazeal, C., Crandall, J. W., Christakis, N. A., Couzin, I. D., Jackson, M. O., Jennings, N. R., Kamar, E., Kloumann, I. M., Larochelle, H., Lazer, D., McElreath, R., Mislove, A., Parkes, D. C., Pentland, A., . . . Wellman, M. (2019). Machine behaviour. *Nature, 568*(7753), 477–486. https://doi.org/10.1038/s41586-019-1138-y

Raveendhran, R., & Fast, N. J. (2021). Humans judge, algorithms nudge: The psychology of behavior tracking acceptance. *Organizational Behavior and Human Decision Processes, 164,* 11–26. https://doi.org/10.1016/j.obhdp.2021.01.001

Repacholi, B., & Slaughter, V. (Eds.). (2003). *Individual differences in theory of mind: Implications for typical and atypical development.* Psychology Press. https://doi.org/10.4324/9780203488508

Roadside bombs 'No. 1 threat' to troops in Afghanistan. (2009, July 9). CNN. https://www.cnn.com/2009/WORLD/asiapcf/07/09/afghanistan.ieds/index.html

Rodrigue, J.-P., Comtois, C., & Slack, B. (2009). *The geography of transport systems* (2nd ed.). Routledge.

Saxe, R. (2013). Theory of mind: How brains think about thoughts. In K. N. Ochsner & S. Kosslyn (Eds.), *The Oxford handbook of cognitive neuroscience: Vol. 2. The cutting edge* (pp. 204–213). Oxford University Press.

Schlund, R., & Zitek, E. M. (2021). Who's my manager? Surveillance by AI leads to perceived privacy invasion and resistance practices. *Academy of Management Proceedings, 2021*(1). https://doi.org/10.5465/AMBPP.2021.11451abstract

Schrage, M. (2016, May 26). What a minor league moneyball reveals about predictive analytics. *Harvard Business Review.* https://hbr.org/2016/05/what-a-minor-league-moneyball-reveals-about-predictive-analytics

Sinha, R. R., & Swearingen, K. (2001). Comparing recommendations made by online systems and friends. In *Proceedings of the DELOS-NSF Workshop on Personalization and Recommender Systems in Digital Libraries.* DELOS. https://www.ercim.eu/publication/ws-proceedings/DelNoe02/RashmiSinha.pdf

Soll, J. B., & Larrick, R. P. (2009). Strategies for revising judgment: How (and how well) people use others' opinions. *Journal of Experimental Psychology: Learning, Memory, and Cognition, 35*(3), 780–805. https://doi.org/10.1037/a0015145

Steiner, C. (2012). *Automate this: How algorithms came to rule our world*. Penguin Group.

Surowiecki, J. (2004). *The wisdom of crowds: Why the many are smarter than the few and how collective wisdom shapes business, economies, societies, and nations*. Doubleday.

Tazelaar, F., & Snijders, C. (2013). Operational risk assessments by supply chain professionals: Process and performance. *Journal of Operations Management, 31*(1–2), 37–51. https://doi.org/10.1016/j.jom.2012.11.004

Tesauro, G., Gondek, D., Lenchner, J., Fan, J., & Prager, J. M. (2013). Analysis of Watson's strategies for playing Jeopardy! *Journal of Artificial Intelligence Research, 47*, 205–251. https://doi.org/10.1613/jair.3834

Treaster, J. B. (2017, February 2). Will you graduate? Ask big data. *The New York Times*. https://www.nytimes.com/2017/02/02/education/edlife/will-you-graduate-ask-big-data.html

Ungar, L., Mellers, B., Satopää, V., Baron, J., Tetlock, P., Ramos, J., & Swift, S. (2012). *The good judgment project: A large scale test of different methods of combining expert predictions*. AAAI Technical Report. (FS-12-06). Association for the Advancement of Artificial Intelligence.

Waytz, A., Gray, K., Epley, N., & Wegner, D. M. (2010). Causes and consequences of mind perception. *Trends in Cognitive Sciences, 14*(8), 383–388. https://doi.org/10.1016/j.tics.2010.05.006

Waytz, A., Heafner, J., & Epley, N. (2014). The mind in the machine: Anthropomorphism increases trust in an autonomous vehicle. *Journal of Experimental Social Psychology, 52*, 113–117. https://doi.org/10.1016/j.jesp.2014.01.005

Yeomans, M., Shah, A., Mullainathan, S., & Kleinberg, J. (2019). Making sense of recommendations. *Journal of Behavioral Decision Making, 32*(4), 403–414. https://doi.org/10.1002/bdm.2118

11

Privacy and Ethics in the Age of Big Data

Sandra C. Matz, Ruth E. Appel, and Brian Croll

Our personal, professional, and social lives have become inseparably interwoven with technology. Social media connects us with our loved ones and allows us to publicly express our views and opinions to millions of people around the world. Search engines provide us with instant access to information from all over the world. And recommendation algorithms help us sift through the enormous amount of online content and find what we are most interested in. Many of our activities that used to be private or shared with only a select circle of friends and close others now create a permanent record of data that can be accessed from anywhere around the world. Instead of sharing an ephemeral conversation with friends, for example, we now leave a durable trace of our social interactions—a digital footprint (Kosinski et al., 2013).

We shape technology by creating increasingly intelligent algorithms and machines. But we are also shaped by technology in ways that we often do not fully understand. As our relationship with technology and data becomes more and more intimate, a major question arises: How can we design ethical technology that uses data generated by our interactions with the digital world responsibly and in a way that benefits—rather than exploits—individuals, and strengthens—rather than undermines—the social fabric of the societies we live in? What is clear from the ongoing debate around the data revolution is that individuals and societies stand to both gain and lose. On the one hand, individuals can benefit from personalized services that make many aspects of everyday life easier than ever before (Matz & Netzer, 2017; Sagiroglu & Sinanc, 2013), and the ability to pool data such as medical records to better understand diseases and ways to combat them holds great promise for the flourishing and health of society as a whole (Tatonetti et al., 2012). On the other hand, Big Data threatens to erode our privacy in unprecedented ways, opening the door for manipulation, polarization, and discrimination (Acquisti et al., 2015; Cohen, 2000; Crawford et al., 2014).

Depending on your own view of technology, you might have had very different reactions to the previous examples. You might have felt that technology is a wonderful enabler of modern life, or you might have felt that technology is increasingly infringing on our autonomy. Research shows that there is a lot of heterogeneity in the ways that people perceive and feel about technologies, both across different individuals and within individuals. Not only are some people generally more concerned about the extent to which technology infringes on their privacy (e.g., Bellman et al., 2004; Fogel & Nehmad, 2009; Korzaan & Boswell, 2008; Sheehan, 1999; Stieger et al., 2013), but there is also considerable variation in the way the same people feel about technology across different situations and use cases (Bansal et al., 2016). For example, individuals are generally more likely to embrace predictive technologies when it comes to health care than they are when it comes to personalized advertising (Popov & Matz, 2016).

Whether or not Big Data and technology are a curse or blessing for modern life largely depends on how they are being used (Matz et al., 2020; O'Neil, 2016)—specifically, the extent to which they are used in an ethical and privacy-preserving way. Consequently, social scientists across many disciplines—including psychologists, sociologists, political scientists, economists, ethicists, and marketing and management scholars—have started to devote their attention to the challenges related to privacy and ethics that are introduced in the age of Big Data. The specific questions they address are manifold, and include the following: How is the Big Data revolution different from previous revolutions in technology (e.g., the advent of the printing press or instant photography) when it comes to privacy and ethics? How much does the average person know and care about potential privacy risks associated with their data? How can we design technology in a way that does not make us addicted to it? How can we leverage data and improve services without infringing on people's privacy?

In this chapter, we discuss the new ethical challenges introduced by the age of Big Data. Although there are many other ethical challenges related to technology and data (e.g., addiction, inequality), we focus our review on the topic of privacy as one of the major challenges associated with Big Data. We first introduce the concept of privacy, briefly discussing its history, universality, and core assumptions that lie at the heart of privacy protections. We then move on to the questions of how Big Data threatens our privacy in unprecedented ways and challenges current approaches to privacy protection. Finally, we shift our focus to potential solutions. We discuss how placing the burden of privacy protection on users alone is misguided and provide a number of potential systemic solutions related to regulation, collaboration, design principles, and technological tools. We finish the chapter with concrete practical guidelines for researchers and practitioners of how to design studies, products, and services that protect individuals' privacy and contribute to a mutually beneficial exchange of data for value.

WHAT IS PRIVACY?

The notion of privacy dates all the way back to the writings of the Greek philosopher Aristotle, who made a sharp distinction between the public sphere (*polis*) and the private sphere (*oikos*; DeCew, 1997). In its simplest form, *privacy* can be defined as the ability of individuals to express themselves selectively, to keep their own matters secret, and to be free from being observed or disturbed by other people. In other words, if an individual considers something private, they usually consider it inherently sensitive and worthy of protection from others.

Historical and anthropological accounts of privacy suggest that the need for privacy is a universal human trait. Behaviors indicative of the desire to protect one's personal life from the interference of others have been observed across time and space, ranging from ancient Rome and Greece (Duby, 1992), over preindustrialized Southeast Asian cultures (Westin, 1984), all the way to our current legal system in the United States and other Western cultures (Warren & Brandeis, 1890). Various religious texts, such as the Koran, the Talmud, or the Bible, hold cues to "the common human quest for privacy" (Acquisti et al., 2015, p. 512). They caution against spying on other community members (Koran), provide instructions on how to build homes that preserve one's privacy in front of the public eye (Talmud), and recount the transition of Adam and Eve to covering their naked bodies in shame before the "prying eyes of God" (Bible).

Although privacy hence appears to be a universal human need, its specific nature, its definition and components, its manifestation in regulations and laws, as well as the threats privacy is facing, are in constant flux. How a particular society thinks about and deals with privacy is constantly evolving as a function of social, political, and economic developments. One of the hallmarks of modern Western conceptualizations of privacy is the highly influential review article "The Right to Privacy" that was published in 1890 in the *Harvard Law Review*. It is considered by many to be the first publication to argue for the right to privacy in the United States. In the article, Samuel D. Warren and Louis Brandeis argue for the "right to be let alone" and argue that the common law needs to adapt to "novel technologies" such as instantaneous photography and the mass distribution of newspapers (Warren & Brandeis, 1890). The principles outlined in the article became a cornerstone in the discussion of modern privacy laws, and in many ways have remained relevant to this date.

However, the introduction of the Internet, and with it the ability to connect to and share information with more than 4.7 billion people around the world (Statista, 2021), has fundamentally changed the privacy landscape. While access to personal information in the early days of the internet was limited, recent years have seen an explosion in human-generated data, with 2.5 quintillion bytes of data being created every day in 2018 (Marr, 2018). That is 2,500,000,000,000,000,000 bytes. In this era of Big Data, traditional conceptualizations of privacy, for example as control of information or secrecy as a human right, exhibit important shortcomings when it comes to explaining why and when individuals care about privacy, and how privacy-preserving regulations could be designed (Nissenbaum, 2010). We explore these shortcomings in more detail later in the chapter and draw on recent theories on privacy to discuss promising new ways to conceptualize privacy.

Notably, the current chapter focuses on *informational privacy*, a term that has been coined to capture the unique challenges posed by the collection, storage, and processing of digital records with regard to data privacy and data protection (Cohen, 2000). In other words, informational privacy is concerned with controlling and restricting the information about a person that can be known and accessed by another person. Although data-driven technologies may also touch on other forms of privacy—for example physical and decisional privacy (Allen, 1987)—the concept of informational privacy is most directly linked to the advancements in technology and data collection methods that we focus on in this chapter. Yet, as we move through different types of data as well as different types of inferences that can be drawn from these data, we will occasionally touch on the other types of privacy.

Privacy Is Not About Secrecy; It Is About Control

To develop a better understanding of how Big Data is changing the privacy landscape and why it requires both researchers and practitioners to adapt their current approaches to privacy, it is important to acknowledge a simple but important distinction: Privacy is not about secrecy. It is not about individuals keeping information from everybody else around them. Privacy at its core is about control. It is about control over who individuals share their information with and what happens with that information once it is shared (Crawford & Schultz, 2014; Tene & Polonetsky, 2012).

Think about the pictures you might take during a holiday. You might be happy to share those on social media so that your friends and loved ones can participate in your life. It is not necessarily a secret that you went to Italy with your family, and making those pictures publicly available and sharing them with others does not violate your privacy. In fact, research shows that disclosing personal information is intrinsically rewarding (Tamir & Mitchell, 2012) and sharing one's thoughts and feelings with others is an important factor in establishing close relationships and creating a joint sense of shared reality (Rossignac-Milon et al., 2021). However, you might feel that your privacy is being violated if you were to learn that the social media platform sells your pictures to third parties or uses them to create a detailed profile about you for the purpose of targeted advertising. Hence, control over one's information can be considered two sides of the same coin. On the one side, freedom of expression is our ability to disclose information about ourselves, ranging all the way from our physical location or appearance to our innermost thoughts and feelings. On the flip side, privacy is our ability to withhold information about ourselves from others. Privacy is hence a constant negotiation about who obtains and maintains control over what information and what they are allowed to do with it.

How Does Big Data Threaten Our Privacy?

Novels such as George Orwell's *1984* and Hollywood movies such as Steven Spielberg's *Minority Report* paint dystopian visions of the future in which every step we take is recorded: Governments know where their citizens are at any given moment and use this information to predict their future actions (e.g., to prevent crime). In *1984*, telescreens and microphones installed in people's homes, at work, and in the streets provide 24/7 surveillance, recording every move the citizens of Oceania make. Similarly, in *Minority Report*, cameras in public places quickly identify the people passing by via retina scans, and billboards update their ads seconds later to display highly personalized messaging.

The era of Big Data has brought us one step closer to these dystopian visions. Today, companies and governments know much more about us than ever before, often extracting highly personal information about us from the behavioral records we leave with every step we take in the digital environment. There is no shortage of scandals showcasing how this power can be abused by those who have it. In 2012, the retail store Target sent a teenager pregnancy-related coupons before her parents knew she was pregnant (Duhigg, 2012). In 2013, the whistleblower Edward Snowden laid bare the surveillance conducted by the U.S. government on their own citizens (Snowden, 2019). And finally, in 2016, the U.K.-based consulting company Cambridge Analytica predicted the psychological profiles of millions of U.S. citizens from their Facebook accounts and targeted them with psychologically customized advertising aimed at discouraging them from voting in the 2016 presidential election (Doward & Gibbs, 2017).

What becomes clear from all these examples is that the data captured by the digital devices and services we use almost 24/7 are highly personal and create extensive records of our personal, professional, and public lives (Kosinski et al., 2013; Matz & Netzer, 2017). Over the past decade, growing numbers of computational social scientists have demonstrated how intimate and invasive these digital footprints can be. They can tell us what you do: what you purchase (e.g., via credit card data), watch (e.g., via video-streaming data), listen to (e.g., via music-streaming data), talk about (e.g., via messenger data), or search for (e.g., via search engine data). They reveal information about where you are and who you are around (e.g., using GPS tracking or facial recognition; Gates, 2011; Michael et al., 2006). In fact, most of us carry our smartphones with us wherever we go, making it possible for service providers and applications to track our location continuously via GPS signals or cell tower connections (Harari, 2020; Langheinrich & Schaub, 2018). On the basis of this information, one can infer which places we have visited, where we live (i.e., the place that our phone is stationary for most nights), where we work, and even who we meet (i.e., the users of the devices colocated with ours). In combination with the millions of cameras installed on many street corners as well as increasingly powerful facial recognition technology, data analysts can create a detailed map of a person's physical whereabouts (for excellent overviews of privacy in the context of mobile sensing and ubiquitous computing, see Harari, 2020, and Langheinrich & Schaub, 2018).

Even more worrying is that this information about people's physical locations and their behaviors also reveals insights into who they are. For example, research shows that by getting access to as little as three credit card transactions or location visits, one can uniquely identify an individual among a large population of people (de Montjoye et al., 2013, 2015). That is because there is likely only one person in New York City who bought a coffee at Starbucks on 72nd Street at 8:52 a.m., then had lunch in a Chinese restaurant in Midtown at 12:15 p.m., and after that took an Uber back to the Upper West Side.

In addition to uniquely identifying individuals based on their data records, people's digital footprints can also reveal intimate information about their sociodemographic and psychological traits (see Chapter 1 in this book for a detailed overview). Personality traits, for example, have been predicted from blogs (Yarkoni, 2010), personal websites (Marcus et al., 2006), Facebook and Twitter accounts (Kosinski et al., 2013; Park et al., 2015), spending records (Gladstone et al., 2019), or people's favorite pictures (Segalin et al., 2016). Beyond personality, digital footprints have been shown to accurately predict a whole range of intimate characteristics, including sexual orientation, mental health (e.g., depression or anxiety), political ideology, intelligence, drug use, personal values, and many more (Kosinski et al., 2013). In addition to the assessment of relatively stable psychological traits, digital footprints can also be used to assess more malleable psychological states. Consumer's emotional experiences and their mood, for example, have been predicted from written and spoken language (AlHanai & Ghassemi, 2017), video data (Teixeira et al., 2012), as well as smartphones and other wearable devices (LiKamWa et al., 2013).

In a nutshell, the inferences that can be made from our personal data challenge our informational privacy by revealing what we do, where we go, and who we are. Importantly, the challenges they pose reach beyond mere informational privacy and into the realm of *decisional privacy* (Allen, 1987), which is defined as the right to exclude others from interfering with our personal decisions and actions (e.g., whom we decide to vote for or whom we would like to marry). For example, a growing number of commercial services use the information about what an individual does, where they go, and who they are to decide whether they get a loan, a favorable insurance plan, or a job. Similarly, the predictions third parties—such as governments or companies—can make about an individual's habits, preferences, and psychological motivations can be used to influence their behavior in ways that lie outside of the individual's control. Matz et al. (2017), for example, showed that companies can increase the likelihood of a consumer engaging with their advertisement and ultimately buying a product if they match their messaging to the consumer's personality traits as predicted from their Facebook Likes. The age of Big Data hence threatens not only our ability to keep intimate details about ourselves and our lives private (i.e., informational privacy) but also our ability to make autonomous and unbiased decisions that are free from external interference (i.e., decisional privacy).

One additional factor makes the age of Big Data particularly challenging when it comes to privacy and privacy protection. The abilities to learn details about a person's life and character and to use this information to influence their decisions are not unique to the digital world. As every individual who has grown up in a small village will be able to attest, our social interactions outside of big settlements and urban areas often involve a high level of intimacy between community members. Just as in our interactions with technology, there are instances in which this intimacy is welcome and others in which it is not. If this is part of our social evolution as a human species, what is all the fuss about when it comes to new technologies and Big Data? A fundamental difference between the ways in which people's privacy is being challenged today is the mere scale at which a single player can observe the lives of millions of people simultaneously with very little effort, and the fact that there are enormous power imbalances between those who observe and those who are being observed (Andrejevic, 2014). There is a growing gulf between the vast majority of individuals around the world who—often unknowingly—provide their data and have become products that are being traded like commodities by a few players who have the power to observe, record, and process these data and benefit financially from it. Today, governments and large tech companies—like Google, Alibaba, Facebook, Apple, Microsoft, Tencent, or Amazon—capture both the data and the attention of millions of people around the world.

This scale and imbalance of power is dangerous because it puts the privacy of everybody in the hands of a few major players (Matz et al.,

2018). Even if Facebook were to do their very best to protect the privacy of their users, there is a very real chance that third parties will try to penetrate the walled gardens of Facebook and attempt to steal user data for nefarious purposes (as happened in late 2018 when the personal accounts of 50 million users were compromised; Isaac & Frenkel, 2018). Similarly, a hack into Gmail would lay bare the lives of billions of users in one single sweep. Of course, even isolated breaches to our privacy can be extremely consequential for individuals (e.g., when person A posts pictures of person B publicly on the internet that person B does not want anyone else to see and cannot easily remove). However, the consequences of such isolated breaches are small compared to the privacy risk that comes from privacy breaches affecting the big tech giants. Because they are impacting the lives of billions of people around the world, they pose a privacy risk for everybody. Simply put, the concentration of data and insights in the hand of a few big players poses a major privacy risk arising from security breaches that are qualitatively different from everything we have experienced before (Matz et al., 2018).

How Does Big Data Challenge Traditional Approaches to Privacy Protection?

Big Data not only poses a threat to people's privacy by providing increasingly intrusive insights into their lives but also challenges the ways in which we think about privacy as well as many of the existing approaches that have been put in place to protect it. That is, traditional conceptualizations of privacy often fail to predict and explain people's privacy preferences and can no longer provide sufficient guidance on how to effectively protect them from privacy violations (Nissenbaum, 2010). Building on the work of prominent privacy scholars (Nissenbaum, 2010; Solove, 2013), social psychological research on distributed responsibility (Bandura et al., 1996; Darley & Latané, 1968), as well as one of our own review articles on privacy in the context of psychological targeting (Matz et al., 2020), we focus on four challenges associated with Big Data that amplify the existing risks for information privacy and that will need to be addressed in an updated framework for privacy protection: (a) blurred lines between public and private information, (b) outdated practices of notice and consent, (c) increasing diffusion of responsibility in tech organizations, and (d) the illusion of data anonymization.

THE BLURRED LINES OF WHAT IS PRIVATE AND WHAT IS PUBLIC

In the age of Big Data, in which content can be generated and consumed by billions of people around the world, it has become increasingly difficult to distinguish the private from the public (Kosinski et al., 2015; Matz et al., 2020; Nissenbaum, 1998). Modern technology allows for new capabilities to collect, store, and evaluate information that enable surveillance and change what is meant by public information in contrast to private information (Nissenbaum, 1998). For example, if someone walks down the boulevard St Giles in Oxford, England, and passes by the Oxford Internet Institute, they are exposed to a higher level of publicity today than they were a few years ago. Today, they can be seen by not just other people passing by, but by everyone with access to the internet because the Oxford Internet Institute livestreams the view of the boulevard (https://www.oii.ox.ac.uk/webcams/). The level of privacy in the public space has thus changed.

A photo of an evening among friends, which could be considered private information, can instantly become public once it is uploaded to social media and shared with the world. Importantly, it is almost impossible to make such information private again after it has been made public on the internet. Even the deletion of a picture could still leave digital traces behind, for example, the state of the internet with the picture saved might have been archived, or other people might have shared the picture in the meantime.

Appropriate regulation of private information is especially difficult because a certain piece of information can affect multiple individuals. For example, the photo of an evening among friends might be considered unproblematic information by one individual, who shares it publicly and then tags their friends in the photo, who might consider the same information about their meeting to be private. Beyond such a direct link, it is also possible that information made public by one individual can indirectly affect what can be known about other individuals with similar characteristics (Barocas & Nissenbaum, 2014). For example, if researchers construct a model to infer political orientation from Facebook Likes, they only need to know the political orientation of a few participants who are willing to share both their Likes and their political orientation in order to predict the political orientation of many more individuals who only shared their Likes publicly. The latter individuals might consider their Likes public information, but not their political orientation.

Informed Consent Is No Longer Enough

Privacy regulation today typically relies on the practice of notice and consent, also known as informed consent. However, privacy experts consider the idea that an individual can fully understand how the data they generate are used and consent to that use as flawed for several reasons (Acquisti et al., 2015; Barocas & Nissenbaum, 2009; Cate & Mayer-Schönberger, 2013; Solove, 2013).

First, today's privacy landscape is much more complex and less transparent than it used to be, making the assumption that individuals can fully understand privacy notices a questionable one (Barocas & Nissenbaum, 2009). One reason for the increased complexity of privacy policies is the high frequency of privacy policy changes and the fact that many privacy policies link to separate third party privacy policies that also need to be reviewed (Barocas & Nissenbaum, 2009).

Second, individuals cannot be trusted to make privacy decisions that are in their best interest (Acquisti et al., 2015). One the one hand, individuals do not possess the knowledge required to evaluate privacy policies critically (Barocas & Nissenbaum, 2009). For example, individuals might mistake location data from their smartphones as innocuous records of their behavior, although this type of information might reveal sensitive information such as their home location, their work location, whether they have regular doctor's visits, or whether they occasionally stay over at other people's homes at night. On the other hand, individuals have cognitive biases that result in suboptimal decision making (Acquisti et al., 2015; Solove, 2013). For example, present-bias (i.e., the tendency to value rewards in the present more than rewards in the future) could encourage individuals to share intimate information like a photo of their newborn child because they feel instantly gratified by the reactions they receive. In the long term, however, the benefit of this instant gratification might not outweigh the privacy risks posed by sharing the information, for example the ability for social media companies to find out the child's date of birth and capture their face from day one.

Third, when individuals are asked to consent to data collection and use, they are not necessarily able to foresee how and in which combinations the data could be used in the future (Barocas & Nissenbaum, 2014; Cate & Mayer-Schönberger, 2013; Solove, 2013). For example, an individual might consent to Facebook sharing their Likes with the public. Later on, they learn that Facebook Likes can be used to predict political orientation via a model that was generated by combining Facebook Likes with study participants' psychological profiles and demographic data. When consenting to Facebook's sharing of their Likes, the individual probably did not foresee the use of Likes to predict political orientation, and might potentially reconsider their consent given this use. Importantly, future and unforeseeable uses of data imply that no matter how much knowledge an individual has and how rational they are, they cannot possibly fully understand what their consent means.

Fourth, the focus on individuals in the context of notice and consent is inadequate because one individual's consent has consequences for other individuals (Barocas & Nissenbaum, 2014). The data collected from consenting individuals could be used to profile nonconsenting individuals with comparable characteristics (Barocas & Nissenbaum, 2014), for example, the prediction of political orientation via Facebook Likes mentioned earlier.

Finally, consent choices usually lack appropriate levels of granularity (Solove, 2013). That is, users often face the choice between either accepting all terms and conditions and uses of their data or leaving a service entirely. Take the example of the social media platform Facebook. Recently, Facebook gave users additional privacy choices like opting out of facial recognition on the platform, but an individual still has to consent to most data use without choice in order to use the platform.

Diffusion of Responsibility

A classic stream of social psychological research shows that people feel less responsible to take action and stand up for others when they are part of a group or surrounded by other bystanders who—in principle—could all take on the responsibility of acting. This phenomenon is known as the diffusion of responsibility or bystander effect (Bandura et al., 1996; Darley & Latané, 1968). Diffusion of responsibility is manifested in many different contexts. For example, individuals are less likely to help a person in need if there are other people around them (Darley & Latané, 1968), and members of a team are less likely to feel personally responsible for the team's outcome as group size increases (Bandura, 1999).

The diffusion of responsibility effect is also observed in workplace settings, where employees often fail to inform managers about concerns when they are part of larger teams that work toward a common goal. This form of moral disengagement has been related to high levels of division of labor, where each individual focuses only on their own individual tasks and fails to consider the wider moral responsibilities of the organization as a whole (Bandura, 1999; Bandura et al., 1996; Trevino, 1986).

Diffusion of responsibility is likely to play a role in the way that the employees of tech companies feel responsible for an ethical application of the products they create or voice concerns about how their product might infringe on users' privacy. Given the complexity of most data-driven products, dividing labor among different parts of the organization is unavoidable. Implementing Big Data solutions successfully often requires big teams. Launching and maintaining products such as Google maps, for example, requires expertise from many different layers and silos within an organization. Each team that is involved contributes to only part of the final product and each team has their unique, localized metric of success, often unrelated to the overall success of the entire product. Engineers and software developers might work on ways in which geolocation data can be collected, stored, and processed, and their success is judged by their ability to deliver a high-quality and fully functional product on schedule. UX designers might think about how to turn these data into insights that are useful for the user, and their success will be judged based on user feedback. Product managers will strategize how to coordinate between teams, and their success will be measured by how well the market receives the product. Each team member is responsible for a small slice of the overall product and is unlikely to have a solid grasp of the entire product pipeline or the effect a product may have when all its pieces are put together. If privacy is not specifically called out as a goal for each individual in every silo, it is easy to diffuse responsibility for protecting user privacy to others. Engineers and developers might assume that this is the task of product managers or UX designers that liaise with the users themselves, while product managers might feel that they do not have the expertise to fully understand the data structures and processes to make informed decisions and therefore rely on engineers and developers to voice concerns.

In a nutshell, the fact that data-driven technology is a complex endeavor that requires different parts of organizations to work together makes it difficult for each individual to embrace their own moral responsibilities. Importantly, this is not necessarily a moral failure, but it often stems simply from the lack of exposure to the bigger picture of what a product or service is going to look like eventually and how users are going to interact with the technology. Consequently, it is critical that organizational leaders visualize the bigger picture for their employees and actively encourage them to engage with ethical issues as part of their daily activities.

The Data Anonymization Illusion

As psychologists, we are trained to respect the privacy of our research subjects by removing all identifiable information—such as names or email addresses—from our data sets. In fact, this procedure of removing personally identifiable information (PII) whenever possible is an integral part of the institutional review board's (IRB) approval process required for scientists in order to conduct and publish research using human subjects. In most traditional psychological research contexts, this could be easily done by deleting variables that could identify a person and replacing names, for example, with unique, anonymous identifiers (codes).

Working with large amounts of user-generated data makes anonymization a difficult and often impossible task. As we outlined before, a few data points are often enough to uniquely identify one person in a large population of people (de Montjoye et al., 2013). Because every person's digital footprint is so unique, it becomes possible to reverse engineer their identity from their digital records, even if we as researchers do our best to remove obvious personal information such as names or email addresses. Such attempts at reverse engineering an individual's identity from their digital records usually takes the form of linkage attacks, where one data set is combined with other data sets to identify unique individuals (Narayanan & Shmatikov, 2008).

There are numerous prominent examples of such attacks. In 2007, Netflix published "anonymized" movie rankings from over 500,000 users in the context of a Netflix challenge that encouraged research teams around the world to help improve their recommendation algorithms. A team of researchers showed that they could identify most of the users by linking the Netflix data to additional data from IMDb (Narayanan & Shmatikov, 2008).

The fact that tech companies like Netflix with hundreds or even thousands of experienced experts in data science failed to adequately anonymize their users' data and foresee privacy concerns arising from publishing "anonymized" data clearly highlights the difficulty of anonymizing data. As social scientists, we often receive no training in what it means to analyze data that has not been collected in traditional laboratory settings or online panels. Even with the purest intentions, researchers are therefore likely to make mistakes under the illusion that they have anonymized their data before making it public.

In fact, there is a growing demand for open science that aims to address the replication crisis by encouraging researchers to make their data and analysis scripts publicly available on open science repositories such as OSF (Open Science Collaboration, 2012). Although this push for more transparency and accountability is certainly desirable, it can lead to unintended consequences and poses risks when it comes to protecting the privacy of research subjects. Researchers, for example, might decide to share data about a person's Facebook profile, removing all PII such as names or email addresses. However, it is likely that information about the unique combination of, for example, Facebook pages a person has followed will allow others to reidentify that person. Consequently, social scientists—including psychologists—will need to receive better training to fully understand when they might put their participants at risk when publicly releasing seemingly "anonymized" data sets. What is needed is a set of agreed-upon guidelines and best practices that they can refer back to and follow closely when designing and sharing their research.

The same is true for IRBs, which are meant to protect researchers from making careless mistakes and to assure the safety of research subjects. However, many IRBs that were trained to consider ethical challenges in traditional social science paradigms and research settings might not possess the knowledge and skills to make an informed and adequate decision on whether a certain data sharing protocol might put the anonymity of research subjects at risk. In addition to training scientists, it is hence paramount to also train the second line of defense—the IRB personnel—on how to evaluate privacy concerns arising from Big Data and to create interdisciplinary teams that include both social and computer scientists.

RETHINKING PRIVACY IN THE AGE OF BIG DATA

As we have outlined, the age of Big Data poses new challenges to traditional approaches to privacy protection. In this section, we argue that in order to identify viable solutions for the future, we need to shift our focus from the traditional focus on *who* collects *what* data, to a new focus of *how* these data are being used (Matz et al., 2020). In which contexts are people's personal data used and for which purposes? There are two frameworks that we believe will prove valuable in thinking about privacy when it comes to new technologies and Big Data. The first one conceptualizes privacy as contextual integrity, and the second one urges us to handle data in the same way we handle radioactive material.

Privacy as Contextual Integrity

The theory of *contextual integrity* was first introduced by the philosopher Helen Nissenbaum, who conceptualized privacy as the appropriate flow of personal information, which is guided by constantly evolving social norms (Nissenbaum, 2010, 2019). According to the theory, whether or not an individual's privacy is upheld depends on the flow of personal information, and the extent to which this flow adheres to informational norms that govern a particular social context. In other words, the specific use of a particular type of information might be upholding a person's privacy in one context, but it might violate it in another. Returning to the notion of "privacy is not about secrecy, but about control," contextual integrity acknowledges that the use of our social media posts, for example, might not be a concern as long as they are used in the way that we intended them to be used (e.g., sharing them with friends and—depending on our picture settings—the world more broadly). However, if the pictures were taken out of context and, for example, put onto a pornographic website or used to target specific advertising at us, the flow of information would no longer adhere to the dominant sociocultural norms by which individuals would generally consent to such a usage. According to the theory of contextual integrity, there are five critical factors that need to be considered when deciding whether or not a particular flow of information adheres to the dominant norms: (a) who sent it, (b) who received it, (c) about whom it is, (d) what types of information are concerned, and (e) which transmission principles (e.g., consent) apply (Nissenbaum, 2019).

According to the theory of contextual integrity, approaches to privacy protection that focus exclusively on solving the problem of *who* collects *what* kind of data are hence insufficient. They are insufficient because they cannot account for the myriads of ways in which the data might be used in the future across different contexts or protect individuals from harmful inferences that can be made based on this data. Acknowledging the complexity of data and the fact that the extent to which individuals are concerned about third-parties accessing and using their data is often highly context-dependent. Nissenbaum's (2019) theory of contextual integrity skillfully shifts the focus to *how* the data are being used (i.e., in which context and for which purpose) and therefore provides a more meaningful theoretical framework of how to conceptualize privacy in the age of Big Data. Importantly, the notion of contextual integrity allows for privacy to dynamically adapt to changing sociocultural norms that develop over time. What our generation might consider unacceptable today could quickly become the norm in the years and decades to come, as younger generations grow up in a tech-enabled, data-driven world and develop new ideas of how technology, individuals, and society at large should interact with one another.

Data as Radioactive Material

Another appealing way of approaching privacy is to think of and deal with data just like we should deal with radioactive materials. In 2008, Cory Doctorow wrote in a *Guardian* article, "We should treat personal electronic data with the same care and respect as weapons-grade plutonium—it is dangerous, long-lasting and once it has leaked there's no getting it back." The idea has been further developed by Brian Croll, one of the authors of this chapter. The reasoning behind this notion is as simple as it is powerful: Just like radioactive material, data can be extremely powerful and valuable to both individuals and society at large. In the same way that X-rays allow us to get insights about the human body that would otherwise be impossible, data can help us better understand the human condition (e.g., the way we interact with one another, our health). However, what is needed are guidelines and international agreements that prevent bad actors from abusing this power in a potentially harmful way. Notably, the same argument we made about the scale of data also holds for radioactive material: The danger of nuclear power increases exponentially with mass. Considering data as radioactive hence builds on the core idea that personal data carry inherent potential harm that gets exacerbated with scale: (a) the collection of data can drive violations of privacy; (b) comprehensive dossiers of individuals fuel manipulation, which weakens democratic institutions; (c) surveillance shuts down dissent and stifles free thought; and (d) massive databases pose a security risk. Because data—just like radioactive materials—are long-lasting, the key to dealing with data is to recognize this inherent harm and treat data accordingly. If appropriate guidelines are put in place, societies will eventually be able to reap the benefits of using personal data while safeguarding their members from the harm that it might cause if left unattended.

Can Individuals Successfully Navigate the Privacy Challenges Posed by Big Data?

Before we outline promising avenues for privacy and data protection, we want to explore a line of reasoning that places the responsibility of privacy and data protection on users (rather than data holders). According to this point of view, users of technologies should be considered self-interested and autonomous decision makers who can be trusted to make adequate decisions about how much information they want to disclose or withhold (Posner, 1981). Instead of placing "crippling" regulations on companies and stymieing innovation, users should decide for themselves what data they want to share with whom. In their excellent review paper, Acquisti et al. (2015) presented various factors that influence individuals' privacy concerns and privacy behaviors and that render individuals ill-equipped to navigate privacy challenges in the age of Big Data. These factors can be grouped into three related themes—the uncertainty of privacy preferences, their context-dependence, and their malleability—which are summarized below. The following sections borrow heavily from Acquisti et al.'s (2015) review to illustrate that placing the burden of privacy protection on users is misguided.

Uncertainty

According to Acquisti et al. (2015), an individual's uncertainty about their privacy preferences means that they are unsure what to share publicly and equally unsure about existing privacy trade-offs. There are two main sources for this uncertainty: (a) asymmetric and incomplete information, and (b) uncertainty about one's privacy preferences.

Asymmetric and incomplete information arises because data are abstract in nature and therefore it is difficult to grasp the extent and consequences of their collection. As Acquisti et al. (2015) pointed out, data collection in the online world is a mostly invisible process, so it often remains unclear which information organizations have and how this information is being used. At the same time, data-driven technologies are often so complex that, even if there were full disclosure and complete visibility over what data is being collected, it is extremely difficult for the average user to determine (a) exactly what is happening with the data and (b) if or how much harm might be caused by third parties obtaining access to their data.

Further, privacy infringements result mostly in intangible costs. While it might be straightforward to measure the damage done by identity theft that results in concrete monetary loss, it is much more complex to estimate the damage incurred by sensitive private information like political orientation or a criminal past becoming public. Adding to the lack of unambiguous information, privacy decisions are hardly clear—they usually involve trade-offs that favor the benefits of sharing data in the here and now (for the broader concept of time discounting or delay discounting, see Frederick et al., 2002). As we have discussed before, a user might be aware that getting personalized search results requires them to sacrifice their personal data (which could have negative ramifications in the future), but they still might be willing to make this trade-off in order to benefit from better search results in the present (i.e., present bias).

There are multiple forces underlying people's general uncertainty about their privacy preferences. First, as becomes evident in the conceptualization of privacy as contextual integrity (Nissenbaum, 2019), context is critically important for people's privacy considerations. This is true for the contexts in which data are being used, but also for the contexts in which people are asked about their privacy preferences and presented with different options for how to protect their data. For example, the endowment effect suggests that people value their privacy more when they currently have it and might lose it, than when they do not currently have it but could gain it. In a clever experiment, people shopping in a mall were randomly assigned to receive a gift card that was either anonymous ($10) or with their personal information being associated with it ($12) for their participation in a survey (Acquisti et al., 2013). They were then offered the opportunity to switch their card for the other one. Among those who were originally assigned the less valuable but anonymous card, about half the individuals decided to keep their card rather than switching to the traceable, higher value card (52.1%). In contrast, only 9.7% of the individuals who had originally received the higher value card decided to switch to the anonymous, lower value card. These findings highlight that the status quo (e.g., default settings) matters: people value their privacy more when they already have it. Third, privacy decision making is not fully rational, but also influenced by emotions, heuristics, and social norms. As we outlined earlier, individuals might be subject to present bias and choose the instant gratification of posting intimate information over the long-term regret of having shared too much (Acquisti, 2004). Finally, humans are naturally inclined to share information about themselves because self-disclosure is inherently rewarding (Tamir & Mitchell, 2012). Digital platforms hence create a tension between the need for self-disclosure and the need for privacy.

The extent to which individuals experience uncertainty about their privacy preferences also provides an explanation for the apparent gap between people's self-declared goals around privacy and the behaviors and actions they take to implement these goals. When you ask people how much they care about privacy, most of them will tell you that they care a lot about it. People do not like the idea of having strangers, businesses, or governments snooping around in their lives—neither in their physical nor in their virtual "homes." A staggering 93% of Americans report that having control over who can access and use their information as important (Madden & Rainie, 2015). Yet, almost no one reads the privacy policies of the services they are using (Solove, 2013), and most people consent to companies accessing their data without giving the decision much thought (e.g., by allowing a calendar app to access their pictures, messages, GPS sensor, and microphone without reading how the data will be used). This phenomenon—the discrepancy between individuals' proclaimed attitudes and intentions about privacy and their observable behaviors and actions—is known as the *privacy paradox* (Acquisti et al., 2015; Athey et al., 2017). The privacy paradox highlights how the uncertainty around what data are being collected, how the data are being used, and how one should adequately respond in different contexts prevents individuals from living up to their own expectations when it comes to privacy.

Context-Dependence

According to Acquisti et al. (2015), *context-dependence* refers to the fact that the same individual can care a lot about their privacy, or not at all, depending on the context in which they find themselves. The uncertainty of privacy preferences implies that individuals look for guidance on their privacy decisions and hence take the cues that their context provides into account. As suggested by the theory of privacy as contextual integrity, which has been discussed in more detail in an earlier section, social norms affect what we consider private and public, and what we are willing to share in a given context.

Contextual influence also depends on one's culture and the behavior others exhibit. For example, reciprocity makes it more likely that individuals share intimate information themselves after their interaction partner shares intimate information about themselves (Acquisti et al., 2012; Moon, 2000). Contextual influence also has a temporal dimension because individuals learn from past experiences. They might become more careful if they experience privacy invasions. However, it is important to note that individuals' behavior might not change permanently after privacy invasions because despite initial behavior change, individuals might get used to increased levels of surveillance and return to their presurveillance behaviors (Oulasvirta et al., 2012).

Context-dependence is especially challenging today since individuals are exposed to mixed contexts. For example, on social media, we might be connected to both our personal friends and our coworkers. While we might have different norms around how much we would like to disclose to these two groups in an offline context, we are forced to follow one "disclosure policy" online, unless the social media platform we are using allows us to select our audience on a more granular level. Further, we do not necessarily know how big our audience truly is nor who will see the content we share. We might have sharing settings on our social media account set to public assuming that no one outside of our friendship circle will be interested in our posts, but one day we might discover that a story went viral and reached a far bigger audience.

Malleability

According to Acquisti et al. (2015), *malleability* refers to the idea that privacy concerns can be influenced by various different factors, and a change in privacy concerns can in turn cause a change in privacy behavior.

For example, if a user creates a social media account, the default privacy settings on the platform will likely shape the settings they choose to adopt.

Malleability can be a result of so-called antecedents that affect privacy concerns (Acquisti et al., 2015). An example of an antecedent is trust in an institution that collects data: This preexisting trust might lead a user to be more willing to share private information with the institution (Joinson et al., 2010). Antecedents can lead to both rational and irrational behavior. While it might make sense to entrust data to an institution that has been trustworthy in the past, there are instances where privacy-related interventions can backfire and lead to unexpected behavior. For example, 62% of participants in a survey erroneously believed that companies posting privacy policies on their website implied that they could not share their personal data with third parties (Hoofnagle & Urban, 2014). Another feature that we have highlighted prominently in the beginning, but that can sometimes produce paradoxical effects, is control. Although giving users control appears to be desirable in most contexts, perceived control can also reduce privacy concerns by creating a false sense of trust. For example, research has shown that individuals who are given additional control over their data might be willing to share *more* data later on (Brandimarte et al., 2013).

Importantly, malleability is not inherently good or bad. Increasing or reducing privacy concerns can be either beneficial or harmful, depending on the context. For example, guaranteeing anonymity might result in increased cheating in one context, but in increased prosocial behavior in another (Acquisti et al., 2015).

The three themes of uncertainty, context-dependence, and malleability highlighted by Acquisti et al. (2015) showed that there is no straightforward way for individuals to protect their privacy because they have uncertain privacy preferences that depend on their current context and are easily influenced. This means that even if individuals care about protecting their privacy, they might not be able to do so effectively, and their stated intentions to protect their privacy might not result in corresponding actions. This insight shows that we need to rethink what privacy means in the age of Big Data and how it could be protected effectively by solutions like privacy by design.

PROTECTING PRIVACY IN RESEARCH AND PRACTICE

Building on the theory of contextual integrity (Nissenbaum, 2010), we outline a number of promising approaches to privacy and data protection. To acknowledge the fact that data and the insights we can draw from them are often extremely valuable to both individuals and society at large, we focus our solutions on those that overcome the false dichotomy of what we call the "privacy versus insight" illusion. What we mean by this is the tendency to portray the choices users face as an "either-or" question: Users can either give up their privacy for convenience and better recommendations (e.g., Netflix recommendations, Google maps), or they can forsake these benefits and enjoy higher levels of privacy. This public perception of privacy as an "either-or choice" is hardly surprising when considering that many of the products and services we use today indeed force their user to make such a decision. For example, we can either use Instagram, Facebook, and Google by agreeing to the terms and conditions they put forward—and by doing so hand over our personal data—or we can give up the connections we maintain through these platforms with our friends and family. Although Facebook does offer users options to tweak their privacy settings, there is often no real path to fully protecting one's privacy while still using the service. However, as we argue below, we believe that this "privacy versus insight" dichotomy is a false one, and that users can enjoy both at the same time.

Importantly, we do not claim that our list of possible solutions is comprehensive. Instead, we have focused on the solutions that maximize the overall utility of both insights and privacy, and try to tackle existing privacy issues by creating win–win scenarios, rather than zero-sum outcomes with unnecessary trade-offs. We hope that readers will consider these suggestions as a foundation for an on-going discourse and reflection on how we can protect users' privacy while reaping the benefits Big Data promises.

Privacy by Design

One potential solution to the privacy paradox that is diametrically opposed to the proposition that each user is responsible for their own privacy is *privacy by design*. Instead of placing the burden of privacy protection on users, privacy by design mandates the integration of privacy protection into the design, development, and application of new technologies and data solutions (Cavoukian, 2011; Monreale et al., 2014). For example, by setting the privacy default to a reasonable level of protection with the requirement for users to opt-in to more data sharing if they desire, the burden of making rational choices that are costly in the moment but beneficial in the long-run would be lifted from users. At the core, privacy by design is about reducing friction on the path to privacy, and about making it easier for users to follow through with their intentions. This shift in responsibility from users to companies is not only promising when it comes to data protection and privacy, it might also bring with it an improvement in products and services. If companies must rely on users' active consent if they want to collect and use more than the data available to them under the privacy-preserving default, they will have to take seriously their promise of using data in order to improve the quality of their product. If the user does not see the value of sharing their data, they will not do so and simply move on to another service. (Interested readers can consult the seven foundational principles of privacy by design [Cavoukian, 2011], which capture the essence of privacy by design and puts forward concrete implementation guidelines.)

Regulation

We believe that regulation needs to be a part of the solution to protect individuals' privacy, as regulation can both send an important signal to companies of what sociocultural norms look like and reorganize incentive structures for companies such that the cost of violating privacy increases. Data-driven technologies are often extremely complex and opaque, making it impossible for the average consumer to understand them and make an informed decision of how to protect their privacy within such ecosystems. Similar to other contexts (e.g., drugs that have to be cleared by a regulator such as the U.S. Food and Drug Administration; Coravos et al., 2019), regulation can help protect citizens who have neither the expertise nor the access and resources to make their own well-informed decisions.

Given that a focus on privacy is often costly for companies, they have little incentive to invest in this domain (other than ethical grounds) as doing so would likely put them at a competitive disadvantage compared to less concerned companies. Hence, government regulation is the only solution that can mandate the same rules for everybody and thus level the playing field by introducing costs for not adhering to certain privacy-related standards.

Privacy regulations vary substantially across countries (and sometimes within countries, as is the case in the United States, where states can implement their own data protection laws), providing an interesting test bed for studying the effects of various levels of stringency. Among the strictest data protection regulations are the General Data Protection Regulation (GDPR) of the European Union and the California Consumer Privacy Act (CCPA), which focus on core principles like transparency and control, including the need to disclose the purpose of data collection, which third parties the data might be shared with, and the right to obtain the data collected about oneself. Notably, the GDPR was also the first to officially address and regulate psychological profiling.

However, although the GDPR and CCPA are the most progressive data protection regulations currently in place, they do not fully deal with the challenges discussed in the context of contextual integrity. For example, future regulations could be improved by not only recommending privacy by design but also putting mechanisms in place to oversee its implementation, and they should aim to protect data across a variety of contexts. This requires protecting data on all levels (including metadata) as well as limiting the insights that can be derived based on them (Nissenbaum, 2019). For example, if location data can be closely linked to health outcomes or other highly intimate personal outcomes, this type of data deserves better protection. In the context of psychological targeting, for example, this would require (a) carefully outlining the data that can be used to create psychological profiles; (b) identifying psychological characteristics that should be protected (e.g., impulsiveness or an addictive personality); and (c) specifying in which contexts psychological targeting can be applied (e.g., marketing of consumer products), and in which contexts it should be restricted (e.g., political campaigning; see Matz et al., 2020). Regulations that are built on contextual integrity and focus on *how* data are being used could greatly enhance trust in governments and organizations, which would have to become more responsible data stewards than they currently are (Cate & Mayer-Schönberger, 2013; Price & Cohen, 2019), and encourage individuals to reap the benefits of Big Data and predictive technology while protecting them from potential harms.

Collaboration

A constructive debate about the key ethical challenges of Big Data and new technologies, as well as promising ways of addressing these challenges, will require collaboration. This collaboration needs to happen both vertically and horizontally, bringing together technology users and creators. Industry leaders, policy makers, and scientists across a range of disciplines—including the computer and social sciences—need to engage with each other (Barocas & Selbst, 2016), as well as the public more broadly. Because technology affects all of us, collaboration is the only way to ensure that different legitimate—but potentially opposing—goals are heard and adequately considered in the solutions that are put forward. For example, psychologists might diverge in their perspective on how acceptable prediction error is when it comes to mental health assessment compared to their computer science peers or industry experts. In contrast, using their computational toolbox, computer scientists might be able to evaluate and uphold principles of "algorithmic fairness" more successfully than psychologists. Finally, industry mantras such as "move fast and break things" are likely to pose a threat to research integrity and the desire to thoroughly vet services and products for their privacy standard, but industry innovations and resources could also speed up the development of technological solutions to privacy protection (see the next section).

Notably, the type of collaboration we envision here is not unprecedented. After the explosion of the first nuclear bomb in 1945 in a remote desert in New Mexico (known as the Trinity explosion) and the nuclear attacks on Hiroshima and Nagasaki less than a month later, nuclear scientists formed the Bulletin of the Atomic Scientists to educate the public about atomic weapons and discuss ethical standards and regulations

(Grodzins, 1963). There are a number of research-focused consortia and think tanks today that address ethical challenges related to Big Data, including AI Now (ainowinstitute.org), the Data Pop Alliance (datapopalliance.org), the Psychology of Technology Institute (psychoftech.org), the Electronic Frontier Foundation (www.eff.org/issues/privacy), Data & Society (datasociety.net/research/privacy-data), or the Life Sensing Consortium (lifesensingconsortium.org). We hope that this conversation continues and will result in improved ethical standards.

Technological Solutions

Many opponents of strict regulations advocate for the use of technology to solve some of the privacy challenges introduced by Big Data. To many psychologists, the idea of solving the problems that have been introduced by technology by using even more technology might seem counterintuitive or even preposterous. However, we argue that the computer sciences have produced a number of compelling technological (algorithmic) approaches to leveraging the power of Big Data while preserving individuals' privacy. Although there are many more solutions that we could cover in this chapter, we would like to highlight two broad categories that we believe are particularly promising. The first one is differential privacy and the second one federated (or collaborative) learning. We will refrain from diving into the underlying math of these solutions and instead discuss the basic ideas behind the two approaches.

Differential Privacy

As we outlined earlier in the chapter, removing personally identifiable information (i.e., PII) from large data sets that contain records of human behavior is often insufficient for guaranteeing the anonymity of the individuals in the data set. This is because one can often combine different data sources via shared variables (e.g., date of birth + zip code), making it possible to reidentify a particular person, even if the data set has been "anonymized" with the intention of protecting individuals' privacy. The fact that sharing data even after PII has been removed poses a risk to individuals' privacy stands in contrast to the desire in many parts of society to gain insights from publicly sharing and pooling data. For example, anonymously sharing health information to see which treatments work best for which parts of the population can be extremely valuable (Tatonetti et al., 2012).

Differential privacy tries to alleviate the risk of reidentification by systematically introducing noise to the information that is shared with others (Dwork, 2008; Dwork et al., 2006). The core idea here is that removing any given individual from the data set would not significantly alter the insights one would draw in response to a particular question. For example, if a data set was meant to reveal whether males or females respond more favorably to a medical treatment, then noise would be introduced in a way that assures that removing any given individual from the data set does not significantly alter the insights. In other words, a process is differentially private if the outcome is basically the same, no matter who is removed from the original data set. The challenge in differential privacy is hence to add just the right amount of noise such that the privacy of individuals is protected, but the insights one can draw from the data are not overly distorted (Dwork et al., 2006). Notably, differential privacy requires sufficiently large data samples to be able to adequately strike this balance between privacy and accuracy. In small data sets, one needs to add a substantial amount of noise to guarantee that an individual cannot be reidentified, which often leads to a stark drop in the accuracy of insights that can be gleaned.

Federated Learning

Differential privacy allows companies and researchers to minimize the risk of third parties reidentifying individuals in a data set while maintaining the ability to draw meaningful insights from data that is pooled across many individuals. However, differential privacy approaches do not address the challenge that our original data is still stored in the same, central place and is potentially vulnerable to attacks and security breaches. In other words, we need a lot of faith in the entity storing our data—whether that is a government, a company, or an individual—to safeguard them. An alternative route to protecting users' privacy while still inferring valuable insights is to leverage the fact that we now have powerful computers in the hands and homes of billions of people around the world: their smartphones and smart devices.

The idea of *federated learning* is to decouple the process of machine learning from the necessity to store all data used in the training and evaluation stages of a predictive algorithm centrally in one place (e.g., on a company server, the cloud; Kairouz et al., 2021; Konečný et al., 2016; McMahan et al., 2017). Federated learning uses mobile devices to perform the process of insight generation locally. Instead of submitting an individual's personal data to a company, all that is sent are insights that help improve the overall predictive performance of the general model. Imagine Netflix's movie recommendations. As of now, a user's preferences are submitted to the Netflix servers where they are processed and used to refine the recommendation algorithm used to provide users with the best experience and movie recommendations. Instead of submitting a person's movie recommendations to Netflix, federated learning loads the existing recommendation model onto the user's device, updates and refines the model based on that user's movie preferences, and then sends an encrypted version of the new model back to Netflix such that all users can benefit from the improved model. Federated learning hence does not require any data to ever leave the user's device, while at the same time provides a way of leveraging the insights gained from each and every user for the greater benefit of all users. As such, federated learning is one technical route to achieving privacy by design.

Addressing the Root Causes, Not the Symptoms

When technologies go awry, the first reaction of regulators and companies is often to focus on quick, immediate fixes to the specific problem at hand. While this might help alleviate some of the potential damage inflicted by technology (e.g., informing customers and implementing short-term fixes after a hack), it is critical to move beyond such short-term fixes and dedicate more energy and resources toward addressing the root causes. The case of Cambridge Analytica provides an interesting example. One of the major storylines was a scientist sharing Facebook data with Cambridge Analytica (Stahl, 2018). Of course, this could be a one-time instance in which a researcher made questionable ethical decisions. Holding that individual accountable is the simplest solution to the problem. However, it might also overlook some more systemic forces that have led to this instance being possible in the first place. In other words, the proximal causes of harmful outcomes based on Big Data (e.g., a scientist illegally sharing data with third parties or a company like Cambridge Analytica aiming to intentionally invade people's privacy), might be difficult to disentangle from deeper structural factors like academic incentive systems or the ad-based business model underlying many of the tech giants' decisions. To address the structural factors that underlie many of the visible privacy breaches, we need to create a value system that breaks down structural barriers to privacy and reassembles the building blocks in a way that allows governments and organizations to create win–win situations. Some of the questions that will need to be addressed are: Which information and contexts do we consider "sacred" and untouchable? In which contexts should the application of predictive technology be limited? Which tradeoffs do we accept when it comes to algorithmic decision making where efficiency/accuracy can come at the cost of justice/privacy? Are there certain business models that are so detrimental to the welfare of societies that they should be restricted by regulation? Do we have the right incentive structures in place to encourage researchers and practitioners to make decisions that are aligned with society's desire for privacy? These questions cannot be answered by any one set of actors. Instead, they require a public debate about the values, norms, and political priorities related to privacy and data use.

RECOMMENDATIONS FOR RESEARCHERS AND PRACTITIONERS

As scientists, we are obligated to apply for ethical approval of our studies with our university's IRB and follow a set of ethical guidelines. These IRB guidelines are based on the *Belmont Report* (National Commission for the Protection of Human Subjects of Biomedical and Behavioral Research, 1979), which identified basic ethical principles that guide research conducted on human subjects. The three core principles can be summarized as follows:

- **Respect for persons:** protecting and upholding the autonomy of research subjects and treating them with courtesy and respect. Requires informed consent and truthful conduct with no deception.
- **Beneficence:** the philosophy of *do no harm* and the goal of maximizing the benefits to research subjects and society while minimizing the risks to all subjects.
- **Justice:** ensuring reasonable and nonexploitative procedures that are administered fairly and equally (e.g., fair distribution of costs and benefits to research subjects).

Although these guidelines apply only to scientists, we argue that they should also be considered foundational among practitioners. The important question that arises from these general principles is how they can be translated to concrete practices in the age of Big Data (for a comprehensive discussion of the Belmont report in the age of Big Data, see Paxton, 2020). How can we as scientists and practitioners uphold and foster the principles laid out in the Belmont report in a data-driven world? Below, we suggest a number of best practices. The list is not exhaustive, and there are likely many additional guidelines that could help researchers and practitioners live up to high ethical standards. We hope that the community will continue to refine them and add additional best practices over time.

- **Be transparent:** Explain what data are being collected and how the data are to be used in a simple way that people can understand.
- **Treat informed consent as a process:** Privacy should not be a matter of just one mouse click. Periodically remind people what data are being collected and how they are used, and allow them to reconsider their privacy choices (Harari, 2020).
- **Push for privacy by design:** Whenever possible, use opt-in mechanisms that make privacy the default.
- **Collect only the data you need:** Think of data as radioactive. Assuming that more data will mean higher costs of providing sufficient protection, do you really need to collect all the data?

- **Share data mindfully:** Be sure to share research data only within the limits of permission, and carefully consider whether anonymity can be guaranteed and the potential reidentification of participants can be avoided (Zook et al., 2017).
- **Take training seriously:** What type of data protection is appropriate for the data type you are working with? How exactly are personally identifiable data defined? What kind of data can be accessed from personal devices or send around via email? Work with your institution's IT and ethics teams to ensure you are aware of important privacy best practices, and make sure that your collaborators are, too.
- **Do the "user check":** Before implementing a data solution, ask yourself how you would feel about a loved one (e.g., your partner, kids, closest friends) using your product or service. If this thought makes you feel uncomfortable, something is off!
- **Set high standards for yourself:** Do not stop at asking, "What is legal?"; ask, "What is ethical?" You might not always be able to live up to your own high standards, but if you do not set them in the first place, you certainly will not.
- **Give back:** If you ask for people's data, give back as much value and insight as possible to create win–win situations. This can mean providing a better service or simply offering users insights into their own data.

CONCLUSION

As we have outlined in this chapter, Big Data can provide tremendous opportunities to improve the lives of individuals and benefit society at large. However, given the considerable challenges that Big Data poses to privacy and the ways in which we approach protecting privacy, it is necessary to identify ways in which we can address the potential downsides of using Big Data while retaining the potential upsides. We argue that the burden cannot be placed on users alone. Instead, we need a combination of design principles, regulations, cross-disciplinary collaboration as well as technical solutions to overcome the false dichotomy of "privacy versus insight." We will have to have an open debate among scientists, policy makers, industry leaders, and the public about what our core values are and how they can be not just upheld and respected, but also nurtured and developed in the age of Big Data. Importantly, this debate needs to carefully weigh different perspectives and solutions. We cannot simply evaluate solutions independently, as if they existed in a vacuum. Instead, we need to consider them vis-a-vis the status quo and available alternatives. For example, a focus on algorithmic fairness might lead people to reject algorithms in the context of recruiting processes, even if the bias introduced by algorithms might be well below that introduced through the stereotypes held by human resources personnel (Cowgill et al., 2020). Hence, we need to ask what the alternatives to data-driven solutions and particular predictive algorithms are, and whether these alternatives serve us better or worse. This also means deciding when it is unethical to experiment with new technology, and when it is unethical not to. The mere fact that a novel approach comes with its own challenges does not mean that it is inferior to a traditional one. Ideal ethics and privacy guidance does not prevent innovation, but it puts the right framework in place so that innovation will be more beneficial than harmful. Like crash barriers for a highway, these frameworks should not limit speed too much, but guarantee an adequate amount of safety.

REFERENCES

Acquisti, A. (2004). Privacy in electronic commerce and the economics of immediate gratification. In *EC '04: Proceedings of the 5th ACM Conference on Electronic Commerce* (pp. 21–29). Association for Computing Machinery. https://doi.org/10.1145/988772.988777

Acquisti, A., Brandimarte, L., & Loewenstein, G. (2015). Privacy and human behavior in the age of information. *Science, 347*(6221), 509–514. https://doi.org/10.1126/science.aaa1465

Acquisti, A., John, L. K., & Loewenstein, G. (2012). The impact of relative standards on the propensity to disclose. *Journal of Marketing Research, 49*(2), 160–174. https://doi.org/10.1509/jmr.09.0215

Acquisti, A., John, L. K., & Loewenstein, G. (2013). What is privacy worth? *The Journal of Legal Studies, 42*(2), 249–274. https://doi.org/10.1086/671754

AlHanai, T., & Ghassemi, M. M. (2017). Predicting latent narrative mood using audio and physiologic data. In *AAAI '17: Proceedings of the Thirty-First AAAI Conference on Artificial Intelligence* (pp. 948–954). AAAI Press.

Allen, A. L. (1987). Taking liberties: Privacy, private choice, and social contract theory. *University of Cincinnati Law Review, 56*, 461–491.

Andrejevic, M. (2014). Big data, big questions| the big data divide. *International Journal of Communication, 8*, 1673–1689.

Athey, S., Catalini, C., & Tucker, C. (2017). *The digital privacy paradox: Small money, small costs, small talk* (NBER Working Paper No. 23488). National Bureau of Economic Research. https://doi.org/10.3386/w23488

Bandura, A. (1999). Moral disengagement in the perpetration of inhumanities. *Personality and Social Psychology Review, 3*(3), 193–209. https://doi.org/10.1207/s15327957pspr0303_3

Bandura, A., Barbaranelli, C., Caprara, G. V., & Pastorelli, C. (1996). Mechanisms of moral disengagement in the exercise of moral agency. *Journal of Personality and Social Psychology, 71*(2), 364–374. https://doi.org/10.1037/0022-3514.71.2.364

Bansal, G., Zahedi, F. M., & Gefen, D. (2016). Do context and personality matter? Trust and privacy concerns in disclosing private information online. *Information & Management, 53*(1), 1–21. https://doi.org/10.1016/j.im.2015.08.001

Barocas, S., & Nissenbaum, H. (2009). On notice: The trouble with notice and consent. In *Proceedings of the Engaging Data Forum: The first international forum on the application and management of personal electronic information.* http://ssrn.com/abstract=2567409

Barocas, S., & Nissenbaum, H. (2014). Big data's end run around procedural privacy protections. *Communications of the ACM, 57*(11), 31–33. https://doi.org/10.1145/2668897

Barocas, S., & Selbst, A. D. (2016). Big data's disparate impact. *California Law Review, 104*(3), 671–732. https://doi.org/10.2139/ssrn.2477899

Bellman, S., Johnson, E. J., Kobrin, S. J., & Lohse, G. L. (2004). International differences in information privacy concerns: A global survey of consumers. *The Information Society, 20*(5), 313–324. https://doi.org/10.1080/01972240490507956

Brandimarte, L., Acquisti, A., & Loewenstein, G. (2013). Misplaced confidences: Privacy and the control paradox. *Social Psychological & Personality Science, 4*(3), 340–347. https://doi.org/10.1177/1948550612455931

Cate, F. H., & Mayer-Schönberger, V. (2013). Notice and consent in a world of big data. *International Data Privacy Law, 3*(2), 67–73. https://doi.org/10.1093/idpl/ipt005

Cavoukian, A. (2011). *Privacy by design: The 7 foundational principles.* https://iab.org/wp-content/IAB-uploads/2011/03/fred_carter.pdf

Cohen, J. E. (2000). Examined lives: Informational privacy and the subject as object. *Stanford Law Review, 52*(5), 1373–1438. https://doi.org/10.2307/1229517

Coravos, A., Chen, I., Gordhandas, A., & Stern, A. D. (2019). We should treat algorithms like prescription drugs. *Quartz.* https://qz.com/1540594/treating-algorithms-like-prescription-drugs-could-reduce-ai-bias/

Cowgill, B., Dell'Acqua, F., & Matz, S. (2020). The managerial effects of algorithmic fairness activism. In W. R. Johnson & G. Herbert (Eds.), *AEA Papers and Proceedings* (Vol. 110, pp. 85–90). https://doi.org/10.1257/pandp.20201035

Crawford, K., Gray, M. L., & Miltner, K. (2014). Critiquing big data: Politics, ethics, epistemology. *International Journal of Communication, 8,* 1663–1672. https://ijoc.org/index.php/ijoc/article/view/2167/1164

Crawford, K., & Schultz, J. (2014). Big data and due process: Toward a framework to redress predictive privacy harms. *Boston College Law Review, 55*(1), 93. https://lawdigitalcommons.bc.edu/bclr/vol55/iss1/4

Darley, J. M., & Latané, B. (1968). Bystander intervention in emergencies: Diffusion of responsibility. *Journal of Personality and Social Psychology, 8*(4, Pt. 1), 377–383. https://doi.org/10.1037/h0025589

DeCew, J. W. (1997). *In pursuit of privacy: Law, ethics, and the rise of technology.* Cornell University Press. https://doi.org/10.7591/9781501721243

de Montjoye, Y.-A., Hidalgo, C. A., Verleysen, M., & Blondel, V. D. (2013). Unique in the crowd: The privacy bounds of human mobility. *Scientific Reports, 3*(1), 1376. https://doi.org/10.1038/srep01376

de Montjoye, Y.-A., Radaelli, L., Singh, V. K., & Pentland, A. S. (2015). Unique in the shopping mall: On the reidentifiability of credit card metadata. *Science, 347*(6221), 536–539. https://doi.org/10.1126/science.1256297

Doctorow, C. (2008, January 15). Personal data is as hot as nuclear waste. *The Guardian.* https://www.theguardian.com/technology/2008/jan/15/data.security

Doward, J., & Gibbs, A. (2017, March 4). Did Cambridge Analytica influence the Brexit vote and the US election? *The Guardian.* https://www.theguardian.com/politics/2017/mar/04/nigel-oakes-cambridge-analytica-what-role-brexit-trump

Duby, G. (1992). *A history of private life: Vol. 1. From pagan Rome to Byzantium.* Harvard University Press.

Duhigg, C. (2012, February 16). How companies learn your secrets. *The New York Times.* https://www.nytimes.com/2012/02/19/magazine/shopping-habits.html

Dwork, C. (2008). Differential privacy: A survey of results. In M. Agrawal, D.-Z. Du, Z. Duan, & A. Li (Eds.), *Theory and applications of models of computation* (pp. 1–19). Springer.

Dwork, C., McSherry, F., Nissim, K., & Smith, A. (2006). Calibrating noise to sensitivity in private data analysis. In S. Halevi & T. Rabin (Eds.), *Theory of cryptography* (pp. 265–284). Springer. https://doi.org/10.1007/11681878_14

Fogel, J., & Nehmad, E. (2009). Internet social network communities: Risk taking, trust, and privacy concerns. *Computers in Human Behavior, 25*(1), 153–160. https://doi.org/10.1016/j.chb.2008.08.006

Frederick, S., Loewenstein, G., & O'Donoghue, T. (2002). Time discounting and time preference: A critical review. *Journal of Economic Literature, 40*(2), 351–401. https://doi.org/10.1257/jel.40.2.351

Gates, K. A. (2011). *Our biometric future: Facial recognition technology and the culture of surveillance*. New York University Press. https://doi.org/10.18574/nyu/9780814732090.001.0001

Gladstone, J., Matz, S. C., & Lemaire, A. (2019). Can psychological traits be inferred from spending? Evidence from transaction data. *Psychological Science, 20*(7), 1087–1096. https://doi.org/10.1177/0956797619849435

Grodzins, M. R. E. (1963). *Atomic age: Scientists in national and world affairs. Articles from the Bulletin of the Atomic Scientists 1945–1962*. Basic Books.

Harari, G. M. (2020). A process-oriented approach to respecting privacy in the context of mobile phone tracking. *Current Opinion in Psychology, 31*, 141–147. https://doi.org/10.1016/j.copsyc.2019.09.007

Hoofnagle, C. J., & Urban, J. M. (2014). Alan Westin's privacy homo economicus. *Wake Forest Law Review, 49*, 261–317.

Isaac, M., & Frenkel, S. (2018, September 28). Facebook security breach exposes accounts of 50 million users. *The New York Times*. https://www.nytimes.com/2018/09/28/technology/facebook-hack-data-breach.html

Joinson, A. N., Reips, U.-D., Buchanan, T., & Paine Schofield, C. B. (2010). Privacy, trust, and self-disclosure online. *Human–Computer Interaction, 25*(1), 1–24. https://doi.org/10.1080/07370020903586662

Kairouz, P., McMahan, H. B., Avent, B., Bellet, A., Bennis, M., Bhagoji, A. N., Bonawitz, K., Charles, Z., Cormode, G., Cummings, R., D'Oliveira, R. G. L., Eichner, H., El Rouayheb, S., Evans, D., Gardner, J., Garrett, Z., Gascón, A., Ghazi, B., Gibbons, P. B., . . . Zhao, S. (2021). Advances and open problems in federated learning. *Foundations and Trends in Machine Learning, 14*(1–2), 1–210. https://doi.org/10.1561/2200000083

Konečný, J., McMahan, H. B., Yu, F. X., Richtárik, P., Suresh, A. T., & Bacon, D. (2016). *Federated learning: Strategies for improving communication efficiency*. arXiv. https://arxiv.org/abs/1610.05492

Korzaan, M. L., & Boswell, K. T. (2008). The influence of personality traits and information privacy concerns on behavioral intentions. *Journal of Computer Information Systems, 48*(4), 15–24.

Kosinski, M., Matz, S. C., Gosling, S. D., Popov, V., & Stillwell, D. (2015). Facebook as a research tool for the social sciences: Opportunities, challenges, ethical considerations, and practical guidelines. *American Psychologist, 70*(6), 543–556. https://doi.org/10.1037/a0039210

Kosinski, M., Stillwell, D., & Graepel, T. (2013). Private traits and attributes are predictable from digital records of human behavior. *Proceedings of the National Academy of Sciences, 110*(15), 5802–5805. https://doi.org/10.1073/pnas.1218772110

Langheinrich, M., & Schaub, F. (2018). Privacy in mobile and pervasive computing. *Synthesis Lectures on Mobile and Pervasive Computing, 10*(1), 1–139. https://doi.org/10.2200/S00882ED1V01Y201810MPC013

LiKamWa, R., Liu, Y., Lane, N. D., & Zhong, L. (2013). Moodscope: Building a mood sensor from smartphone usage patterns. In *MobiSys'13: Proceeding of the 11th Annual International Conference on Mobile Systems, Applications, and Services* (pp. 389–402). Association for Computing Machinery. https://doi.org/10.1145/2462456.2464449

Madden, M., & Rainie, L. (2015). *Americans' attitudes about privacy, security and surveillance* [Report]. Pew Research Center. https://www.pewresearch.org/internet/2015/05/20/americans-attitudes-about-privacy-security-and-surveillance/

Marcus, B., Machilek, F., & Schütz, A. (2006). Personality in cyberspace: Personal web sites as media for personality expressions and impressions. *Journal of Personality and Social Psychology, 90*(6), 1014–1031. https://doi.org/10.1037/0022-3514.90.6.1014

Marr, B. (2018, May 21). How much data do we create every day? The mind-blowing stats everyone should read. *Forbes*. https://www.forbes.com/sites/bernardmarr/2018/05/21/how-much-data-do-we-create-every-day-the-mind-blowing-stats-everyone-should-read/?sh=28d4fbe260ba

Matz, S. C., Appel, R. E., & Kosinski, M. (2020). Privacy in the age of psychological targeting. *Current Opinion in Psychology, 31*, 116–121. https://doi.org/10.1016/j.copsyc.2019.08.010

Matz, S. C., Kosinski, M., Nave, G., & Stillwell, D. J. (2017). Psychological targeting as an effective approach to digital mass persuasion. *Proceedings of the National Academy of Sciences, 114*(48), 12714–12719. https://doi.org/10.1073/pnas.1710966114

Matz, S. C., & Netzer, O. (2017). Using big data as a window into consumers' psychology. *Current Opinion in Behavioral Sciences, 18*, 7–12. https://doi.org/10.1016/j.cobeha.2017.05.009

Matz, S. C., Rolnik, G., & Cerf, M. (2018). Solutions to the threats of digital monopolies. In G. Rolnik (Ed.), *Digital platforms and concentration* (pp. 22–30). Stigler Center. https://promarket.org/wp-content/uploads/2018/04/Digital-Platforms-and-Concentration.pdf

McMahan, B., Moore, E., Ramage, D., Hampson, S., & Agüera y Arcas, B. (2017). Communication-efficient learning of deep networks from decentralized data. In A. Singh & J. Zhu (Eds.), *Proceedings of the 20th International Conference on Artificial Intelligence and Statistics (AISTATS)* (PMLR 54; pp. 1273–1282). http://proceedings.mlr.press/v54/mcmahan17a.html

Michael, K., McNamee, A., & Michael, M. G. (2006). *The emerging ethics of human-centric GPS tracking and monitoring*. IEEE Explore. https://doi.org/10.1109/ICMB.2006.43

Monreale, A., Rinzivillo, S., Pratesi, F., Giannotti, F., & Pedreschi, D. (2014). Privacy-by-design in big data analytics and social mining. *EPJ Data Science, 3*, 10. https://doi.org/10.1140/epjds/s13688-014-0010-4

Moon, Y. (2000). Intimate exchanges: Using computers to elicit self-disclosure from consumers. *The Journal of Consumer Research, 26*(4), 323–339. https://doi.org/10.1086/209566

Narayanan, A., & Shmatikov, V. (2008). Robust de-anonymization of large sparse data sets. In *2008 IEEE Symposium on Security and Privacy (SP 2008)* (pp. 111–125). IEEE. https://doi.org/10.1109/SP.2008.33

National Commission for the Protection of Human Subjects of Biomedical and Behavioral Research. (1979). *The Belmont report: Ethical principles and guidelines for the protection of human subjects of research*. U.S. Department of Health, Education, and Welfare. https://www.hhs.gov/ohrp/regulations-and-policy/belmont-report/read-the-belmont-report/index.html

Nissenbaum, H. (1998). Protecting privacy in an information age: The problem of privacy in public. *Law and Philosophy, 17*(5–6), 559–596. https://www.jstor.org/stable/3505189

Nissenbaum, H. (2010). *Privacy in context: Technology, policy, and the integrity of social life*. Stanford Law Books.

Nissenbaum, H. (2019). Contextual integrity up and down the data food chain. *Theoretical Inquiries in Law, 20*(1), 221–256. https://doi.org/10.1515/til-2019-0008

O'Neil, C. (2016). *Weapons of math destruction: How big data increases inequality and threatens democracy*. Broadway Books.

Open Science Collaboration. (2012). An open, large-scale, collaborative effort to estimate the reproducibility of psychological science. *Perspectives on Psychological Science, 7*(6), 657–660. https://doi.org/10.1177/1745691612462588

Oulasvirta, A., Pihlajamaa, A., Perkiö, J., Ray, D., Vähäkangas, T., & Hasu, T. (2012). Long-term effects of ubiquitous surveillance in the home. In *UbiComp '12: Proceedings of the 2012 ACM Conference on Ubiquitous Computing* (pp. 41–50). Association for Computing Machinery. https://doi.org/10.1145/2370216.2370224

Park, G., Schwartz, H. A., Eichstaedt, J. C., Kern, M. L., Kosinski, M., Stillwell, D. J., Ungar, L. H., & Seligman, M. E. P. (2015). Automatic personality assessment through social media language. *Journal of Personality and Social Psychology, 108*(6), 934–952. https://doi.org/10.1037/pspp0000020

Paxton, A. (2020). The Belmont Report in the age of big data: Ethics at the intersection of psychological science and data science. In S. E. Woo, L. Tay, & R. W. Proctor (Eds.), *Big data in psychological research* (pp. 347–372). American Psychological Association. https://doi.org/10.1037/0000193-016

Popov, V., & Matz, S. C. (2016). *Trust and predictive technologies 2016* [Report]. University of Cambridge, The Psychometric Centre/Edelman. https://www.slideshare.net/EdelmanInsights/trust-predictive-technologies-2016

Posner, R. A. (1981). The economics of privacy. *The American Economic Review, 71*(2), 405–409. https://EconPapers.repec.org/RePEc:aea:aecrev:v:71:y:1981:i:2:p:405-09

Price, W. N., II, & Cohen, I. G. (2019). Privacy in the age of medical big data. *Nature Medicine, 25*(1), 37–43. https://doi.org/10.1038/s41591-018-0272-7

Rossignac-Milon, M., Bolger, N., Zee, K. S., Boothby, E. J., & Higgins, E. T. (2021). Merged minds: Generalized shared reality in dyadic relationships. *Journal of Personality and Social Psychology, 120*(4), 882–911. https://doi.org/10.1037/pspi0000266

Sagiroglu, S., & Sinanc, D. (2013). Big data: A review. In *Proceedings of the 2013 International Conference on Collaboration Technologies and Systems (CTS)* (pp. 42–47). IEEE. https://doi.org/10.1109/CTS.2013.6567202

Segalin, C., Perina, A., Cristani, M., & Vinciarelli, A. (2016). The pictures we like are our image: Continuous mapping of favorite pictures into self-assessed and attributed personality traits. *IEEE Transactions on Affective Computing, 8*(2), 268–285. https://doi.org/10.1109/TAFFC.2016.2516994

Sheehan, K. B. (1999). An investigation of gender differences in on-line privacy concerns and resultant behaviors. *Journal of Interactive Marketing, 13*(4), 24–38. https://doi.org/10.1002/(SICI)1520-6653(199923)13:4<24::AID-DIR3>3.0.CO;2-O

Snowden, E. (2019). *Permanent record*. Picador/Macmillan.

Solove, D. J. (2013). Privacy self-management and the consent dilemma. *Harvard Law Review, 126*, 1880–1903. http://papers.ssrn.com/abstract=2171018

Stahl, L. (2018, September 2). Aleksandr Kogan. The link between Cambridge Analytica and Facebook [Interview]. *CBS News: 60 Minutes*. https://www.cbsnews.com/news/aleksandr-kogan-the-link-between-cambridge-analytica-and-facebook-60-minutes/

Statista. (2021). *Global digital population as of January 2021*. https://www.statista.com/statistics/617136/digital-population-worldwide/

Stieger, S., Burger, C., Bohn, M., & Voracek, M. (2013). Who commits virtual identity suicide? Differences in privacy concerns, internet addiction, and personality between Facebook users and quitters. *Cyberpsychology, Behavior, and Social Networking, 16*(9), 629–634. https://doi.org/10.1089/cyber.2012.0323

Tamir, D. I., & Mitchell, J. P. (2012). Disclosing information about the self is intrinsically rewarding. *Proceedings of the National Academy of Sciences, 109*(21), 8038–8043. https://doi.org/10.1073/pnas.1202129109

Tatonetti, N. P., Ye, P. P., Daneshjou, R., & Altman, R. B. (2012). Data-driven prediction of drug effects and interactions. *Science Translational Medicine, 4*(125), 125ra31. https://doi.org/10.1126/scitranslmed.3003377

Teixeira, T., Wedel, M., & Pieters, R. (2012). Emotion-induced engagement in internet video advertisements. *Journal of Marketing Research, 49*(2), 144–159. https://doi.org/10.1509/jmr.10.0207

Tene, O., & Polonetsky, J. (2012). Big data for all: Privacy and user control in the age of analytics. *Northwest Journal of Technology and Intellectual Property, 11*(5), 239–273. https://scholarlycommons.law.northwestern.edu/cgi/viewcontent.cgi?article=1191&context=njtip

Trevino, L. K. (1986). Ethical decision making in organizations: A person–situation interactionist model. *Academy of Management Review, 11*(3), 601–617. https://doi.org/10.5465/amr.1986.4306235

Warren, S. D., & Brandeis, L. D. (1890). The right to privacy. *Harvard Law Review, 4*(5), 193–220. https://doi.org/10.2307/1321160

Westin, A. (1984). The origins of modern claims to privacy. In F. D. Schoeman (Ed.), *Philosophical dimensions of privacy: An anthology* (pp. 56–74). Cambridge University Press. https://doi.org/10.1017/CBO9780511625138.004

Yarkoni, T. (2010). Personality in 100,000 words: A large-scale analysis of personality and word use among bloggers. *Journal of Research in Personality, 44*(3), 363–373. https://doi.org/10.1016/j.jrp.2010.04.001

Zook, M., Barocas, S., boyd, d., Crawford, K., Keller, E., Gangadharan, S. P., Goodman, A., Hollander, R., Koenig, B. A., Metcalf, J., Narayanan, A., Nelson, A., & Pasquale, F. (2017). Ten simple rules for responsible big data research. *PLOS Computational Biology, 13*(3), e1005399. https://doi.org/10.1371/journal.pcbi.1005399

12

The Psychology of Technology: Where the Future Might Take Us

Moran Cerf and Sandra C. Matz

"Prediction is very difficult. Especially about the future." This proverb, attributed to (among others) physicist Neils Bohr, has never been truer than in the realm of the future of technology. Technology has repeatedly shown exponential growth, and its trajectories and scales have continued to surpass the most optimistic predictions. In fact, the suggestion that people *underestimate* the technological advances of the year ahead (and *overestimate* that of decades into the future) has been validated time and again. Betting against a technophobe who suggests that "AI will never [*fill in the blank*]" has been a consistent winning gamble.

Similarly, looking at future interactions between psychology and technology is likely to stretch our imagination, given the number of uncertainties that each discipline encompasses. Yet, tasked with the challenge of making predictions about such a future, we propose a number of domains and research questions that are likely to dominate the conversation on

---

https://doi.org/10.1037/0000290-013
The Psychology of Technology: Social Science Research in the Age of Big Data, S. C. Matz (Editor)
Copyright © 2022 by the American Psychological Association. All rights reserved.

how technology will impact our psychology. The aim of this somewhat eclectic mix is to generate new ideas for psychologists who want to work the field of the *psychology of technology*. While each of the chapters in this book summarizes future directions in their respective fields, this short final chapter is meant as a complementary blue-sky-thinking exercise. There is no common thread running through the ideas we discuss over the following pages other than that we believe these topics will be of growing interest to psychologists and the general public over the next decade. As a reader, you can think of this as an invitation to join us in envisioning what the future of technology might hold, and how psychologists might be able to use these advances as fertile ground for cutting-edge research.

1. HOW WILL TECHNOLOGY SHIFT OUR SOCIAL NORMS AND ETHICAL STANDARDS?

Advances in technology require us to constantly update and refine our social norms and ethical standards. For example, "Is it appropriate to look at our phone while having dinner with a friend?" (a simple conundrum that experts in etiquette guidelines have weighed in on; Zimmerman, 2012), or "How do we deal with the fact that every online conversation we have may be recorded by one of the participants and used against us later?" More complex examples include questions such as whether it is acceptable to keep eating animal meat when advances in technology afford us plant-based alternatives that are indistinguishable in taste and potentially better for the environment. It is not unlikely that very soon—say a generation ahead—we will have to justify ourselves to our kids with regard to how we could have lived in a society that murdered animals when viable alternatives existed. Extrapolating this idea further, we can imagine advances in technology that will challenge aspects of our existence. What is currently regarded as ethically controversial may well become mainstream in just a decade from now. For example, advances in genetic engineering might stir a conversation in which a future child accuses her mother of careless parenting: How could she fail to optimize the daughter's biological making by not removing the potentially harmful *BRCA* gene? Or how could she carry the daughter in a biological womb instead of a toxin-free and pollution-free incubator? This shift in social judgment could be considered akin to our generation's horrified view of parents who drank or smoked during pregnancy. Being asked to alter our social norms and ethical guidelines in accordance with new technologies is likely to require a psychological shift in how we think about the world and our place within it. Social psychologists are uniquely suited to study such shifts by trying to understand and predict how the introduction of new technologies might challenge fundamental moral assumptions and change the way we interact and judge the behaviors of those around us.

2. HOW WILL TECHNOLOGY CHANGE THE WAY WE THINK?

Instead of perfecting machines to become more similar to us, one of the unfortunate consequences of technological advances is that humans often adapt their thinking to the limitations of the devices they use. At the cusp of the millennium, when online searches originated, a person new to technology might have searched for a picture of the U.S. President on a computer using a phrase like: "Computer, please find me a picture of President Clinton next to the White House." Today, the same search is more likely to be phrased: "Clinton image white house." That is, instead of computers adapting to our nuanced prose, we have learned to simplify our language to adhere to their systematic queries. It is not unimaginable that such changes to our vocabulary and phrasing will take a toll on our thinking and psychology. If language permeates into one's thinking, then forcing ourselves to become more structured and poignant in our conversations may mean that technology will alter the way we interact, not just with machines but also with one another. Extrapolating this idea further, we can imagine a world where instead of integrating the best of human communication and empathy into machines, we would further internalize their limitations into our psychology.

Technology might not just change our cognitive processes but also the biological hardware underlying these processes. Our psychology is a manifestation of the neural circuits that govern our thinking. Hence, the cognitive changes driven by technology are likely to alter our brain's physiology. For example, research has shown that using virtual reality in fear-inducing scenarios (e.g., walking on a plank high above ground in a virtual environment) leads to a change of our neural processing (Krupić et al., 2021). People with fear of height may overcome the anxiety using such exposure and negate the innate feelings of danger. The feedback from the virtual experience changes centuries of evolutionarily developed fear that may have been prewired into our psychology for good reasons (Small et al., 2020). Other examples of how technology is altering our neural circuits include changes to areas of the brain responsible for navigation (less dense hippocampi—structures implicated with navigation in the brain—due to increased use of GPS instead of our innate compass; Maguire et al., 2000), logic and math (less use of regions dedicated to computation now that a calculator is always in our pockets; Greenfield, 2009), memory (the date of the French Revolution is available within a call to "Alexa"), or reward systems (the expectations of immediate gratifications in our current fast-paced environment lead to frequent disorders of attention). The impact of technology on our cognitive and neural systems provides a fascinating playground for cognitive and neuro psychologists to study how human cognition is evolving and how we can build technologies in a way that complement rather than undermine our cognitive strengths.

3. WILL TECHNOLOGY DECREASE OUR OPENNESS TO UNKNOWNS?

While some technologies allow individuals to be exposed to experiences that may not be easily accessible otherwise (e.g., virtual reality tours of nearly unchartered locations), the technological advances aimed at making our lives more convenient and our experiences more personalized often minimizes people's openness to new experiences by virtue of exploration. Decades ago, a search for an entry in the encyclopedia was likely to expose an individual to different ideas and terms in passing (looking for "Bhutan" may have led to the unplanned discovery of "Botany"). Today, search engines give you exactly what you have been searching for, leaving little space for accidental stumbling upon a novel and intriguing concept. Similarly, navigating the streets of Nashville without a map may have led to an unplanned discovery of an artisan's shop in a small alley—whereas, today, mobile phone maps will ensure that you do not deviate from your given path. Gone are the days of heading to the airport without knowing where we would choose to fly to. As technology takes over the planning of our lives—making it more convenient and easier—the psychological traits that govern exploration and openness to new experiences may suffer a notable decline. As our preferences are no longer truly devoid of those offered to us by technology, we may need to relearn what our core desires and interests really are and actively create room to explore the world and learn about our preferences and ourselves. Personality psychologists may contribute unique insights to this debate by studying collective personality change over time and individual personality change as a function of technology use.

4. HOW WILL TECHNOLOGY CHANGE THE MEANING OF WORK AND LEISURE?

As technology becomes prominent in *replacing* human workers (e.g., replacing assembly workers in many factories) it is possible that work will become superfluous for large parts of the population. Currently, most people's identity is intricately tied to their participation in the workforce and their professions. In addition to providing for our livelihoods, work gives us meaning and purpose. When people lose this source of meaning and purpose (e.g., through unemployment), their physical and mental health suffer (Murphy & Athanasou, 1999). The current debate about technology in the workplace focuses on how advances in technology are likely to shatter people's sense of meaning and lead to a demoralized society. However, one could imagine a more positive approach to the same problem, where instead of considering the growing use of machines as a "loss of work for humans," we might consider it a gain of "human leisure time." Instead of labelling a person who is unemployed "between jobs," we may reframe the situation by defining work time as "between leisures." Such reframing surely will also manifest in a new psychological view of work and its place in our lives, as well as society at large. At the same time, new professions may emerge that will require humans to "work for machines" in order to make them more valuable to society. For example, as the need for training data increases with the growth of machine learning capabilities, new professions might emerge that employ people as machine learning supervisors (akin to a nanny helping a baby learn to walk) or data generators (e.g., creating training data for movies with a happy ending, so that the machine can distinguish between "good ending" and "bad ending"). Organizational psychologists are in a unique position to help us understanding how people adjust to the new world in which machines and human coexist and how we might have to collectively rethink the meaning of work.

5. CAN TECHNOLOGY IMPROVE THE WAY WE TACKLE MENTAL HEALTH PROBLEMS?

As machines are becoming better at processing spoken language and communicating seamlessly with humans (repeatedly passing Turing tests, which were regarded the hallmark of human performance just decades ago), the potential of using technology for clinical care *instead of* humans is becoming real. Whereas in 1966, a simple language processing software, ELIZA, showed impressive yet limited ability to help people communicate psychological challenges when they found it hard to talk to humans, decades later the language processing capabilities of machines have been greatly improved and may become a reliable source of help for people who need care but cannot get it from a human (e.g., because they cannot afford it, because they prefer to not communicate with a human). In addition, machine learning is becoming increasingly adept at diagnosing mental health problems via passively collected records of human behavior (e.g., language on social media, GPS signals). Leveraging technology in the assessment and treatment of mental health problems is likely to expand the realm of clinical psychology beyond a small group of therapists. This technological expansion will make it possible to offer support to more people in need and provide the personalized care necessary to battle the growing mental health pandemics observed in large parts of the world. Clinical and health psychologists are likely to play a critical role in making sure that new technologies are not just driven by a desire for technological innovation but grounded in centuries of research in the field of mental health. Their knowledge of the antecedents and expressions of mental health problems, as well as ways in which these problems can be tackled effectively, will be critical for developing effective technologies that can truly make a difference.

6. WILL TECHNOLOGY EVENTUALLY BECOME PART OF US?

One of the most sought-after advances among technology companies in recent years has been the integration of technology in our body and brain. Wearable devices and neural implants are at the forefront of current technological conversations with public tech figures promising grandiose advances within less than a decade. While the timeline of such an integration of brain and machines is unclear, the efforts conducted by those technology giants suggest that the endeavor is likely to materialize in the near future. With the embedding of neural implants into our brains, an alteration of our psychological making seems inevitable. Not only will our thinking evolve and change, but so will our priorities, our identity, and, potentially, our personality. The ability to outsource thoughts to an artificial "thinker" within our brain will elevate us to new levels of mental capacity and allow us to perform operations that would have previously been impossible. Similarly, neural implants might be developed to control and regulate our emotions in a way that lets us tune them up and down as we please.

The advanced abilities afforded to us by neural implants are, in many ways, the source of fantastical science-fiction speculation, but even the conservative projections suggest a potential change to our perception (e.g., we will be able to see more of the reality surrounding us with devices that sample a richer spectrum of light or sound), our thinking (e.g., faster processing), our memories (e.g., being able to use external storage, not being certain if memories we carry are true or implanted), and potentially even our locality (e.g., being able to see through cameras feeding information from outside our eyes or sense temperature outside our geography). It is challenging to imagine what such advances may do to our identity and psyche, but it is safe to assume that it would lead to a major shift in our experiences of ourselves as well as our environments and social interactions. For example, the integration of technology and the human brain might alter the very notion of what it means to be "us." Our perception of who we are, of what we call "me," is effectively the sum of all our neural modules (e.g., the visual system, the reward system, our memory) operating in concert. Once neural implants are introduced into the brain, our existence as a coalition of modules will become salient (e.g., one would be able to, say, turn off the implant that governs "solving math problems") and the fact that we are a collection of many parts (some hidden from our conscious awareness) will become notable. This understanding will likely challenge our current view of our identity as a singleton and might have profound implications for our concept of self.

Besides integration of technology into our brain, one can imagine a similar integration into our body that will alter our perception, health, and sensations even further. Take, for example, the finding that superficial changes to the face muscles due to the use of botulinum toxin (i.e., Botox) results in changes in our psychological response to affective situations (Davis et al., 2010)—making us more emotionally numb. Extrapolate that to a world where technology will govern much of our physique (e.g., what postures we use when we interact with devices, or one of many similar aspects of embodied cognition) and one can imagine a set of conditions that will alter our mental projection of the body entirely. In particular, as many people's image of themselves is tied to their physical appearance—their height, weight, or skin color—changes to what our bodies look like and are able to do will likely alter our mental image of ourselves. Recent evidence for the negative effect of technology on our sleep, vision, or muscles should be seen as precursors to the types of changes that lie ahead, as those effects are intricately linked to psychological consequences (e.g., lack of sleep affects mood and behavior).

Finally, a more speculative—yet based on recent scientific evidence (Vrselja et al., 2019)—change to our physical existence in the wake of technological advances is the amendment of the concept of death. If death becomes optional or superfluous in the future, we will have to reinvent what it means to be human. If we become immortal, how should we think about the meaning of life, our own existence, and our place in the world?

Understanding how the integration of technology into the physical and mental self will impact the human psyche is too big a task for any one psychological subdiscipline and will require collaboration not just among psychologists but also other adjacent disciplines. What happens to people's identities and self-concepts? How does being one with a machine change the way we interact? What might conceding control to an external entity do to people's mental health? These questions can only be answered collectively by psychology as an integrative discipline that is concerned with the human experience as a whole.

The list above provides a snapshot of how technology could impact our psychology—the way we think about ourselves, the way we interact with others, and even the biological foundations underlying who we are—in the years to come, and how these developments provide exciting new opportunities for psychological research. Importantly, all of these suggestions are the product of a mind that only knows today's technology and that is bounded by today's psychology. Simply, our brain is unable to wonder outside its own boundaries. If this chapter had been written by a machine, it might already have made superior projections. In 2019, for the first time, a machine generated a mathematical formula that the human programmers could not solve themselves; this may well be true for writing about psychology's future. Put differently, the growing ability of artificial intelligence to start projecting ideas into the future may provide new psychological opportunities (and challenges) that are not even in the realm of our limited human imagination.

Decades ago, when asked about the technology that would be used during a possible World War III, Albert Einstein acknowledged that he could not predict what types of weapons would be used. Yet, Einstein suggested that World War IV would use arrows and daggers again. With this remark, he reminded his peers of the dangers of rogue technology and the fact that our use of it may eventually lead to humanity's doom. The world we live in today manifests this notion. As prior chapters suggested, technology carries with it some remarkable advances that stand to improve our abilities, our research, and the quality of our lives. However, with those advances comes a risk that the same technologies could alter our psyche in ways that are unforeseeable and uncontrollable. Once the technology genie is out of the bottle, it is nearly impossible for us to put it back. It is up to us—researchers, thinkers, citizens—to consider whether we choose to chase the next shiny advanced technology in a pursuit of godlike abilities, or whether we prefer to, collectively, contain some of those tempting trajectories. This is true as long as our psychological making affords us the choice and the ability to the tame our desires—as long as we are better in understanding our psychology than the machines.

REFERENCES

Davis, J. I., Senghas, A., Brandt, F., & Ochsner, K. N. (2010). The effects of BOTOX injections on emotional experience. *Emotion, 10*(3), 433–440. https://doi.org/10.1037/a0018690

Greenfield, P. M. (2009). Technology and informal education: What is taught, what is learned. *Science, 323*(5910), 69–71. https://doi.org/10.1126/science.1167190

Krupić, D., Žuro, B., & Corr, P. J. (2021). Anxiety and threat magnification in subjective and physiological responses of fear of heights induced by virtual reality. *Personality and Individual Differences, 169*, 109720. https://doi.org/10.1016/j.paid.2019.109720

Maguire, E. A., Gadian, D. G., Johnsrude, I. S., Good, C. D., Ashburner, J., Frackowiak, R. S. J., & Frith, C. D. (2000). Navigation-related structural change in the hippocampi of taxi drivers. *Proceedings of the National Academy of Sciences, 97*(8), 4398–4403. https://doi.org/10.1073/pnas.070039597

Murphy, G. C., & Athanasou, J. A. (1999). The effect of unemployment on mental health. *Journal of Occupational and Organizational Psychology, 72*(1), 83–99. https://doi.org/10.1348/096317999166518

Small, G. W., Lee, J., Kaufman, A., Jalil, J., Siddarth, P., Gaddipati, H., Moody, T. D., & Bookheimer, S. Y. (2020). Brain health consequences of digital technology use. *Dialogues in Clinical Neuroscience, 22*(2), 179–187. https://doi.org/10.31887/DCNS.2020.22.2/gsmall

Vrselja, Z., Daniele, S. G., Silbereis, J., Talpo, F., Morozov, Y. M., Sousa, A. M. M., Tanaka, B. S., Skarica, M., Pletikos, M., Kaur, N., Zhuang, Z. W., Liu, Z., Alkawadri, R., Sinusas, A. J., Latham, S. R., Waxman, S. G., & Sestan, N. (2019). Restoration of brain circulation and cellular functions hours post-mortem. *Nature, 568*(7752), 336–343. https://doi.org/10.1038/s41586-019-1099-1

Zimmerman, E. (2012, March 10). Smartphones should know their place at work. *The New York Times.* https://www.nytimes.com/2012/03/11/jobs/etiquette-for-using-personal-technology-at-work-career-couch.html

Index

Aaltonen, I., 314
Academic incentive systems, 410
Acquisti, A., 398–400
Actionable intelligence, 269
Activity-tracking applications, 276
Addiction-related scales, 202
Addiction to social media, 213, 218–221
Adolescents, mental health of, 199
Advertising, targeted, 29
Affective polarization, 253, 255–256
Ahn, S. J., 173
AI. *See* Artificial intelligence
AI Now, 407
Alessandretti, L., 139
Algorithm Appreciation (Logg), 351, 358–359, 363–366
Algorithmic advice, 356, 367–368, 371
Algorithmic Hiring (Logg), 363
Algorithms, 93, 349–373
    accuracy of, 355–356, 367–368, 371
    bias in, 27, 291–293
    black box, 359
    building, 370, 372
    capabilities of, 357
    complexity of, 31–32
    framing as threat or tool, 357
    future research, directions for, 367

    human judgment vs., 367–371
    human response to, 371–372
    improvement of, 368
    and last mile problem, 350–351, 353–354, 358, 372
    power of, 354–358
    psychology of, 358–359
    and theory of machine, 360–367
    as tools, 371–373
    in the workplace, 281, 283, 293
Alibaba, 387
Allcott, H., 215
Allport, G. W., 63
Amazon, 88, 387
Amazon Mechanical Turk, 170, 285
Amazon Redshift, 98, 99
Amazon Web Services, 99, 118
American Psychological Association, 274, 288
Anonymity, of online communication, 251
Anonymization of data, 270, 393–395
Anthropomorphism, of machines, 369, 371
Apache Hadoop, 98
Apache Hive, 99
Apache Spark, 99, 118
APA Ethics Code, 294

INDEX

Apple, 387
Application programming interfaces (APIs), 95–96
Aquinas, Thomas, 49
AR (augmented reality), 180–181
Archives of Scientific Psychology (journal), 274
Arena, D., 156
Aristotle, 381
Arm gestures, 311
Artificial intelligence (AI)
  chatbots, 74
  early days of, 13
  "explainable," 32
  future research, directions for, 354
  and inference of human intentions, 329–330
  in the workplace, 281, 319, 321, 325, 328–331
Attribution theory, 223–224
Augmented reality (AR), 180–181
Autism simulations, 173–174
Autoencoder models, 113
Automated psychological assessments
  accuracy of, 23–24
  capturing real behavior in naturalistic contexts, 21–23
  future research, directions for, 31–35
  and messiness of data, 25–26
  and privacy, 28–30
  reliability/validity of, 26–28
Automated text analysis, 277
Avatars, 158, 164–165, 174–175
Aviation, 356

Badger robots, in retail settings, 318
Bailenson, J. N., 156, 159, 174, 175, 179, 180
Balcetis, E., 135
Barker, R. G., 10, 21
Bayesian machine learning, 117–118
Bayesian optimization, 113

BBC, 97
Beck Depression Inventory, 214
Behavior, in real vs. virtual environments, 168
Behavior-based linguistic traits (BLTs), 63–64
Behavior change, 176–177
Behaviorism, 21
"Being there," sense of (VR), 160, 161
Belmont Report, 410–412
Belonging, sense of, 223
Beneficence, in human subjects research, 411
Bergen Facebook Addiction Scale, 219
BERT model, 60
Bias
  in Big Data, 27
  linguistic positivity, 20
Bible, 382
Big Data, 4, 87–120
  and Big N/Big V/Big T, 89–91
  data collection using, 94–97
  and data preprocessing, 100–108
  and data storage, 98–100
  definitions of, 88–89
  ethical considerations with, 294
  explanatory vs. predictive research using, 91–94
  modeling using, 108–115
  need for guidelines governing use, 397
  predictive modeling for, 111–115
  psychology of, 353, 358–359
  resource considerations with, 115–119
  statistical inference for, 109–111
  as term, 88
Big Data and privacy, 380–414
  core assumptions, 381–384
  potential solutions, 395–411
  public vs. private data, 388–395
  threats and challenges, 384–389

432

INDEX

Big Data in the workplace, 267–295
    applications of, 276–291
    characteristics of, 268–269
    and diversity/inclusion, 291–293
    and employee turnover, 288–289
    future of, 293–295
    in performance management, 283–287
    and privacy, 269–272
    for recruitment/training, 280–283
    sources of, 272–276
    validity/reliability of, 274–276
Big Five personality traits, 16, 22, 62–64, 279
Big N, 89–90, 92, 109
"Big O notation," 116
Big T, 89–92
Big V, 89–92, 113
Biocca, F., 160
Biological sex differences, and VR, 166
Black avatars, 174–175
Black box algorithms, 359
"Black-box" models, in Big Data, 93
#BlackLivesMatter movement, 241, 254–255
Blame, for social media harms, 223–224
Blascovich, J. J., 167
BLTs (behavior-based linguistic traits), 63–64
Bluetooth, 15
Blurring of real and virtual, in children, 180
Boase, J., 210
Body movement, 311
Body ownership illusion, in VR, 163, 164
Body transfer, in VR, 162, 163
Böhmer, M., 132
Bohr, N., 421
Bot-generated data, 101–102, 254
Botox, 428
Brain
    integration of technology and the, 427–428
    technology and changes in physiology of, 423–424
Brandeis, L., 382
Breast cancer, prediction of, 356
Brexit, 68
Bruges, Belgium, 179
Bulletin of the Atomic Scientists, 406–407
Busa, R., 49
Business models, ad-based, 410
Bystander effect, 391

California Consumer Privacy Act (CCPA), 271–272, 405
Cambridge Analytica, 96, 385, 409, 410
Campbell, R. S., 52
Campion, M. C., 279
Cancel culture, 241, 253
Cancer, prediction of, 356
Candy Crush, 125–128, 132, 137, 138, 140, 143
Carbonell, X., 219–220
Cardiovascular disease, 64–65, 356
Castlight Health, Inc., 290
CATA (computer-aided text analysis), 48, 277
Cattell, R. B., 89, 211
Cave automatic virtual environment (CAVE) systems, 156–157
CBOW (continuous bag-of-words) model architecture, 57
CCPA (California Consumer Privacy Act), 271–272, 405
Cell tower connections, tracking of, 385
CEOs, female, 71
CG (computer-generated) experiences, 157–158
Chamorro-Premuzic, T., 275
Chan, E. T. H., 135
Chatbots, potential of, 74
Chaulet, J., 133
Children
    perspective-taking of, 176
    and VR, 178–180

433

INDEX

Chomsky, N., 45
Circumplex model of valence and arousal, 312
CITI training, 294
Civil Rights Act of 1964, 290
Civil Rights Act of 1991, 292
Classroom, VR in the, 178–180
"Clickers," 213
Cloud computing, 99, 101, 118
Clusters (of servers), 118
Cobots (collaborative robots), 317
Code optimization, 119
Cognition, technology and changes in, 423–424
Cognitive bias
 in humans, 390–391
 in robots, 213
Collaboration, between technology users and creators, 406–407
Collaborative robots (cobots), 317
Collective action, and moral emotions, 254–255
Colorblindness simulations, 172–173
Columbia Business School, 4
Commercial spaces, robots in, 314
Common Rule, 269, 294
Compactness, maximizing, in AI models, 329
Companion robots, 310–311
Computational complexity, 116–117
Computer activity, employees, monitoring of, 269
Computer-aided text analysis (CATA), 48, 277
Computer-generated (CG) experiences, 157–158
Computer processing power, 117–118
Concept drift, 28
Confederates, in VR research, 167
Connectedness, personal, 137–138, 207
Conscientiousness, 16–18, 23
Consumer data, 71
Contagion, emotional, 247, 248

Content algorithms of social platforms, 246
Content words, in sentiment analysis, 50–51
Context-dependence, of privacy concerns, 400–401, 403
Contextual integrity, privacy as, 395–397, 403
Continuous bag-of-words (CBOW) model architecture, 57
Control and realism trade-off, in VR, 167–168
Corporate culture, 69–70
COVID-19 pandemic, 97, 276–277, 283, 285
CPU, 117
Creative work, robots in, 319
Credit card records, as digital footprint, 16
Croll, B., 397
Cross-sectional studies, of social media use, 209, 211–212
Cross-validation, 92, 111–112, 114
Crossword puzzles, 33
Crowd dynamics, complex, 168
Culture
 assessment of, 287–288
 geographic differences in, 225
 organizational, 287–288
 and social media use, 224–226
Cummings, J. J., 159
Curriculum, embedding of VR into, 178
Customer service interactions of robots, 313
Cybersickness, 166

Darabi, H., 328
Data. *See also* Big Data
 anonymization of, 270, 393–395
 capture, in VR, 170
 collection of, 169–171
 disclosure policies, 401
 leaks, 290

434

INDEX

liaison, 352
missing/sparse, 100, 103–105
online collection of, 398
preprocessing of, 100–108
privacy regulations, 270–272
as "radioactive" material, 397
volume/variety/velocity/veracity of, 268, 286
"Data box" heuristic, 211
Data collection, 94–97
Data.gov, 274
Data liaisons, 353–354
Data models, 91, 92
Data Pop Alliance, 407
Data & Society, 407
Data storage, 98–100
Death, changes in concept of, 428
Decisional privacy, 386–387
Decision making
human–robot collaborative, 324–330
and last mile problem, 353–354
and theory of machine, 363–367
Dede, C., 181
Deduplicating, 103
Deep neural networks (DNNs), 14, 328
Deindividuation, 251
DELETE (HTTP command), 95
Democratic Party, 14
Depersonalization, in social media, 247
Depression
Facebook use and prediction of, 18–19
and ostracism, 223
prediction of, 66
and social media use, 207–208, 213–214
Diagnosis
of addiction to social media, 220–221
of mental and physical disorders, from digital footprint, 64–67

Diagnostic training, nonimmersive VR for, 177
Dictionaries, pre-defined, 52–54, 73, 75
Dictionaries, prespecified, 277
Dienlin, T., 209
Diermeyer, D., 278
Dietvorst, B. J., 366
Differential privacy, in PII, 408
Diffusion of responsibility, in the workplace, 392
Digital footprints
automated psychological assessments from, 13–15, 21–30, 62–64
diagnosis of mental and physical disorders from, 64–67
future research, directions for, 31–35
geographic differences in interpretation, 26
Likes as, 11–13
non-social media and, 15–20
and privacy, 385–386
quantity of data in, 11
Digital media, integration into daily life, 211
Digital platforms, profit-generating motive of, 27–28
Dimensionality, 90
Dimensionality reduction, 113, 115
Disability, virtual embodiment of, 163
Disinformation, 72–73, 253–254
Disinhibition, and social media, 251
Displacement/reinforcement, 209
Displays, head-mounted, 156–157
Dispositional attribution, in social dynamics, 224
Distributed file systems, 99
Distributed responsibility, 388
Diversity
lack of, in social media research, 225
in the workplace, 291–293

435

DNNs (deep neural networks), 14, 328
Doctorow, C., 397
Document-term matrix, 53
Domestic chores, robots for, 310–311
Dunleavy, M., 181
Dunning, D., 135
Dystopian visions of future, 384

EBRD (expectations-based reference dependent) loss aversion, 323
EDriving, 285
Education
  robots in, 178, 307, 310
  VR in, 164, 177–182
EEG studies, 90
Eggermont, S., 208
Egypt, 356
Eichstaedt, J. C., 65–66
Einstein, A., 429
Ekman, P., 312
Elastic net, 93
Elder care, robots for, 310
Elderly, perspective-taking of, 176
Election of 2016, 253, 385
Electronic Frontier Foundation, 407
Electronic performance monitoring (EPM), 273, 281, 283–287, 290
Elements of Statistical Learning (Hastie et al.), 117
Eli Lilly and Company, 288
ELISA, 426
Ellis, D. A., 211
Email, 70, 269, 287
Embedded images, in online communication, 249
Embodiment, in VR, 162–165
Emotion, as research topic, 239
Emotional contagion, 247, 248
Emotional expression
  of robots, 311–312
  on social media, 239
Empathy, in VR, 163, 165, 171–177
Empirical stimuli, 167–169

Employees
  acceptance of monitoring, 284–285
  behavioral data of, 281
  cell phone data of, 273
  monitoring and type of job, 286
  performance management of, 273, 281, 283–287, 290
  recruitment of, 280
  remote working by, 276–277
  as research participants, 269–270
  selection of, 274–275, 277–280
  training of, 280
  turnover of, 288–289
Engl, S., 134
Ensemble models, in Big Data, 93
Entertainment, VR applications, 171
Entertainment robots, 310–311
Environmental presence, 160
EPM (electronic performance monitoring), 273, 281, 283–287, 290
ESM (experiencing sampling method), 141
Ethical Principles of Psychologists and Code of Conduct (APA), 294
Ethical standards, technology and future of, 422–423
Ettema, D., 134
European Union (EU), 271
Exclusion criteria, in data pre-processing, 102
Expectations-based reference dependent (EBRD) loss aversion, 323
Experiencing sampling method (ESM), 141
Explanatory research, 91–92, 107–108
Extraction of distribution parameters, 107
Extroverts, social media use by, 139–140, 218
Eye gaze, 179, 311

INDEX

Facebook, 88, 129
  addiction to, 219
  APIs, 95–96
  deactivation of, 215
  digital footprints left on, 10–15
  fastText models, 60
  international presence, 225
  Likes on, 11–15, 24, 26, 29–30, 33, 62, 390, 391
  and marketing, 387
  mobility data, 97
  natural language use on, 46–47
  and place, 129
  and prediction of depression, 18–19, 66
  and prediction of psychological traits, 11–12, 14–15, 24–30, 33, 62, 279, 385–386
  and privacy, 388, 394, 403, 409
  status updates, 18–19, 27, 62
  usage volume, 10
  and user well-being, 207–208
  volume, 88, 109
  within-person research, 64
Face-to-face interactions, 209
Facial recognition technology, 385, 391
"Fake news," 72–73, 253–254
Faking, in employee assessments, 275–276
"False memories," in children exposed to VR, 180
Farman, J., 130
Fast, N. J., 363–364
Fastlink package (R), 103
Fauville, G., 179–180
Fear of missing out (FOMO), 208
Feature engineering, 100, 105–108
Feature extraction, in text analysis, 278
Feature importance scores, 114
Feature selection, 92
Federalist Papers, 50
Federated learning, 408–409

Feedback
  algorithmic, 363
  automated, in the workplace, 283
Felnhofer, A., 165
Female CEOs, 71
Female technology self-efficacy, 165
Festinger, L., 207
Fidelity, of online emotion representations, 247, 248
Field trips, virtual, 179
Figeac, J., 133
Filtering, of Big Data, 101–103
Financial transactions, 16–17
Fire
  emergency evacuations, virtual, 168
  simulations, 177
First-person perspective, in VR, 171
Firth, J. R., 57
Fitbit, 273
Flash Airlines, 356
Flesch–Kincaid Grade Level, 53
FOMO (fear of missing out), 208
Football simulations, 178
Footprints, digital. *See* Digital footprints
Formosa, N. J., 172–173
Foster, M. E., 314
Fox, C. R., 213
Fox, J., 156
Frequency-based analysis, 49–56, 75
Friemel, T. N., 134
Frison, E., 208
Function words, in sentiment analysis, 51
Future, dystopian visions of, 384
Future use of data, 391
Fuzzyjoin package (R), 103
Fuzzy merging algorithms, 103
Fuzzywuzzy package (Python), 103

Galileo, 21
Game playing, with robots, 321
Gamification, 281–282
Gamma distributions, 102

INDEX

"Garbage in, garbage out," 368
GDPR (General Data Protection Regulation), 271, 405
Gender, and response to VR, 165–166
General Data Protection Regulation (GDPR), 271, 405
Generalizability, 93, 115
Genetic engineering, 422–423
GET (HTTP command), 95
Gill, D., 323
Glasgow, T. E., 135
Glassdoor.com, 353
GloVe vectors, 60
Gmail, 388
Goodhart's law, 276
Google, 27, 88, 95, 274, 387
Google BigQuery, 98–100
Google Books, 28
Google Cloud Platform, 99, 118
Google Maps, 403
Gorilla, taking on role of, 179
GPS, 15, 65, 102, 103, 181, 273, 385, 424
Gradient boosting, 93, 112, 114
Granularity, observational, 90–91
Groom, V., 175
Group identities, 68–69, 250–252
Growth curve modeling, 91
Guardian, 397

Hacking, 290
Hadoop Distributed File System (HDFS), 99, 101
Harari, G. M., 273
Harford, S., 328
Harley Davidson, 11, 12, 16
Harvard Law Review, 382
Hasler, B. S., 174–175
Hastie, R., 117
HDFS (Hadoop Distributed File System), 99, 101
Head-mounted displays (HMDs), 156–157
Health care, VR applications, 171

Health information, 408
Healthy behaviors, tailored persuasive appeals for, 34
Heart disease, early diagnosis of, 65–66, 356
Hedge-maze environment, virtual, 168
Hello Kitty, 12
Herrera, F., 176
HFT (high frequency trading), 357
Hierarchical latent tree analysis (HLTA), 54
High frequency trading (HFT), 357
High-performance computing cluster, 118
Hiroshima, nuclear attack on, 406
HLTA (hierarchical latent tree analysis), 54
HMDs (head-mounted displays), 156–157
Hoffman, M., 166
Homelessness, perspective-taking of, 176
Hospitality applications, for robots, 318
HRI. *See* Human–robot interaction
HTML, 96
HTTP (hypertext transfer protocol), 95
Human activity, robots' perception of, 309
Human "coders," of text, 47–48
Human–robot interaction (HRI), 305–332
  attitudes towards robots, 323, 327
  challenges in, 306–309
  collaborative decision making, 324–330
  competition, 321–324
  in the home, 310–313
  in public spaces, 313–315
  in the workplace, 315–332
Humans and robots, trust between, 312
Humphreys, L., 130
Hyman, I. E., Jr., 133

Hyperband, 113
Hyperparameters, 92, 112–113, 117
Hypertext transfer protocol (HTTP), 95
Hypotheses, 90, 91

ICUs (intensive care units), 358
IEDs (improvised explosive devices), 358
Iliev, R., 28
Image data, 90
Image quality (VR), 159
IMDb (Internet Movie Database), 394
Immersion, in VR, 158–159
Immersive virtual classrooms, 179
Immersive virtual environments (IVEs), 156–158, 167, 169–170, 179
Impairment simulation, 172–173
Impression management, 247
Improvised explosive devices (IEDs), 358
Incremental reliability of data, 275
Industry collaborations, for data collection, 97
Inference statistical models, 111, 114
Informational norms, 396
Informational privacy, 383
Informed consent, 30, 270, 390
Ingroup/outgroup perceptions, 174, 246, 252, 255–256
Innovation, and algorithms, 369
In-place contexts, of mobile technology, 127–132
Input, algorithmic, 360–361
Institutional review boards (IRBs), 393, 395, 410
Intensive care units (ICUs), 358
Intentional misrepresentation, 25
Intergenerational attitudes, 175
Intergroup motives, 250
Interindividual analyses, 211–212
Internalization, 244
Internet Movie Database (IMDb), 394
Interpersonal disposition, in social dynamics, 224

Interpersonal motives, 250
Interpupillary distance (IPD), 166
Interrater reliability, 48
Intraindividual analyses, 211–212
Invasions of privacy, 401
Invasiveness of employee monitoring, 284
Involuntary employee turnover, 288
Involuntary turnover, 288
IPD (interpupillary distance), 166
IPO social presence questionnaire, 161
IRBs. *See* Institutional review boards
Ireland, M. E., 67–68
Isolation, 207
IVEs. *See* Immersive virtual environments

Jandura, O., 129
Job descriptions, 281
Job interview, virtual, 175
Johnson & Johnson, 349
JSON, 95
Junco, R., 213
Justice, in human subjects research, 411

Kahneman, D., 213
Kanda, T., 314
Karim, F., 328
Karnowski, V., 129
Kellerman, A., 139
Kern, M. L., 70
K-fold cross validation, 111, 113
Kilteni, K., 164
Kinateder, M., 168, 169
Klippel, A., 179
Koran, 382
Kosinski, M., 15, 96
Kshirsagar, A., 323
Kudenov, P., 133
Kulkarni, V., 63
Kuno, Y., 314

INDEX

Language. *See also* Text analysis
  and personality psychology, 61–64
  and personality traits, 19–20
  and politics, 68–69
  and prediction of depression, 18–19
  and romantic attachments, 67–68
  and social psychology, 67–69
  universality/pervasiveness of, 45–46
Lasso, 93, 112, 113
Last mile problem, 350–351, 353–354, 358, 372
Latent Dirichlet Allocation (LDA), 19, 51, 54–56, 107
Latent semantic analysis, 105, 278
LCD screens, 157
LDA. *See* Latent Dirichlet Allocation
Learning
  federated, 408–409
  in immersive VR, 178
  machine, 13–15, 32–33, 87
  reinforcement, 245
  supervised, 106, 111–115
  unsupervised, 107
Lee, C.-j., 216
Lee, K. M., 160–161
Leisure, technology and changes in, 425–426
Li, X., 134–135
Life satisfaction, 207–208
Life Sensing Consortium, 407
Likes, 11–15, 24, 26, 29–30, 33, 62, 206, 243, 390, 391
Linear models, 92, 109
Linear regression, 110
Ling, R., 210
Linguistic Inquiry and Word Count (LIWC), 52, 277–278
Linguistic positivity bias, 20
Linkage attacks, 394
LinkedIn, 11, 275, 280
LIWC (Linguistic Inquiry and Word Count), 52, 277–278

Location data, 385–386
Logg, J. M., 351, 358, 359, 363
Logical schemas, 98
Loneliness, 206–207
Longitudinal data, 91
"Loop of interaction," 354
Luca, M., 368
Luo, M., 135
"Lurkers," 213

Ma, X., 170
Machine learning, 13–15, 32–33, 87
Majumdar, S., 328
Malleability, 401–402
Malls, robots in, 314
Manufacturing, 171, 317
MapReduce, 99
Maps, AR-enhanced, 181
MAR (missing at random), 103
Marginalized groups, perspective of, 173–176
Marketing, 71–72, 387
Markov Chain Monte Carlo algorithms, 117–118
Markowitz, D. M., 179
Marksberry, P., 278
Matrix factorization, 113
Maxhall, M., 173
Mayer, R. E., 178
MCAR (missing completely at random), 103, 104
Measurement, of personal networks, 205–207
MEC Spatial Presence Questionnaire, 161
Medical records data, 380
Medical volunteer simulations, 177
Meehl, P. E., 356–357
Mehrotra, A., 129
Memory (computer), 118–119
Mental health
  in adolescents, 199
  early diagnosis of disorders, 65–67

and mobile device use, 139
and privacy concerns, 34–35
technology and changing
  approaches to, 426–427
Mentor (app), 285
Merging, of Big Data, 102–103
Meta-analyses, 198–205
Metadata, 405
#MeToo movement, 254
Microsoft, 387
Microsoft Azure, 99, 118
Microsoft Hololens, 180
Microsoft Teams, 276
Mindsets, and social media use, 221–222
Minority Report (film), 384
Mirror strategy, in VR, 174
Misinformation, 72–73
Misrepresentation, intentional, 25
Missing at random (MAR), 103
Missing completely at random (MCAR), 103, 104
Missing data, 103–105
Missing not at random (MNAR), 103
Mobile devices and technologies, 125–144, 180, 199, 409.
  See also Smartphones
  future research, directions for, 142–144
  in-place use of, 127–132
  methodological approaches with, 140–142
  on-the-go use of, 132–136
  and user–device connection, 137–140
Model complexity, 114
Model fitting, in data processing, 108
Modeling of data, 108–115
Model performance, assessment of, 108
Mokhtarian, P. L., 138
Monitoring, by employers, 284–286
Monte Carlo cross validation, 112
Mood, and social media, 221

Moral character judgments, 252
Moral disengagement, 392
Moral emotions. *See* Social media and moral emotions
Moral transgressions, 245–246
Morcno, R., 178
Mosteller, F., 50
Motivation, for social media use, 208
Moxi robots, in hospitals, 318
Mundane realism, 167
Museums, robots in, 314
Music apps, 129
Mussakhojayeva, S., 315

Nacke, L. E., 134
Nagasaki, nuclear attack on, 406
Nass, C. I., 159
National Football League, 349
Natural environment, 163
Natural language processing (NLP), 46–47, 90, 279
Navy Seals, 349
Netflix, 359, 394, 403, 409
Neural networks, 93, 112, 114
The New York Times, 28
Ng, A., 74
Niemelä, M., 314
1984 (Orwell), 384
Nissenbaum, H., 395–396
NLP. *See* Natural language processing
Nonverbal behaviors, in robots, 309
Nonverbal social cues, 242
Normativity, 244
Nuclear weapons, 406–407
Numeric inputs to algorithms, 369
NVivo, 277

Obama, Barack, 11, 12
Object representation, 247–249
Occupational Informational Network, 274
Ocean acidification simulations, 179–180

INDEX

Ocean bottom simulations, 179
Odbert, H. S., 63
Offline violence, 241
Oh, S. Y., 175
OLS regression, 112
One Boy's Day: A Specimen Record of Behavior (Barker and Wright), 10, 21–22
Online payments, 16
Online reviews, 73–74
Online social interactions, 239
On-the-go contexts, of mobile technology, 132–136
Openness to ideas, 16, 17, 33
Open science, 394
Open Science Framework (OSF), 274, 394
Orben, A., 197, 199
Organizational culture, 287–288
Organizational email, 269, 287
Organizational psychology, 69–71
Orwell, G., 384
OSF (Open Science Framework), 274, 394
Ostracism, social technology-related, 223–224
Oulasvirta, A., 141
Output, algorithmic, 361, 362
Overfitting, 13, 90, 92, 108–110, 112, 113
Oxford Internet Institute, 389

PAD emotional models, 312
Panova, T., 219–220
Parallelization (parallel computing), 117
Pareto optimization, 292
Park, G., 62–63, 279
Parong, J., 178
"The Party" (VR film), 173–174
Patient, perspective-taking of, 176
Paul, G., 33
"Paycheck vulnerability," 269
Penalized regression models, 115

Pennebaker, J. W., 51, 52
Perceived roles, in human-robot collaboration, 327
Perception of human activity by robots, 309
Perishability, of data, 269
Permutation importance, 114
Personal connectedness, 207
Personality psychology, language in, 61–64
Personality questionnaires, 21–23
Personality traits
  automated feedback of, 34–35
  Big Five, 22, 62–64, 279
Personalized medicine, 33
Personalized psychological interventions, 33–35
Personally identifiable information (PII), 393–395, 407
Personal presence, 161
Person–job fit, 70
Personnel data, 272–273
Perspective taking, in VR, 171–177
Peterson, H., 285
p hacking, 92
Phubbing, 223
Physical disorders, early diagnosis of, 64–66
Physical impairment simulations, 172–173
Physical movement, monitoring of, 269
Physical presence, 160
Physical rehabilitation, VR applications, 171
Physiological activity monitoring of, 269, 284
PII (personally identifiable information), 393–395, 407
PLSA (probabilistic latent semantic analysis), 54
Pluralistic ignorance, and content algorithms, 246
Plutchik, R., 312

INDEX

Pneumonia, diagnosis of, 32
P-O fit (person–organization fit), 69–70
Pokémon Go, 130, 131
Police brutality, videos of, 241
Policymaking, 353
Political discussions, in online platforms, 253
Political news, 73
Positivity bias, linguistic, 20
POST (HTTP command), 95
Posting of photos on Internet, 389
Power imbalances, 387–388
Predictive modeling
    for Big Data, 111–115
    LDA in, 55–56
Predictive research, 92–94, 108
Pregnant women
    advertising targeting, 29
    workplace discrimination against, 290
Premier League soccer teams, 349
Preprocessing, data, 100–108
Presence, in VR, 155, 160–166, 180
Primaries, 14
Priming, in questionnaires, 213
Privacy
    and automated psychological assessments, 28–30
    and Big Data. *See* Big Data and privacy
    challenges, successful navigation by users, 398
    concerns, technological solutions to, 407–409
    as contextual integrity, 395–397, 399
    and control, 383–384
    decisional, 386–387
    decision-making about, 395–402
    by design, 403–405
    differential, 408
    and Facebook, 388, 394, 403, 409
    informational, 383
    infringements, 399, 404
    policies, 390, 391, 400, 402
    preferences, users' uncertainty about, 398–400
    regulation related to, 389, 404–405
    in research, 403
    and self-disclosure, 400
    and VR, 170–171
Privacy paradox, 400
"Privacy vs. insight" illusion, 403
Private information, public vs., 29
Probabilistic latent semantic analysis (PLSA), 54
Process, algorithmic, 361–362
Processing power, 117–118
Profit-generating motive, of digital platforms, 27–28
"Protecting Human Subjects Training" (NIH), 294
Proteus effect, 164
Prowse, V., 323
Proxemics, 315
Przybylski, A. K., 199
Psychological assessments, automated. *See* Automated psychological assessments
Psychological constructivism, 240
Psychological contract, 286, 289–290
Psychological distance, 250
Psychological profiling, 29–30, 405
Psychological targeting, 388
Psychological traits, prediction of, 11–12, 14–15, 24–30, 33, 62, 279, 385–386
Psychological well-being. *See* Social media use and psychological well-being
Psychology
    of algorithms, 358–359
    of big Data, 353, 358–359
    value of, 293–294
"Psychology beats technology," 226

443

Psychology of technology, as field, xii, 3–5
Psychology of Technology Institute, 407
Psychometrics, 28
Public data, 274
Public information, private vs., 29
Publicly available data, 279
Purchase decisions, 73–74
Purpose of employee monitoring, 284
PUT (HTTP command), 95
PySpark, 103
Python (programming language), 95, 96, 99, 101, 103, 105, 106, 118, 119

Questionnaires
  design of, 209, 212–214, 220
  personality, 21–23
  self-report, 61–62, 64

R (programming language), 95, 96, 99, 101, 103, 105, 106, 118, 119
Racial bias, in VR, 174–175
Racial stereotyping, 163
"Radioactive" material, data as, 397
Random forests, 93, 112, 114
Raveendhran, R., 363–364
Ravid, D. M., 283–284
RDBMSs (relational database management systems), 98–101
Reading apps, 129
Recruiting of minority candidates, 291
Reddit, 46
Reeves, B., 159
Reference models, 93
Regularization, data, 90, 115
Regulation, privacy and, 404–406
Reichow, D., 134
Reinforcement learning, 245
Relational database management systems (RDBMSs), 98–101
Reliability, data, 274
Religiousity, 12

Rendering, 157
Replication crisis, 90
Replication of experimental conditions, 169
Repositories, open science, 394
Representational state transfer (REST) APIs, 95–96
Representative sampling, 169–170
Republican Party, 14
Research, VR as tool for, 167–171
Resource requirements, with Big Data, 115–119
Respect for persons, in human subjects research, 411
Responsible data stewardship, 290
REST (representational state transfer) APIs, 95–96
Retail, robots in, 318
Reviews, online, 73–74
Reward systems, 424
"Rich get richer" hypothesis, 218
Ridge, 93, 113
"The Right to Privacy" (Warren & Brandeis), 382
Robo-Coaching (Logg), 363
Robotics. *See also* Human–robot interaction
  and changes in work/leisure, 425–426
  future research, directions for, 331–332
Role reversal, 176
Rollercoaster, virtual, 166
Rui, J. R., 209
Rule-based sentiment analysis, 50–51

Sample size, 23–24, 89–90, 109
Sandstrom, G. M., 131
Sandygulova, A., 315
Scharkow, M., 211
Schema, in RDBMSs, 98
Schizophrenia simulations, 172
Schlund, R., 364

INDEX

Schnall, S., 135
Schneier, J., 133
Schrage, M., 350
Science and scientific method, xi–xii, xiv
Screenomics, 215
The Secret Life of Pronouns: What Our Words Say About Us (Pennebaker), 51
Security breaches, 388
Segovia, K. Y., 180
Self-disclosure
    as human trait, 384, 400
    and robots, 314, 319
    on social media, 208
Self-esteem, and social media, 207–208
Self-presence, in VR, 161
Self-reported data, 140–141, 220
Semantic analysis, latent, 103
Semantic structure, of sentences, 53
Sensorimotor correspondence, in VR, 163
Sentiment analysis, rule-based, 50–51, 239
Sharing settings, on social media accounts, 401
Shi, C., 314
Shmueli, G., 93–94
Shrinkage models, in Big Data, 93
Signal correspondence, in VR, 162
Simulations. *See* Virtual reality (VR)
Singleton, P. A., 134
Singular value decomposition (SVD), 56
Situational attribution, in social dynamics, 224
Six degrees of freedom systems (6DoF), 157–158
Skip-gram model architecture, 57
Slater, M., 159, 174, 215
Slepian, M. L., 131
Smartphone addiction, 219–220
Smartphones, 15–16, 65, 126, 127, 137, 140–142, 210–211, 385, 390
Smartphone sensing, 90

Smithsonian Museum, 314
Snowden, E., 385
Social communication, in robots, 315
Social comparison, 207, 208
Social compensation, 208
Social connection, 223
Social cues, lack of, in online communication, 247
The Social Dilemma (docudrama), 195–197, 225–227
Social environment, 179
Social exclusion, 207
Social feedback buttons, 243
Socially anxious people, 218
Social media, time spent on, 209–210
Social media and moral emotions, 239–256
    amplification, consequences of, 253–256
    communication via symbolic representation, 246–249
    large social networks, 249–253
    social feedback delivery systems, 242–256
Social Media Test Drive, 216
Social media use and psychological well-being, 195–227
    across populations, 202–204, 224–226
    and active vs. passive use, 201–202, 213
    and addiction, 218–221
    and blame for harms, 223–224
    causal link between, 204–205
    cultural factors, 224–226
    individual differences in, 217–218
    mapping of relationship, 205–209
    measurement of, 202, 210–211
    meta-analysis overview, 198–205
    methodological issues, 209–216
    and mindset, 221–222
    overall effect, 198–199
    over time, 199–200

445

INDEX

Social movements, 253
Social navigation, of robots, 315
Social network mapping, 287
Social networks
  size of, 250
  VR applications, 171
Social norms, technology and future of, 422–423
Social praise, 244
Social presence, 161
Social reward, 244
Social-sensory abilities, 179
Social structure, 205–207
Social support, 206–207
Social work and therapy, robots in, 318
Society for Industrial and Organizational Psychology (APA), 288–289
Software sharing, 169
Sparse data, 105
Spatial abilities, 165–166
Spatial contexts, of mobile technology use, 127–137
  in-place use, 127–132
  on-the-go use, 132–136
Spatial presence, 160–161, 165
Spending decisions, 17–18, 30
Spielberg, S., 384
Splink package (PySpark), 103
Spotify, 11, 12, 359
SQL (Structured Query Language), 98–100
Sri Kalyanaraman, S., 172
Srivastava, S. B., 69–70, 287
Stachl, C., 16
Stanney, K., 166
Startups, as data collectors, 97
Static nature, of online emotion representations, 247
Statistical inference, for Big Data, 109–111
Statistical modeling, 90–91
Statistics Netherlands, 218
STEM virtual environments, 178
Stereoscopic glasses, 157

Stereoscopic views, 157
Sterling, J., 68
Streamlined feedback, 243
Stroke simulations, 172–173
Structured Query Language (SQL), 98–100
Student motivation, 178
Substance use disorders, 219, 220
Supervised learning, 106, 111–115
Support vector machines (SVMs), 114
Surgical simulations, 177
Surveillance, 389
SVD (singular value decomposition), 56
SVMs (support vector machines), 114
Symbolic representation, 246–249
Synchronicity, of employee monitoring, 284
Synchrony, 162

Tai Chi simulations, 178
Tailored marketing, 387
Taipale, S., 139
Talmud, 382
Target (retailer), 29, 385
Targeted advertising, 29
Tarr, M. J., 168
Teacher role simulations, 179
Teachers, eye gaze of, 179
Technically mediated social cues, 242
Technological determinism, 197
Technological Skinner box, 244
Technology
  advances in, xi, 421
  and changes in cognition, 423–424
  human perception of, 380
  integration of, into ourselves, 427–430
  and mental health, 426–427
  and openness to unknowns, 424–425
  pervasiveness of, 3
  shaping humans, 379–380
  and social norms, 422–423
  and work/leisure, 425–426

Technology-enhanced employee training, 281–283
Technology-mediated contagion, 249
Technology-mediated social learning, 246
Telepresence, 160
Temporal data, 90–91
Tencent, 387
Test scores, 92
Text analysis, 45–75, 119
   in consumer psychology, 71–74
   frequency-based methods, 49–56
   future research, directions for, 74–75
   in health and clinical psychology, 64–67
   in organizational psychology, 69–71
   in personality psychology, 61–64
   in social psychology, 67–69
   topic modeling, 54–56
   word-embedding-based models, 57–61
Theories, 90, 91
Theory of machine, 351–353, 360–367
Theory of mind, 352
Therapy robots, 309
Thinking, technology and changes in, 423–424
3D virtual instructor, 178
Three degrees of freedom systems (3DoF), 157–158
360-degree videos, 173–174
Threshold, for moral emotion expression, 245
Thrun, S., 314
Time magazine, 97
Time series regression, 91
Time travel simulations, 179
Title VII of Civil Rights Act of 1964, 290
Tokens, 53
Tomczak, D. L., 284–285
Topic modeling, 54–56, 73, 75

Toyota, 278
Tracking
   of learner growth, 281
   of movement in immersive virtual environments, 157–158
Training
   definition of, 280
   monitoring effectiveness of, 281
   VR for, 177–178
Training needs analysis, 281
Transit stations, robots in, 313
Transparency, of employee monitoring, 284
Trinity explosion, 406
Trump, D., 68
Truncated singular value decomposition, 105
Tübingen, Germany, 168
Turnover, employee, 288–289
Twitter, 279
   bots "users" on, 102
   and emotional information, 243, 245
   and mortality rates, 65–66
   natural language use on, 46–47
   and political beliefs, 68, 72
   and prediction of personality traits, 20, 25, 70, 279, 386
   usage volume, 11
   virality of posts, 72–73
   volume of posts, 48
Type I errors, 90, 92
Type II errors, 90

Uber, 285–286
Ultrasonic wristbands, 273
UMAP, 113
Underrepresented populations, in psychological research, 109
Unemployment rate, 288
Uniform resource identifiers (URIs), 95
University of Cambridge, 4
Unknowns, technology and openness to, 424–425

INDEX

Unsupervised learning (machine learning technique), 107
URIs (uniform resource identifiers), 95
URL, 96
U.S. Department of Labor, 274
U.S. Food and Drug Administration, 404
"Using Algorithms to Understand the Biases in Your Organization" (Logg), 368
UX, 392–393

Validated measures, 209–211
Validity of data, 274–275
Value-based decision-making, 243
Van Veen, H. A., 168
Variable interval reinforcement, 244
Variable ratio reinforcement, 244
Variables, in Big Data, 90
Variable selection, 110, 113, 115, 117
Variety of data, 268, 286
Velocity of data, 268, 286
Veracity of data, 268, 286
Verduyn, P., 216–217
Verkasalo, H., 128, 132
Victims, perspective-taking of, 163, 176
Videoconferencing, 276
Videos, 360-degree, 157
Violence, offline, 241
Viral content, on social media, 72–73
Virtual co-learners, 179
Virtual embodiment, 163
Virtual field trips, 179
Virtual interview software, 276
Virtual memory, 119
Virtual mobilities, 139
Virtual reality (VR), 155–182
 applications of, 171–181
 and changes in brain physiology, 424
 and children, 177–181
 in education, 177–181
 embodiment in, 162–165
 empathy and perspective taking in, 163, 165, 171–177

exposure therapy utilizing, 161
gender and response to, 165–166
immersion and presence in, 158–162
as psychological phenomenon, 158–166
as research tool, 167–171
as term, 156
in training, 281–282
types of, 156–158
Virtual reality perspective-taking (VRPT), 171–177
Virtual schemas, 98
Virtual training, 276
Volume of data, 268, 286
Volumetric video (VV), 158
Voluntary turnover, 288
Vosoughi, S., 72–73
Voter efficiency improvement, in AI models, 329
Voter sentiment, 68
VR. *See* Virtual reality
VRPT (virtual reality perspective-taking), 171–177
VV (volumetric video), 158

Wallace, D. L., 50
Ward, G., 68
Warehouse logistics and materials handling, robots for, 317–318
Warren, S. D., 382
Warren, W. H., 168
The Washington Post, 54
Wearable devices, 273, 287
Web-scraping, 96–97
WEIRD populations, 24, 227
Well-being. *See also* Social media use and psychological well-being
 cultural differences in, 225
 and VR applications, 171, 196
Wheel of emotions, 312
Wide data sets, 354

Wilbur, S., 159
Withdrawal, from social media, 221
Within-person analyses, 67
Word2vvec, 51, 57–60, 71
Word choice, 48–49
Word co-occurrence, 278
Word embeddings, 57–59, 71, 75
Word frequency distributions, 90
Word vectors, 57, 59
Work, technology and changes in meaning of, 425–426
Workplace
    algorithm use in the, 281, 283, 293
    Big Data in the. *See* Big Data in the workplace
    diffusion of responsibility in the, 392
    diversity in the, 291–293
    email in the, 70, 269, 287
    future of, 425–426
    human–robot interactions in the, 315–332
World Meteorological Society, 326
Wright, H. F., 10, 21

XML, 95
X-rays, 32

Yee, N., 174, 175

Zarsky, S., 284–285
Zhang, S. M., 134–135
ZipRecruiter, 280
Zitek, E. M., 364
Zoom, 46, 276

About the Editor

Sandra C. Matz, PhD, is the David W. Zalaznick Associate Professor of Business at Columbia Business School. As a computational social scientist, she studies human behavior using a combination of Big Data analytics and experimental methods. Her research explores how psychological characteristics influence real-life outcomes in different business-related domains (e.g., financial well-being, consumer satisfaction, or team performance), with the goal of helping businesses and individuals make better and more ethical decisions.

Dr. Matz's research has been published in the world's leading scientific journals and is frequently covered by major news outlets around the world. She has won numerous awards, including Data IQ's most influential people in data-driven marketing, *Pacific Standard's* 30 top thinkers under 30, and *Poets & Quants's* 40 under 40 best business school professors.